CAMBRIDGE LIBRARY COLLECTION

Books of enduring scholarly value

Life Sciences

Until the nineteenth century, the various subjects now known as the life sciences were regarded either as arcane studies which had little impact on ordinary daily life, or as a genteel hobby for the leisured classes. The increasing academic rigour and systematisation brought to the study of botany, zoology and other disciplines, and their adoption in university curricula, are reflected in the books reissued in this series.

Journal kept by David Douglas during his Travels in North America 1823-1827

David Douglas (1799–1834), the influential Scottish botanist and plant collector, trained as a gardener before attending Perth College and Glasgow University. His genius for botany flourished and his talents came to the attention of the Royal Horticultural Society. With the society's backing he went to North America in 1823, beginning his life-long fascination with the region's flora. He discovered thousands of new species and introduced 240 of them to Britain, including the Douglas fir. Douglas continued to explore and discover plant species until his death in the Sandwich Islands (present-day Hawaii) in 1834. This remarkable journal, which remained unpublished until 1914, describes his adventures in North America during 1823–7. It also includes extracts from his journal of his explorations of Hawaii during 1833–4. The appendices include a listing of the plants Douglas introduced to Britain, and contemporary accounts of investigations into the mysterious circumstances of his death.

Cambridge University Press has long been a pioneer in the reissuing of out-of-print titles from its own backlist, producing digital reprints of books that are still sought after by scholars and students but could not be reprinted economically using traditional technology. The Cambridge Library Collection extends this activity to a wider range of books which are still of importance to researchers and professionals, either for the source material they contain, or as landmarks in the history of their academic discipline.

Drawing from the world-renowned collections in the Cambridge University Library, and guided by the advice of experts in each subject area, Cambridge University Press is using state-of-the-art scanning machines in its own Printing House to capture the content of each book selected for inclusion. The files are processed to give a consistently clear, crisp image, and the books finished to the high quality standard for which the Press is recognised around the world. The latest print-on-demand technology ensures that the books will remain available indefinitely, and that orders for single or multiple copies can quickly be supplied.

The Cambridge Library Collection will bring back to life books of enduring scholarly value (including out-of-copyright works originally issued by other publishers) across a wide range of disciplines in the humanities and social sciences and in science and technology.

Journal kept by David Douglas during his Travels in North America 1823-1827

DAVID DOUGLAS

CAMBRIDGE UNIVERSITY PRESS

Cambridge, New York, Melbourne, Madrid, Cape Town,
Singapore, São Paolo, Delhi, Tokyo, Mexico City

Published in the United States of America by Cambridge University Press, New York

www.cambridge.org
Information on this title: www.cambridge.org/9781108033770

© in this compilation Cambridge University Press 2011

This edition first published 1914
This digitally printed version 2011

ISBN 978-1-108-03377-0 Paperback

JOURNAL

KEPT BY

DAVID DOUGLAS

1823—1827

PRINTED BY
SPOTTISWOODE AND CO. LTD., COLCHESTER
LONDON AND ETON

David Douglas F.L.S. 1798-1834
enlarged from a pencil drawing   ætat 30   by his niece Miss Atkinson

JOURNAL

KEPT BY

DAVID DOUGLAS

DURING HIS TRAVELS IN

NORTH AMERICA

1823—1827

TOGETHER WITH A PARTICULAR DESCRIPTION OF
THIRTY-THREE SPECIES OF AMERICAN OAKS
AND EIGHTEEN SPECIES OF PINUS

WITH APPENDICES

CONTAINING

A LIST OF THE PLANTS INTRODUCED BY DOUGLAS
AND AN ACCOUNT OF HIS DEATH IN 1834

*PUBLISHED UNDER THE DIRECTION OF THE
ROYAL HORTICULTURAL SOCIETY*

LONDON
WILLIAM WESLEY & SON
1914

PREFACE.

MANY causes have contributed to the delay which has taken place in the publication of the Diaries of David Douglas's journeys on behalf of the Royal Horticultural Society :—

1. The handwriting is nowhere easy to read, and in places is most difficult, occasionally almost if not quite impossible.

2. In the course of nearly one hundred years the ink has faded and become in places very hard to decipher.

3. After the Diary of his journey in North Western America had been prepared for the press and set up in type, a second manuscript was discovered which was at first sight taken to be a duplicate, but which on closer examination was found to contain a great deal of additional information. It had therefore to be compared word for word with the Diary and the additions inserted in their proper places.

4. All the botanical names mentioned have been very carefully looked up by Mr. H. R. Hutchinson and the name given to the plant in *Index Kewensis* or by other later authority is quoted at the bottom of the page with a reference to the author responsible for it.

5. Last, but by no means least, the work was entrusted to the Secretary of the Society and to Mr. Hutchinson, the Librarian, each of whom had already more work in hand than he could conveniently compass, and only occasional spare time could be given to a work which had already waited nearly one hundred years for publication before it was committed to their charge.

* * * * * * * *

I desire to express my very great indebtedness to Mr. Hutchinson for the extraordinary carefulness with which he has assisted me both in the deciphering of the difficult writing and also in the identification of the plants mentioned, which latter work adds enormously to the value of the publication and to the ease with which it may be consulted by modern Botanists and Horticulturists.

W. WILKS, *Sec. R.H.S.*

Westminster, August 18, 1914.

CONTENTS

JOURNAL

KEPT BY

DAVID DOUGLAS

DURING HIS TRAVELS
IN NORTH AMERICA

ON BEHALF OF THE ROYAL HORTICULTURAL SOCIETY, 1823
AND FOLLOWING YEARS

1823

June 3rd.—LONDON : left Charing Cross by coach for Liverpool. Morning very pleasant ; had rained through the night, country very fine for seventeen miles from the metropolis ; found during time of changing horses *Conferva egerops* (Ball) *confer.* Beautiful fields at Woburn Abbey tastefully laid out and divided by hedgerows in which are planted Horse-chestnuts (*Aesculus Hippocastanum*) at regular distances, all in full flower ; had a very imposing appearance. *Menyanthes nymphoides*,[1] for the first time I ever saw it in its natural state. Northampton at 2.30 o'clock P.M., rested 25 minutes ; reached Lancaster quarter to 10 P.M., took supper, started again half past 10, rain during the night ; very cold. Arrived at Liverpool 4 o'clock afternoon. After calling on Messrs. Monal & Woodward and learning that the *Ann Maria* of New York was to sail the following morning, in which a passage had been taken for me, I arranged my business as to my departure and made for Botanic Garden. Mr. Shepherd received me in the most handsome manner ; showed me all his treasures (of which not a few were from North America). I was particularly pleased with *Ferns* ; the luxurious appearance of *Sarracenia adunca*,[2] *rubra, flava, purpurea* was rather weak. *Ranunculus aconitifolius*, which I learn is rare ; Mr. Munro says that he never saw it save in the collection of Mr. Don of Forfar. *Andromeda Catesbaei*[3] var., large flowers of a dirty white. *Pinguicula grandiflora*, growing in a small pot of peat earth, in flower ; *Ranunculus parnassifolius*, in a border of a north exposure, in the same ; *Dentaria bulbifera*,[4] *Lilium pyrenaicum* not in

---

[1] *Limnanthemum nymphoides*, Griseb. in DC. Prod. ix. p. 138.
[2] *S. variolaris*, S. Wats. Bibl. Ind. N. Am. Bot. p. 40.
[3] *Leucothoë Catesbaei*, A. Gray, Syn. Fl. N. Am. ii. I. p. 34.
[4] *Cardamine bulbifera*, Ind. Kew. fasc. ii. p. 733.

flower, *Anemone alpina*, *Claytonia sibirica*, *Arenaria paniculata*, *Inula alpina*,[1] *Anthericum edule* [sic]. Of Scotch rarities *Menziesia*, *Azalea*, *Arbutus alpina*,[2] *Arbutus Uva-ursi*,[3] *Tofieldia palustris*, *Cherleria sedoides*, all in an excellent state of cultivation. The plants in the hot-houses look well, particularly *Scitamineae*, *Panax fragrans*, *Saxifraga ligulata* (two plants in good condition). After spending a little time with Mr. Shepherd and returning thanks, I made my way for town again. I had an interview with Capt. Tair, who commands the *Ann Maria* : he informed me that he intended to sail the following morning.

5th.—Went on board at 6 A.M., when to the great mortification of the Captain he could not clear the river, but was truly goodness to me. I then came on shore again and called at Botanic Garden a second time. I had thus full scope of seeing it in perfection. Mr. S. received four species of *Tillandsia* from Trinidad as also about thirty good plants of *Arracacia*, the same as I saw at the Society's Garden, only in a further advanced state, and one of the species of *Tillandsia* was remarkably woolly.

He had some other specimens received last year from same source, doing well nailed on the back wall of his stove.

6th.—On board at 9 o'clock A.M. in tow of two power steam-boats, which left 15 miles down the channel; we made but little progress, wind being rather contrary.

7th.—All day tossing in the channel, made little or nothing ; few of the passengers were exempted from sickness. I felt perfectly comfortable, only a headache which was occasioned by cold when on my way to Liverpool.

8th.—Thick rainy weather and strong gales.

9th.—Strong heavy gales and tremendous sea towards noon ; a fine pleasant breeze.

10th.—This was the first good morning we had ; most of passengers still sick. Clouds of sea-fowl continue to surround the vessel ; Welsh coast in sight.

11th and 12th.—Winds averse ; Isle of Man on the right, Isle of Anglesea on the left, at a great distance rocky shores of Wales in view ; sun obscured.

June 13th.—Light airs and cloudy. Put on allowance of water, two quarts to each individual.

14th.—Same as preceding day. Lat. 47.33 N.

15th and 16th.—Very hazy weather, thick, foggy ; sun obscured.

17th.—Light airs during forenoon and calm towards even. I was amused by enormous shoals of porpoises surrounding the ship.

18th and 19th.—Perfect calm ; a ship in company passed, a small vessel bound eastward.

20th.—A brig and a sloop in company, which afforded pleasure. Lat. 44.19.

21st and 22nd.—Calm.

---

[1] *I. grandiflora*, Boiss. Fl. Orient. iii. p. 187.
[2] *Arctostaphylos alpina*, A. Gray, Syn. Fl. N. Am. ii. i. p. 27.
[3] *Arctostaphylos Uva-ursi*, A. Gray, *loc. cit.*, p. 27.
[4] *Heteropanax fragrans*, C. B. Clarke, in Hook f. Fl. Brit. Ind. ii. p. 734.

23*rd*.—Thick weather, a ship in company. Lat. 43.35.

24*th*.—Light breeze from the north. Lat. 42.53.

25*th and* 26*th*.—This being my birthday (I completing my twenty-fourth year) and the market day of my native place, I could not help thinking over the days that were gone. Light airs of wind; making little progress.

27*th and* 28*th*.—Morning, thick fog; sun obscured all day.

29*th*.—At 8 o'clock A.M. had a delightful view of Flores [one of the Azores Islands], distance about fifteen leagues; appears to be very rocky. Pleasant. Shot four sea-fowl, but I could not pick them up, the current being strong. Lat. 39.34; Long. 36.12.

30*th*.—At 4 o'clock A.M. had Flores within three leagues. North of the island, observed small shrubs with Captain's glass; prepared the boat for going on shore.

July 1*st*.—Wind favourable, which prevented the Captain sending the boat on shore, which was a disappointment to me. Peak of Pico in view, distance nine leagues.

2*nd*.—Rather foggy; lost sight of the islands. About midday the sun shone, and towards evening the sky was beyond description. Lat. 40.2 N.

3*rd*, 4*th*, *and* 5*th*.—Nothing particular; almost a perfect calm. I found my Spanish grammar a great relief, and sometimes I took a book of 'Linnean Transactions.' Sun obscured.

6*th*, 7*th*, *and* 8*th*.—Dull, heavy weather; making little progress.

9*th*.—At 4 A.M. heavy shower. I could not but observe how the dogs eagerly licked the decks. Some of passengers washing their clothes. Lat. 43 N.

9*th and* 10*th*.—Light breezes, squally, rainy, very changeable; sun obscured.

11*th and* 12*th*.—Thick fogs, accompanied with rain, &c.

13*th and* 14*th*.—Very unpleasant fogs on the west coast of Newfoundland.

15*th*.—Strong winds and very heavy sea, tore the sails; we look all pitiful indeed; at 8 evening more moderate, and at midnight pleasant.

16*th*.—Cloudy; wind averse.

17*th*.—This morning was pleasant; towards midday a good breeze; afternoon foggy.

18*th and* 19*th*.—Pleasant in the morning; towards night squally. Water very bad at tea. Sun obscured.

20*th*.—Pleasant, making little way.

21*st*.—Strong breezes; midday, moderate wind from the north.

22*nd and* 23*rd*.—Foggy.

24*th*.—Light airs and pleasant weather; 45 fathoms water on George Bank; 12 o'clock night, 25 fathoms of water.

25*th*.—West end of George Bank; cleared the Nantucket Shoals with good prospect of making land soon; passed crowds of fishing-vessels. Sun obscured.

26*th*.—The Captain sent on board two fishermen and procured fresh mackerel, which was a very great delicacy.

27th.—At 10.30 got sight of Cape Cod ; a pleasing spectacle, distance six leagues. Lat. 41.38.

Monday, July 28th.—Strong breeze and heavy sea. Lat. 40.48.

29th.—Every person on board now became uneasy. The ship's crew were out of tobacco, and many of the passengers who 'found' for themselves were out of provisions ; nothing but passengers buying, bartering, &c., which made good sport ; sailors chewing their tobacco, drying it, and then smoking the same. Ten P.M., 12 fathoms of water ; foggy.

30th.—Fine light breeze. At 12 o'clock made Block Island ; I now felt myself nearer.

31st.—Long Island in sight ; I cannot express the satisfaction I feel ; shores sand and partly rock. Four vessels in company.

August 1st.—Fine wind ; at midday squally ; views of Long Island.

Saturday, 2nd.—The ship this morning was all in an uproar, in consequence of a horse, which one of the passengers had, being looked on as dying ; it cost him £200 in England, and after troubled passage the poor man lost his horse. At 12 o'clock saw light at Sandy Hook.

Sunday, August 3rd.—Four o'clock A.M. saw more of the New World. Every face seemed to feel glad, and at 7 A.M. took a pilot on board ; at 10 passed the floating light lately erected, the Captain of which came on board of the *Ann Maria* ; 4 o'clock passed the *Nourain waspe*, and the other forts on the right and left ; half-past 4 cast anchor and considered ourselves at land ; 5 o'clock boarded by the Health Officer, who signified that fourteen days of quarantine was requisite in consequence of small-pox ; at 6 o'clock went on shore on Staten ; returned to the vessel at 7.

Monday, August 4th.—This day presented nothing but bustle ; every person seemed anxious to get on shore. I was permitted to go on shore at midday for two hours, and returned to the vessel again.

August 5th.—On application to the Medical Officer at quarantine ground as to my going to the city, I was informed that I could not possibly go with any of my clothing which was on board the vessel, which induced me to purchase clothing and go up ; but had to return down in the evening. Nothing was yet permitted to leave the ship. Messrs. Kennedy & Maitland being out of town, I then called on Dr. Hosack, but unfortunately I did not see him either. I then made my way through the town again, and by steamboat got to Staten Island. I felt the heat very oppressive. Thermometer in the shade 96° Fahrenheit.

August 6th.—This morning can never be effaced ; it had rained a little during the night, which cooled the atmosphere and added a hue to Nature's work, which was truly grand—the fine orchards of Long Island on the one side, and the variety of soil and vegetation of Staten on the other. I once more thought myself happy. I went to the city in the afternoon to see what steps I should take as to the progression of my business ; finding that I could not get my luggage for two days to come, I began to feel uneasy after having such a long passage, and then to be perplexed. I had an interview with Dr. Hosack ; the cordial manner in which I was received by this gentleman made an impression upon me. I called on Dr. Torrey whom I found an intelligent botanist, an agreeable person, and much disposed to aid me. I landed again at Staten Island.

August 7th.—To-day I got my luggage on shore, and through the Custom-house. The after part of the day I devoted to a trip through the island; observed the oaks and maples growing spontaneously. *Lirioden-dron*, I observed, *Pyrola maculata*[1] and *P. umbellata*[2] and *Mitchella repens* growing one mile west of marine hospital. On the hill behind the church I observed two species of *Stellaria*, one of which was small, like *S. scapigera* of Britain. On this small hill I found asbestos in great abundance; three species of *Rubus* growing by the roadside; of *Eupatorium* I saw three species, one *E. perfoliatum* fine, on which was growing *Cuscuta americana*; *Vernonia, Verbena hastata*, all growing on dry gravelly soil; two species of *Smilax*, and one of a native vine, twisting themselves round the trees.

August 8th.—To-day was occupied in getting my boxes to town; the after part of the day I spent with Dr. Hosack.

9th.—I waited on Mr. Hogg, who expressed himself glad to see a person from London; on his knowing my object, he readily offered to aid me and in the meantime would, on the recommendation of Dr. Hosack, accompany me to all principal parts of the city, till I should get a little acquainted.

Sunday, August 10th.—Mr. Hogg accompanied me to some of his neighbours and religious friends; had a glimpse of their gardens, &c.

Monday, August 11th.—Early this morning I went to the vegetable market, the *Fulton*. It had a beautiful appearance, beet of superior variety and fine carrots, raised in this country; I observed a very great deficiency of cauliflower, indeed they were miserably poor; onions were fine, mostly red; the immense supply of melons and cucumbers—the latter of which, however, were not so fine as may be expected and appeared for the most part to be the same as the short prickly ones cultivated in England —the melons were fine. An abundant supply of early apples, pears, peaches—the two former were fine, but the peaches looked rather bad, being ripened immaturely and the trees being sickly; immense varieties of squashes or gourds, plums, early damsons, a great supply of pineapples from the West Indies, and cocoanuts. I observed a fine head of *Musa sapientum* which weighed 40 lb. At 8 o'clock this morning we set off for Flushing and visited the establishment of Mr. Prince. I found him a man of but moderate liberality; he has some good specimens of *Magnolia*, of *Berberis Aquifolium*, a few European plants, common shrubs and herb plants. Indeed on the whole I must confess to be somewhat disappointed, for his extensive catalogue and some talk had heightened my idea of it; but most of his ground is covered over with weeds. I was much pleased with the beautiful villas on the banks of the Sound; saw people employed in preparing their operations for diving to the *Hussar*, a British frigate taken during the late war.

August 12th and 13th.—Early this morning we crossed the River Hudson and visited some of the finest fruit-orchards of that place. I saw one par-

[1] *Chimaphila maculata*, A. Gray, Syn. Fl. N. Am. ii. I. p. 45.
[2] *Chimaphila umbellata*, A. Gray, *loc. cit.*, p. 45.

ticularly fine, belonging to an old Dutch farmer, consisting of twenty acres
(English). Of peaches he had about twenty-four varieties in fine health,
only four years old. The stones he sows in the month of April about four
feet between the rows and six or ten inches between the seeds in the rows ;
they generally make their appearance towards the middle of May, after
which he takes care to keep them clear of weeds, which he does by means of
a plough drawn by one horse. As soon as any lateral shoots appear, they
are pinched off, suffering the plant to run up, having only a top. About
the end of August the same season they are inoculated [budded], headed
off at the usual time the following spring, and frequently the second season
after budding they produce fruit. Plums are frequently raised from
seed. The whole of this day was occupied in the gardens, &c., and was
devoted chiefly to ornamental plants in the swamps: we observed
Sarracenia purpurea ; a species of *Chironia* and its varieties ; *Blitia tuberosa*
[sic] [1] fading ; two of *Rubus* ; in the swamps *Lycopodium dendroideum* and
L. clavatum, and two other species, one of a *Rose*, growing in the margins
of pools, appears to have had a great show of blossoms in form of a
corymb. *Catalpa* along the roadsides, in conjunction with *Salix babylonica*,
Populus balsamifera, *P. tremella*, *Liriodendron*, made an agreeable appear-
ance. In the pools *Nymphaea odorata*, *Nuphar advena*, and *N. luteum*,
Kalmianum (? ?) — is this the same as *N. minimum* of Britain ?—
Orontium aquaticum in abundance.

  Thursday, 14*th.*—This day I visited the Botanic Garden, which is now,
I am sorry to say, in ruins ; one of the hot-houses is taken down, one
stripped of the glass, and the greenhouse still in a sort of form. It has
been well chosen, in point of soil and situation, there being a great
variety of both ; some good trees of *Magnolia cordata* and *M. macrophylla* ;
all the herbarium plants are gone from the greenhouse, and have been given
to the lunatic asylum. In the afternoon I called on a Mr. Codie, who had
a good fruit orchard and garden ; he promised me some fruit in the
autumn and seeds of melon, in which he excels.

  Friday, 15*th.*—Through the medium of Dr. Hosack I learned of a fine
plum named ' *Washington*,' a name which every product in the United
States that is great or good is called. This gentleman kindly sent me
four which weighed and measured as follows : largest, $3\frac{3}{4}$ oz. avoir., cir-
cumference $7\frac{3}{8}$ in. ; second, $3\frac{1}{2}$ oz., circumference $7\frac{1}{2}$ in. ; third, $3\frac{1}{4}$ oz.,
circumference 7 in. ; fourth, 3 oz., circumference $6\frac{3}{4}$ in. Form of a
greengage ; colour somewhat between cream and sulphur ; flavour very
delicious, like greengage ; the stone remarkably small, the skin thin. Purchased by a Mrs. Miller about thirty years since out of the flower-market.
After standing in her garden for five years it was during a thunderstorm
cleft nearly to the bottom, which caused its death so far as was rent ; next
spring it sent up suckers and the great Wm. Bolmer, Esq., obtained one of
them, which he planted in his garden and in a few years produced fruit
without any grafting ; the fruit has improved every succeeding year,
the taste being the best. Soil, pure red sand ; the original was removed

---

[1] (?) *Bletia tuberculosa*, now *Phaius tuberculosus*, Blume, Mus. Bot. Lugd. Bat. ii. p. 182.

and three cartloads of good soil from a cultivated field taken in and the tree-plant given a little decayed vegetable matter as a manure. He lays its roots totally bare during the winter months. I put the above fruit in spirits.

Saturday, August 16th.—Wrote to England early in the morning, and then went to Staten Island and crossed the bay for Long Island. They attend little to any cabbage, broccoli, and beet—carrots are objects of more attention.

Sunday, August 17th.—After attending divine service in the forenoon, I dined with Dr. Hosack and spent the evening at his house.

Monday, 18th.—I was engaged rectifying my lists, &c.

Tuesday, August 19th.—In company with Mr. Hogg we went on board a steam packet to New Brunswick, thence by coach to Trenton by Kingston, Princeton, Lawrenceville. The whole country appeared well cultivated; of a rich light soil; for upwards of ten miles it appeared a garden. Slept at Trenton.

Wednesday, August 20th.—From Trenton to Burlington, went by steamer, distance fifteen miles. I made a point of calling on W. Coxe, Esq., who used me very hospitably. I saw his vast and extensive orchards. He gave me such fine descriptions, &c., it was quite a new place. His trees are all in fine health, and he was just on the eve of commencing his cider harvest; he has in his garden a choice selection of peaches, apples, pears, &c. He gave me to understand that he would send the Society a collection of keeping fruit in the fall, after his checking off what he considered to be new, and recommended me to deal with a Mr. Smith, with whom at one time he was in company. I took my departure for the inn. All his family were poorly, except one of his daughters.

Thursday, August 21st.—Four miles east of Burlington on the pine barrens were *Epigaea repens, Pyrola* two species, *Rubus cuneifolius,* two *Euphorbia marylandica* [sic], and two species more, *Asclepias tuberosa* and *A. syriaca,* in Mr. Coxe's field ; *Rhexia virginica, Monarda punctata, Sarracenia, Laurus Sassafras,*[1] &c. At 5 o'clock P.M. left Burlington by steamboat and arrived at Philadelphia 7 evening.

Friday, August 22nd.—After waiting on E. Collins, Esq., a friend of Dr. Hosack's, who is a botanist of distinction in that quarter, who recommended me to go to see the place of the venerable John Bartram, and Mr. Lisle of Woodlands, Henry Pratt, Esq., and the principal nurseries. I am truly obliged to this gentleman for his kind attention and his willingness to forward the views of the Society. I called on Mr. Will Dick, janitor of the university, who also used me kindly. The garden he has established partly on his own account ; offers to aid me by presenting *Euphorbia variegata*[2] from Arkansas ; *Donia ciliata,*[3] figured in Hooker, 'Exot. Fl.' i. (1823) t. 45, a most singular plant belonging to the natural order *Compositae*[4] ; root

---

[1] *Sassafras officinale,* Meissn. in DC. Prod. xv. i. p. 171.
  E. marginata, Boiss. in DC. Prod. xv. ii. p. 63.
[3] *Aplopappus ciliatus,* A. Gray, Syn. Fl. N. Am. i. II. p. 125.
[4] In MS., *Cucurbitaceae* instead of *Compositae,* an error probably arising from comparing the root of the *Donia* with that of *Bryonia dioica.*

large, resembling *Bryonia dioica* ; has a taste like quassia ; gives a liquor like porter by fermentation ; from the Rocky Mountains (Major Long's Expedition, 1822) ; *Aster graveolens*,[1] Arkansas ; *Ribes* sp. differs from *Ribes aureum* or any preceding American species ; *Verbena bipinnatifida*, from Arkansas ; *Fumaria* sp.—a species of vine from Long's Expedition, differs from any in cultivation ; *Salix* sp., beautiful, not in Britain ; *Helianthus specularius* [?]. I am also in hopes that I shall obtain some seeds in the fall. I made a journey to Mr. McMahon, which is three miles north of the city. I did not find him at home ; I looked round the garden, and after a patient search found *Maclura*, two plants, height about seventeen feet, bushy and rugged ; they had a few fruits on the trees ; it is well described in Pursh's Preface of his 'Flora Amer.' Then I called at Bartram's old place, but found no person at home.

Saturday, August 23rd.—In the morning I visited the vegetable market, &c. I found the supply finer than that of New York, and the produce so likewise. The peaches, apples, &c., look superior in every respect ; they have not that sickly appearance which is found among the fruit of New York. I made a visit to Landreth, near the city, who has got a great many fine plants : a rose which originated in the Southern States, called the Champneya rose ; also three fine good varieties ; a fine tree of *Magnolia cordata, auriculata*,[2] *macrophylla* ; a single plant of *Berberis Aquifolium*, rather sickly. In front of his house he had his greenhouse plants ; oranges were particularly fine indeed. I observed fine plant of *Lagerstroemia indica*, imported from China, in flower. I here saw a fine plant of *Maclura*, about twenty feet high, very rustic, leaves large, ovate, at the stalk of which is a large thorn, a few fine fruits on the tree ; stands perfectly well the winter in a poor, light, sandy soil ; the shoots of this year are in length five feet. He has a few stout plants propagated from cuttings last season, two of which I feel glad to have the pleasure of carrying to England from this place. Grapes thrive well here, running up poles or on trellises ; some of the native species are cultivated.

Sunday, August 24th.—To-day I was shown the principal places in and near town. *Centaurea americana* I saw at a small garden in town ; plants are cultivated in this city with a good deal of taste.

Monday and Tuesday, August 25th and 26th.—This morning at 4 o'clock I set off in company with Mr. Hogg to Chester, State of Delaware, fifteen miles from Philadelphia, said to be good for the neatness of its gardens, &c. From thence to Wilmington, Newcastle, and Newport. We saw nothing different from what we had already seen, only in fine order in point of management : *Rubus cuneifolius*, on the banks of the Creek Brandywine, among *Asclepias tuberosa* and *A. syriaca*, and a species of *Eupatorium*. On the morning of Tuesday we got to town again, called at Woodlands four miles from town, where I saw what might be said to be the finest American establishment. Mr. Lisle we did not see ; the whole place has the appearance of nicety. South of the navy-yard about three-quarters

[1] *A. oblongifolius*, Torr. and Gray, Fl. N. Am. ii. p. 144.
[2] *M. Fraseri*, S. Wats. Bibl. Ind. N. Am. Bot. p. 29.

of a mile, on the right-hand side of the road, grows the famous *Cyamus luteus* [1] in conjunction with *Sagittaria obtusa*,[2] *Nymphaea odorata*, and *Nuphar advena*.

Wednesday, August 27th.—We set off by steamboat at 11 o'clock from Philadelphia to Bordentown. Here stands the house of Joseph Bonaparte, a most splendid mansion, fields well cultivated, pleasure grounds laid out in the English style; there are many fine views. We then took stage and came to Amboy during the night, where we slept all night.

Thursday, August 28th.—Took steamboat at half-past 4 o'clock and passing up past Elizabethtown, Staten Island, landed at New York at half-past 10 P.M. As soon as possible I had the plants from the office where they had been left, took them to the son of Mr. Hogg, and had them planted and secured. I cannot but consider myself happy at meeting with Mr. Hogg; he carefully attends to the little treasures during my absence.

Friday, August 29th.—This morning I put the Osage apple in spirits; afterwards I waited on Mr. Floy for the purpose of selecting specimen trees from his grounds, &c.

Saturday, 30th.—I made in the morning a visit to the market and during the forenoon went round to a Mr. Wilson, a market gardener, where I saw good vegetables, particularly celery, and a good stock of trees; he is building a large, elegant greenhouse, &c.

31st.—I went to Flushing this morning; talked to Mr. Prince of the plants. I found only the son at home, who is a great pedant; I returned about 2 P.M. The remaining part of the day I was employed with my lists and catalogues.

Monday, September 1st.—In the morning, wrote to Mr. Sabine and sent my despatch off at 10 A.M. I prepared to leave town early by steam-boat; owing to a change of boats I was prevented for the present. I set my dry plants to rights in the afternoon, and attended in the evening a committee meeting of the New York Horticultural Society, for the purpose of offering their assistance to me during my residence in this city. M. Hoffman, Esq., the President, is a man of reputation, being a wealthy merchant here. He uses me with all possible attention imaginable; invited me to stay at his house all night, which I did. They will assist me materially in the way of my selection.

Tuesday, September 2nd.—I visited, at the advice of Dr. Hosack, a friend of his, seven miles from town who is a fruit grower. I obtained a curious sort of bean of South American origin. I saw twenty trees of *Seckel* pears, loaded to the ground; fruit is this season smaller than usual. I rode in the afternoon with Dr. Hosack to some of his friends, fourteen miles from town.

Wednesday, September 3rd.—The greatest part of this day I spent with a Mr. Shaw, a private gentleman of this city, who is a fruit grower of more than ordinary merit. His garden is about two acres English, occupied solely with peaches and grapes. His peaches are in excellent state of

[1] *Nelumbium luteum*, S. Wats. Bibl. Ind. N: Am. Bot. p. 37.
[2] *S. sagittifolia*, Micheli, in DC. Monog. Phan. iii. p. 67.

bearing, only four years old ; he has a plant of Isabella grape, trained up over the veranda of his house, covering a space of about 75 feet in length, and has a very weighty crop, thought to prove a valuable thing in wine or probably dessert.

Thursday, September 4th.—At 5 o'clock this morning I went on board the steamboat *James Kent* and proceeded up the River Hudson towards Albany. The scenery was particularly fine on the west side : the perpendicular rocks covered with wood gave it an appearance seldom to be met with. About forty miles from New York, in the highlands, many pleasant villas are seen from the river. West Point is still pointed out to strangers, being the place where the unfortunate, but good, Major André paid the debt of nature in his country's defence. The approach to the Fort and Military School on the north has a beautiful effect. At half-past 1 o'clock I went ashore at Governor Lewis', eighty miles from New York. Calling at his house I found that he and his family were on a visit to the western part of the State. I then made towards an inn, to obtain refreshment, which was three miles distant. In the afternoon I took a walk by the side of the Hudson. I found nothing interesting except a rose which was growing out of a crevice of rock ; no flowers on it, forty capsules in a corymb.

Friday, September 5th.—At 5 o'clock this morning I was aroused from bed by a James Thomson, Jun., of Elerslie, who is neighbour to Governor Lewis. He very kindly took me from the tavern and afforded me his house during my stay there. I received great attention from Mrs. Thomson, which was thankfully acknowledged. Mr. Thomson, Sen., I did not see, he being engaged in the Supreme Court at New York. I saw a fine collection of vegetables, fruits, &c., in their garden. Peaches in particular in great vigour and health, neither pruned nor get any manure ; apples, the usual choice ones peculiar to America ; pears, &c. In his woods I found on a point a quarter of a mile west of his house *Gerardia flava* in great perfection, growing in dry gravelly soil, partially shadowing another species of *Gerardia*, small and starved. On the outskirts of the wood, *Eupatorium* four or five species, *Inula*, *Solidago*, and *Aster*. The whole of this place seems in cultivation, &c., like England. Four large oaks of different species, acorns of which I am to get on my return. I stayed here all day ; went to bed early, as I intended to take the steamboat for Albany, which passed early the following morning.

Saturday, September 6th.—I embarked in the *Richmond* steamboat, on her way to Albany, at 1 o'clock A.M. Very little difference in point of scenery, &c., from that which I had seen two days before. At 11 o'clock I arrived at Albany. I waited on His Excellency Governor Clinton, who showed me attention, desiring me to call the following day at 12 o'clock and he would consult with me as to my route. I left with him my instructions and withdrew. I went to the vegetable market in the afternoon, where there was an abundant supply of fruit, &c. A small native plum is brought to market from the opposite banks of the river, where it grows plentifully. I here observed the cabbages and beans were superior to those I had already seen.

Sunday, September 7th.—This morning at 5 I was awakened with tremendous peals of thunder. On my looking out of the window the streets were quite inundated; the town standing on a gentle declivity, the water rushed with great rapidity. I waited at 12 on His Excellency the Governor, who said that his opinion was for me to proceed to Canada without delay, the season being far advanced, and particularly as the steamboat *Superior* was to sail from Buffalo on Saturday next. After giving me letters of introduction to all the places of science or influence on my line of journey, accompanied with a small guide and verbal instructions, I prepared to leave Albany, calculating to see it perfectly on my return. I wrote to Jos. Sabine, Esq. Left Albany at 4 o'clock for Schenectady, where I arrived at 9 o'clock of the same evening; the rain fell in torrents all the way.

September 8th.—At 3 this morning I pursued my journey in the stage towards Utica. This morning was cool, the rich verdure of Nature, the lofty mountains on the right hand, the fertile fields, and the Mohawk gliding down on the left, gave to the country an appearance, fine beyond description. All the farmers here have orchards cultivated: seldom more than ten or twelve varieties of apples; pears are scarce. They make their own cider. In every village or cottage I passed stood a cider-mill, casks, and men busily employed preparing for their cider harvest. Two or three varieties of plums are in cultivation. At 2 o'clock I came to Little Falls, 70 miles from Albany. The bad road and the jolting reduced me so much that I was obliged to give up that mode of travelling. The line of coaches and canal-boats being in company, I took the canal-boat at this place. Here is a beautiful, elegant bridge across the Mohawk, consisting of three arches, built of granite, serving as an aqueduct and a bridge, built in the space of two months and two days, dedicated to De Witt Clinton, &c., Commissioner of Canada. I arrived at Utica about 9 o'clock at night.

Tuesday, September 9th.—At 8 o'clock this morning I left Utica by canal-boat for Rochester; sixty miles from here without a lock.* The fields on its banks are rich and fertile, and in general well cultivated. Passed Rome at 12. On the right, about one mile further on, two fine seats on an eminence. Passed some large swamps of fir, walnut, ash, oak, elm. I here, to my astonishment, found a *Magnolia* ten feet high, leaves large, smooth, ovate, and acute. I saw no more but itself; no seed on it; had no appearance of having flowered. The boats are fitted up on good principles: accommodation for twenty-four ladies in one cabin and as many in the men's. I slept on board, but was much disturbed passing the locks,* &c. To-day presented nothing different from yesterday. I arrived at Rochester at 6 o'clock on the 11th.

Thursday, September 11th.—After breakfast at 8 o'clock I took the mail to Avon, twenty-four miles south, with the intention of getting the coach on the western road to Buffalo. In this movement I was disappointed, and had to stay all night. In the afternoon I went to see the celebrated sulphur-springs known by the name of *Avon Springs*; seem to partake of magnesium. Many people are here for the benefit of their health.

Friday, September 12th.—Left early in the morning by stage from

* *Sic* in MS.—ED.

Buffalo. Breakfasted at Caledonia, a settlement of Scotch people. Country generally flat; by the roadsides stand settlements in infancy, just clearing the ground by burning. Batavia a neat little village. The country here presents a fine appearance, the fields are rich and well cultivated. Left here at 4 o'clock, passed through for the most part wood and marsh, a wild and desolate place. Buffalo, 12 at night.

Saturday, September 13th.—Early in the morning I wrote to Jos. Sabine, Esq., and then called on Oliver Forward, Esq., a gentleman of considerable wealth and friend of Governor Clinton. I took a walk round the town and returned to breakfast. Went on board the steamboat at 9 o'clock A.M., and sailed at 10, and after a pleasant passage of sixty hours landed at Amherstburg. As soon as I got my trunk on shore I waited on Mr. Briscoe, from whom I received great kindness; he readily pledged his exertions for the furtherance of the Society's objects. I felt sorry to learn the loss of his birds. I spent the evening with him, and should to-morrow prove a good day we intend to make an excursion in the woods.

Amherstburg, Tuesday, September 16th.—This morning at daylight, I called on Mr. Briscoe, who on my making my appearance said he had waited a long time for me. We took breakfast at 6 A.M. and set out north-east of Amherstburg. He took his dogs and gun and proved himself, before we had been out two hours, a marksman of the first sort. This is what I might term my first day in America. The trees in the woods were of astonishing magnitude. The soil, in general, over which we passed was a very rich black earth, and seemed to be formed of decomposed vegetables. On the south-east of the town, five miles, near the margin of Lake Erie, the soil is of a reddish cast and produces fine crops of Indian corn; and now for the last four years, I learn they have cultivated tobacco with great success for the Montreal market, and according as the general opinion goes it will form an article of great importance to our Canadas and at no distant period. The woods consisted of *Quercus* (several species, some of immense magnitude); *Juglans cathartica*,[1] *J. nigra* (immensely large), *J. porcina*[2]; in dry places *Fagus*, and on its roots a species of *Orobanche*; also a species under oaks very different from the former one. Four miles east of Amherstburg I observed a species of rose of strong growth, the wood resembling *multiflora*, having strong thorns. All the tender shoots and leaves were eaten off by cattle or sheep, which prevented me from knowing it.[3] I gathered seeds of some species of *Liatris*, which, along with *Helianthus, Solidago, Aster, Eupatorium,* and *Vernonia* form the majority of which I had an opportunity of seeing in perfection. In a field east of Amherstburg grew spontaneously *Gentiana Saponaria* (?) and *crinita,* and I secured seeds. Towards mid-afternoon the rain fell in torrents, urging us to leave the woods drenched in wet.

Wednesday, September 17th.—This morning I made a visit to a small island in the River Detroit, opposite to Amherstburg. It is about one mile long and three-quarters at its greatest breadth; it appeared to be a spot worthy of notice, as I found before evening. The whole island is low, and

---

[1] *J. cinerea,* C. DC. in DC. Prod. xvi. ii. p. 138.
[2] *Carya porcina,* C. DC. *loc. cit.,* p. 144.   [3] Cf. Oct. 22nd.

I was told frequently overflowed by the water from the upper lakes. The soil is very rich black loam, covered with trees of large size. *Quercus* sp. 44 Herb., trees from 50 to 70 feet high, 40 feet without branches. With a shot from my gun I cut some branches, leaves and acorns. They seem to me to be fine and different from any which I had seen before. With a few shots more I secured specimens and paper of seeds. I did not observe any trees but of great magnitude, rough bark, leaves deeply lobed, acorns large and almost enveloped in the cup. *Quercus* sp. 45 Herb. (soil the same as the preceding) tree is also large, but not so handsome as *Quercus* sp. 46 Herb. On the south end of the island the trees are much smaller than the others; it probably is not a distinct species, as it was growing on sand on the margin of the river. How glad I am to see my rose of yesterday growing in perfection!—very strong growth, leaves large serrated and strong veined, no flowers, seed-vessels in clusters; the soil was very rich, mostly all decayed leaves. I lost no time in securing plants and seeds. *Crataegus* sp. 12 Herb., a beautiful large shrub, or more properly a tree, fully 30 feet high, having a large round top, growing in every part of the island. *Crataegus* sp. 13 Herb., a tree of very large size, leaves large and slightly serrated, fruit uncommonly large, almost like crab-apples. This genus seems to attain a much greater size in Canada than in the United States; this last species grew in abundance and appeared to be constant in its characteristics. *Crataegus* sp. 14 Herb., a small tree and very different from the former, the fruit small but very numerous; they must have a fine effect in the early part of summer. *Allium* sp. 1, growing under the trees in rich soil, roots small but very acrid, stem 14 inches high, the leaves were dead. In the middle of the island, in rich soil, I found *Lonicera* sp. 57 Herb. : the leaves were almost yellow, very large, woolly underneath, the wood strong; the birds had devoured all the berries, which prevented me from having it in my power to carry it in that state. I then secured plants of it; all plants grew luxuriantly, but by no means common. I hope it is *Caprifolium pubescens* [1] of Hooker; I feel anxious to see them in a garden in London. *Hordeum* sp. on the margin of the river, *Lobelia inflata* in open places where it was a little wet. I crossed the river and secured the plants of *Rosa* and *Lonicera*, and put away my specimens and seeds. In the evening I called on Mr. Briscoe for the purpose of soliciting his advice as to the most likely places of affording a harvest. He strongly recommended a visit to the River Thames, as also did his friend Rob. Richardson, M.D., physician for the Indian department. (Dr. R. knew Captain Sabine when he was at Stamford on the Niagara.) He kindly offered to take me in his car as far as Sandwich, which is fifteen miles from Amherstburg. As there was no time to lose, I proposed to start in the morning. These proposals being agreed to, we parted at 10 o'clock P.M.

Thursday, September 18th.—I set out in the morning in company with Dr. Richardson for Sandwich, on the east bank of the River Detroit. For three miles up from Amherstburg the ground is thickly covered with wood

---

[1] The authority for *Caprifolium pubescens* is Goldie, and it is so figured by Hooker, Exotic Flora, vol. i. t. 27; but is referred to *Lonicera hirsuta*, by A. Gray, Syn. Fl. N. Am. i. II. p. 17.

consisting of *Quercus, Acer, Juglans, Fagus, Fraxinus*; there are no pines here; the underwood is *Crataegus, Pyrus, Rhus*; also on the side of the river a species of vine, the fruit small, scarcely ripe—in fact the birds eat them as soon as ripe—there were some leaves large, the shoots small. I took seeds of it. Passed a long swamp intersected by natural ditches in which grew *Nuphar advena, N. Kalmianum, Eriocaulon* (a small species and much like our British one on Loch Sligachan, Skye). *Sagittaria sagittifolia* grew spontaneously and seemed to contend with the *Nymphaea odorata* for a prior right of habitation. On passing the swamp just mentioned, which is fully four miles long, we came to what is called ' The French Settlement.' The fields are well cultivated and divided by fences; attached to each house is a neat garden laid out and kept with taste. The cultivated apples comprise about eight or ten varieties; they are known by black, red, white, &c. There are a few pears, which are scarcely seen in the western part of the State of New York; probably the emigrants took them from France at the first settling of the country. They have a few peaches, in appearance the same as in the States. The climate seems to be particularly favourable for them. I am informed they ripen in ordinary seasons; they have not that sickly appearance which they have in the States, occasioned probably by excess of heat. Four miles from Sandwich the ground rises of a brown loam on gravel. I saw some fine species of *Solidago* in flower, but no seed ripe. On my arrival at Sandwich I found many friends of Mr. T. Mason.

Friday, September 19th.—I made a visit to a wood four miles northeast of the town. *Quercus* sp. 41 Herb., 50 to 60 feet high in sandy peat soil, the leaves lobed, fruit small, cup covered with scales. *Quercus* sp. 40, 30 feet high, poor soil; this is near one which I saw on the island at Amherstburg. *Crataegus* sp. 11, on gravelly soil on the side of wood, a small tree. In a small marsh grew *Chelone glabra*, var. *alba*, a fine plant; it had not seeds, I therefore took plants. In the same place *Coreopsis* sp. 1 Herb., flowers a fine bright yellow, growing in water; a Manong, a species of *Bidens*, which had no seeds. On my way home, two miles from Sandwich, among underwood grew a *Phlox* 18 inches high, leaves linear, opposite, and having flower large in proportion to the plant; soil, sandy peat or nearly all sand. I was much pleased with this, it had no seeds; I of course took plants. On the same place *Gerardia* sp. looked like *quercifolia*, plenty of seeds but no flower. I wish they may grow. *Neottia* or *Satyrium* in peat. *Cypripedium* in sandy peat soil among bushes. I here observed a crab-apple larger than the common crab, flat and well tasted. The trees are small, leaves large, spurs or thorns very large. I here got three different species of *Liatris*.

Saturday, September 20th.—Early in the morning got a car and hired a man to conduct me up the country. I set out slowly, moving along the riverside, picking up anything which presented itself. After passing through a country of twelve miles from Sandwich, well cultivated, came then to a morass of about two miles long, succeeded by a passage of sand along Lake St. Clair; rendered the horse so weak that I had to stop for a day. I was glad to do so, as there seemed to be a good field; accordingly towards midday I set off with the man I had taken with me. Here I found

Crataegus sp. 1 Herb., in rich soil close by the side of the lake, a low tree but spreading; sp. 2 growing in the same place; *Crataegus* sp. 3, a very fine tree, fruit large and of a bright scarlet colour, leaves serrated; *Euonymus* sp. 1 Herb., a small tree in rich soil, fruit yellow, warted, different variety from common; a wild plum of a brown and yellow colour, tartish tasted, with red specks, on the margins of the lake very abundant; *Fraxinus* sp. 1, in rich soil, from 40 to 60 feet high, not of great thickness; *Quercus* sp. 42 Herb., soil black, rich, so like one on the island of Amherstburg but not so deeply lobed, fruit very large, mossy cup and fringed, from 40 to 60 feet in height, 2 feet to 2½ thick. *Quercus* sp. 43: during my day's labour I had the misfortune to meet with a circumstance which I must record as it concerns not only my business, but also my personal affairs. I got up in the oak 43 for the purpose of procuring seeds and specimens; the day being warm, I was induced to take off my coat and in that state I ascended. I had not been above five minutes up, when to my surprise the man whom I hired as guide and assistant took up my coat and made off as fast as he could run with it. I descended almost headlong and followed, but before I could make near him he escaped in the wood. I had in my pockets my notes and some receipt of money, nineteen dollars in paper, a copy of Persoon's ' Synopsis Plantarum,' with my small vasculum. I was thus left five miles from where I had left the car, in a miserable condition, and as there was no remedy that could be taken to better myself, I tied my seeds in my neckcloth and made to my lodging. I had to hire a man to take me back to Sandwich as I could not drive; and the horse only understanding the French language, and I could not talk to him in his tongue, placed me in an awkward situation. I had to borrow a coat as there was no tailor to make me one. On my getting to Sandwich I remonstrated with the man who recommended my assistant to me, but he said that he never did so to his knowledge, and so on. However I found my guide was a runaway Virginian.

Sunday, September 21st.—I returned to Sandwich as I have said; stayed the remainder of the day, and on Monday proceeded again to Amherstburg.

Tuesday, September 23rd.—I made an excursion across the river to the Michigan Territory, at which place I found several species of *Liatris*, a *Smilax* of a curious appearance, a species of *Elymus* very strong in the marshes, *Lobelia inflata* and *syphilitica* in wet places; in a dry part of the wood among dead leaves *Botrychium, Arum triphyllum,*[1] *Pothos foetidus.*[2] These plants struck me as singular; they, without fail, in most cases, frequent moist places. I have not seen *Sarracenia* in the Upper Province. *Trillium* is also scarce. I have in the neighbourhood found a rose cultivated which agrees with *pendulina* of Pursh, and which he suspects to be of European origin, which is probable. I am informed that it was taken by the French to Canada at the first settling of the country. Mr. Briscoe to-day received orders to remove to Kingston by the first steamboat, which news sealed my disappointments: first a long passage, the loss of

[1] *Arisaema triphyllum,* Britton and Brown, Ill. Fl. N. U. St. i. p. 361.
[2] *Symplocarpus foetidus,* Engler, in DC. Monog. Phan. ii. p. 212.

my coat and money, bad weather—all these combined made me glad to relinquish the idea of Canada at such a late period of the season. It certainly is a fine field and would afford an abundant harvest.

Wednesday, September 24th.—To-day it rained till 4 o'clock P.M., at which time I made a few steps down the river-side towards the lake. *Quercus* 44, very tall and thick, and growing in sand on the beach on the shore of Lake Erie. I went back through the fields and secured two other species—one in wet poor soil, a small tree; the other was in light brown loam and larger. Several *Liatris* on fine marl (2 Herb.), tall, colour bright red, almost scarlet; a specimen of *Lilium canadense* in sandy peat soil. I here gathered a promiscuous group of several things but my time is so short that I cannot insert them.

Thursday, September 25th.—I packed my gleanings of plants and seeds and specimens, and stood in readiness for the steamboat from Detroit, which came in the afternoon. In company with Mr. Briscoe and family during the whole of the passage; we experienced a motion that could not be surpassed in an ocean. It blew a tempest the whole of the time. Towards midnight on Sunday (28th) we reached Buffalo, one of the wheels of the boat being swept away, and otherwise disabled.

Buffalo, Monday, September 29th.—In the morning I wrote to Jos. Sabine. At 10 o'clock I set out in company with Mr. Briscoe for Niagara. We went on the American side for two miles and crossed to the Canadian side. The weather was very cold and there was a snow-shower which lasted for three-quarters of an hour. We had a good deal of difficulty in gaining the Canadian side; the wind blew from the west. Along the side of the River Niagara the soil in general is rich, of black and brown loam. We travelled slowly, stopping frequently; half-way from Buffalo to the Falls we passed a wood, principally *Ulmus*, which in many places is scarce. Along the margin of the river grows *Crataegus* such as I saw at Amherstburg and Sandwich. One species of *Euonymus*. *Hamamelis* in blossom; it had a fine appearance, being destitute of leaves and with abundance of fruit. Oliver Forward, Esq., of Buffalo, advised me to make a call on a Mr. Clark who lives six miles from the Falls, having a taste for fruits, &c. He has two large orchards of apples, consisting of about twelve or fourteen varieties; he had also a crop of Indian corn in the orchards. He obtained his fruits from New York. He had only two different pears, and they were both bad : one was large, which I think is called pound pear; the other was smaller, and coarse also. He has also a good many peaches planted out as standards; the trees were in good health and had not been pruned. Three varieties of plums—*Magnum bonum* (or Egg Plum, as he called it), *Blue Orleans*, and *Washington*. He has lately got from Europe some vines, *Black Prince* and *Hamburgh*, &c. At 5 o'clock in the afternoon we reached the Falls.

Tuesday, September 30th.—This morning before daylight I was up and at the Falls. I am, like most who have seen them, sensitively impressed with their grandeur. Out of the cliffs of the rock grow Red Cedar, *Juglans amara*,[1] and *Quercus*. On the channel of the river I picked up an *Astra-*

---

[1] *Carya amara*, C. DC. in DC. Prod. xvi. ii. p. 144.

galus and beside it a *Viola*, both in seed ; the *Viola* grew in sand and its seed-pods were buried among it. I crossed below the Falls to the American side, and then to the island called Goat Island. It is partly covered with woods of large dimensions ; the soil is variable, part rich and part sand and gravel. The sugar maple, *Acer saccharinum*, on the brink of the rocks grew very large ; they had all been tapped or bled and still seemed uncommonly vigorous. There were a few pines of two species, but had no cones. *Botrychium*, two species in shady parts of the wood in decayed leaves ; two species of *Orobanche*, in dry places also among leaves. *Trillium* seemed to be plentiful, but the leaves being decayed, I could not get as many as I would like. *Arum triphyllum*,[1] *Dracontium* sp., and *Pothos foetidus* [2] : I was not a little surprised to see *Pothos* in a dry place ; they had perfected seeds. *Rhus Vernix* in conjunction with some species of *Smilax*, and another species of *Rhus* clad the trunks of the large trees. On the south side of the island there is very good limestone and a good kind of gypsum.

Wednesday, October 1st.—I went across the river again four miles down from the Falls, where is great diversity of soil. Opposite what is called the Whirlpool grew three species of *Quercus* on barren rocks, with narrow serrated leaves, acorns small and olive-shaped ; they are certainly different from any in my possession. Among the roots of the trees in crevices of rocks grew *Pteris atropurpurea*,[3] *Asplenium*, and *Polypodium*—of all of which I took plants and seeds, and specimens of the oaks. On my returning to the inn I found that Mr. Briscoe was to leave in the afternoon. The weather being bad, and the seeming approach of winter, made me anxious to get to New York as soon as possible. We accordingly set out and passed Stamford, where Mr. Briscoe informed me Captain Sabine lived. The face of the country is variable, with some beautiful rising eminences. At 4 o'clock we reached Queenston, where I parted from Mr. Briscoe. I am under great obligations to this gentleman. Crossed the river at Queenston for Lewiston on my way to Lockport. I got a box, packed my plants, and then went to bed.

Thursday, October 2nd.—At 4 o'clock I started for Bucks, where I arrived at 7 o'clock. I left the stage and went in the country for four miles to Lockport, for the purpose of calling on David Thomas, Esq., chief engineer for the western district of the Erie Canal. Governor Clinton spoke of him in the highest terms of respect. My meeting with him showed him to be a gentleman of great attainments. Mr. Thomas is, in America, looked on as a mineralogist of the first standing. He possesses a knowledge of birds and botany. In the afternoon we walked along the canal and saw the operation of forming and building.

Lockport, Friday, 3rd.—In company with Mr. Thomas, who kindly offered me the '*use of his person*,' as he said, I visited woods north of Lockport which were almost all beech ; we found two specimens of *Corallorrhiza*—one like *innata* of Britain, the other a small one. I secured plants of them ;

---

[1] *Arisaema triphyllum*, Britton and Brown, Ill. Fl. N. U. St. i. p. 361.
[2] *Symplocarpus foetidus*, Engler, in DC. Monog. Phan. ii. p. 212.
[3] *Pellaea atropurpurea*, Christensen, Ind. Fil. p. 478.

the soil was dead leaves, very dry. Along with them grew *Triphora*,[1] also a *Cypripedium* in partly shaded parts, soil dry sandy peat ; *Viola* sp. leaves round, large, flowers small, white, striated. *Mitchella repens*, *Pyrola maculata* [2] and *P. umbellata* [3] formed a carpet on the ground. In a swamp east of Lockport *Cypripedium* on high tufts of decayed grass, and in the same place *Orchis* sp., tall, in seed. On a rising bank having a dry soil, consisting mostly of decayed leaves, with a southern aspect, two species of *Hydrophyllum, virginicum* and *canadense* ; I know it is called a salad, but on the loss of my coat at Lake St. Clair I lost my note of it ; I have tasted it cooked, it is good when dressed as spinach. *Orchis orbiculare* [4] (sic) in a dry part of the wood shaded on decayed wood ; *Satyrium*, two species, small ; *Neottia pubescens* [5] in great plenty, on detached pieces of rock near the canal ; this likes a dry, light soil. On our way home on Friday observed *Phlox* sp. (?) on the outskirts of the woods, small, hardly four inches high, had no flower, leaves small, round, or nearly so ; I thought it might be *ovata*, still it looks very different from that species. On a decayed root of pine an Orchid, very curious, two narrow leaves, ovate, and acute at the point, of a reddish colour and black warted ; this is certainly not a plentiful plant. On my taking it up, Mr. Thomas observed that he had only seen it once before. At 2 P.M. I got my things packed and prepared to return to Bucks to catch the stage. Left Lockport at 4 o'clock and got to Bucks at 6, where I remained for the night.

Saturday, 4th.—To-day I went by stage to Rochester, where the line of canal-boats starts from. Fifteen miles of this route is through a swamp and trees laid crossways, which rendered it very unpleasant ; a portion of the country is clear of wood and well cultivated.

Sunday, October 5th.—In the morning visited the Falls of the Genessee River, one mile from Rochester, and started at 9 o'clock for Utica by canal-boat. The forests are now seen to advantage, all the tints imaginable. At 9 o'clock in the morning of the 7th I got to Utica; I called on Alexander Coventry, M.D., a friend of Governor Clinton's, but found that he was away from home and would not return for some time. Left Utica at 11 A.M. by canal-boat for Little Falls, where I took the stage for Albany. Slept at Palatine and started again for Albany next morning at 4 o'clock; from cold I was seized with rheumatism in my knees, which alarmed me a little. At 2 o'clock got to Albany. This day is the celebration of the opening of the Western Canal. The town was all in an uproar—firing of guns, music, &c. Governor Clinton's situation prevented me from seeing him for the day. I had considerable difficulty in getting lodgings in the inn. I unpacked my seeds, arranged, and put in fresh papers.

Thursday, October 9th.—At 6 o'clock in the morning I had the pleasure of meeting Dr. Hosack, who had come to Albany to participate in

---

[1] *Pogonia*, Benth and Hook. f. Gen. Pl. iii. p. 616.

[2] *Chimaphila maculata*, A. Gray, Syn. Fl. N. Am. ii. I. p. 45.

[3] *Chimaphila umbellata*, A. Gray, loc. cit., p. 45.

[4] (?) *Orchis orbiculata = Habenaria orbiculata*, Britton and Brown, Ill. Fl. N. Un. St. i. p. 46.

[5] *Goodyera pubescens*, Ind. Kew. fasc. iii. p. 304.

yesterday's proceedings. We breakfasted and then called on the Governor, who, with his usual way, received us kindly. His duties occupied his time ; he suggested to me to devote the day in the neighbourhood of the city. 1 left them, having an invitation to call and spend the evening. Being informed when at Lockport, by David Thomas, Esq., from whom I had much valuable information, that *Pterospora andromedea* was to be found near Albany, and, if he was rightly informed, south of the town— information which I hailed with pleasure—accordingly I set out visiting every place which was likely for it. After a search of seven hours I had the fortune to find it in a small ravine two miles south of Albany. On the right hand, going up the channel of a small rivulet for half a mile, where it branched to the right and left, on the angle, stood a large tree of *Pinus alba*,[1] and under its branches *Pterospora*, 10 feet above the level of the rivulet ; soil, light blackish-brown loam and so dry that every other vegetation refused to grow, it looked as if no rain had fallen for the summer. Whether it is annual or perennial I am unable to say. I am of opinion Mr. Nuttall's description is very correct, from 14 to 24 inches high. Being late in the season, it was of course out of flower : a rusty stem covered with a glutinous substance ; I counted ninety-seven capsules on one stem. How glad Mr. Munroe and Mr. Lindley would be to see it. The root is much like *Corallorrhiza innata*, only smaller, and it seems to be a sort of parasite like *Monotropa* or *Orobanche*. .On going a few steps to the left hand on the north aspect of the valley, I found it in a situation entirely different from the former, excluded from the sun, soil stiff, wet, and covered with *Hypericum* and *Jungermannia*. The plants here were stronger than in the former place. I have no doubt but it will cultivate. A gentleman (Mr. Tracy), Governor Clinton has just informed me of, is very fond of botany and from him I hope to have information as to it. I propose in the morning to sow seeds of it.

Albany, Friday, October 10th.—I waited on Mr. Tracy, who Governor Clinton said could and would feel glad to aid me. In the first place, he invited me to look at his herbarium, which was extensive and in a good state of preservation, arranged according to the Linnean system. In it he had *Pterospora*. He informed me that a friend of his who was fond of plants but possessed no knowledge of botany, in his rambles through the adjoining woods found it three years since and placed plants in his garden in June. They continued to thrive throughout the season. The following year they made their appearance and flowered all the summer. This season it sprung again and equally vigorous, flowered, and ripened seed. This information made me stare. I set out again for plants as they affirm it to be perennial. Every plant which I could see I took up, but found only one which had the least appearance of being perennial ; however this had not flowered, therefore it is not perennial but either biennial or annual. On my calling on Mr. Tracy again, he said that he would take me to see his friend's plants ; accordingly, on our examining the spot, we found the roots dead. It was[2] the last year in different places of the garden. The soil was light loam, at the foot of a wall, north aspect,

---

[1] *Picea alba*, Mast. in Journ. R. Hort. Soc. xiv. p. 221.   [2] MS. obscure.—ED.

Did it rise from seed ? (Yes. D. D.)—Mr. Tracy informed me that
the *Erythraea* which was found on the Missouri by Mr. Nuttall had
been lately found 12 miles from Albany shortly since. We set out for
it, but on our arriving at the spot we had the mortification to look
at the spot but unable to get to it. It was on a rising spot in the middle
of an extensive swamp which was completely inundated by the late rains.
Mr. Tracy expressed his extreme regret, but pledged himself to transmit
plants to Mr. Hogg at New York for the Society next year on its showing
itself. On our way home we observed in a wood near the city, in light red
soil, a species of *Corallorrhiza* different from those at Lockport, *Pyrola
maculata*,[1] *P. umbellata*[2] and *P. secunda*, and a small species of annual
Polygala. On the dry places grew *Epigaea repens*. *Quercus* sp. 18, a
stately tree almost everywhere, foliage large and entire, fruit small and
yellow. The trees were covered with pigeons pecking the fruit ; on rocky
soil. Called on Stephen van Ransaleer,[3] Esq., who is the most wealthy man
in the United States. He has a large garden and orchards, and a fine range
of hothouses, chiefly filled with vines. The grapes were all cut and hung
on strings in a fruit-room. Mr. van Ransaleer, being of Dutch extraction,
has many friends on the continent of Europe, who furnish him with different
kinds of fruits ; there were *Black Prince*, *Hamburgh*, *White Sweet Water*,
Grizzly Frontignan, and *Malmsey*. No attention is paid to the native vines
of North America. His apples and pears are much the same as at New
York and Philadelphia ; plums thrive much better. He has a large space
of ground occupied as pleasure or flower garden, which is a novelty in
America, as little attention is paid to anything but what brings money or
luxury for the table. His flower garden is kept in good order, under the
direction and management of his daughters, with much taste. Roses from
France, herbaceous plants from Germany, grace the plots, with annuals,
&c., from London. I had a letter of introduction to him from Governor
Clinton. Mr. van Ransaleer is a man of taste. He used me with kindness
and invited me to breakfast should I make it convenient.

Albany, Saturday, October 11th.—Early in the morning I called on
Jesse Bull, Esq., a friend of Governor Clinton's, and formerly printer for
the State of New York. He has now retired from a pressing business with
an ample fortune, to his pleasing pursuit of farming and gardening, of which
he is very fond and shows skill in both. His garden is yet in infancy, but
laid out with taste and utility. His farm is large and all divided by
hedges of *Crataegus Oxyacantha* from Britain. Hedging is a thing unknown
in a general sense. Mr. Bull offered to send me some things to New York
as my time was limited. He was kind and very affable for the short
space of four hours' acquaintance. On my way to Albany I called on Paul
Clark and saw *Columba migratoria* in a domesticated state. I waited
on Governor Clinton, who was to see me at New York after furnishing
me with a letter of introduction to James Thomson, Esq., of Elerslie, at
whose house I was already used very politely, unknown to Mr. Clinton.
After making acknowledgment for the very great interest which he had

---

[1] *Chimaphila maculata*, A. Gray, Syn. Fl. N. Am. ii. i. p. 45.
[2] *Chimaphila umbellata*, A. Gray, *loc. cit.*, p. 45.   [3] (?) Rensselaer.

taken for the furtherance of the Society's views, I went on board the steamboat and at 10 at night arrived at the house of General Morgan Lewis.

Sunday, October 12th.—General Lewis in his country stands very justly, high; his house is open and frequented by all denominations of people, and particularly by foreigners. We went to church to the forenoon service and returned at midday.

October 13th, 14th, 15th, 16th.—When at Albany I was seized with rheumatism in my knees, which almost reduced me from being able to do anything. Here I became lame for two days. I can never forget the attention paid to me by General Lewis and family. On Wednesday and Thursday I was able to crawl about a little. I went with Mr. Lewis over the greater part of his estate. He pays great attention to agriculture and gardening; he has all the newest modes of tillage and many of the newest and most improved implements. His garden is about two acres English, of a light gravelly soil. He has fine apples, pears of the finest varieties; peaches do not do so well as in many other places. They have been, within a few years, seized with disease occasioned by a sort of worm, which makes them drop their fruit before being ripe. He has a collection of grapes, consisting of such as are generally cultivated; they do well. In October he digs in well-fermented dung, and lays them down on the surface of the ground and covers them with earth for protection. Some of them are trained on espaliers, some on poles. They have been tried on southern aspects of walls, but were unsuccessful; probably the intensity of heat during the summer months hurt them, as the fruit, before ripening, became shrivelled and dropped from the trees. North of his house, *Quercus* sp. 16, tree tall, but not thick; soil, rich black loam; fruit very long—*Q. Castanea,*[1] I think. The squirrels eat the fruit of this with eagerness; I am informed they prefer them before any sort of nuts, *Corylus* excepted. Two species of pine, one with large cones like *sylvestris*; the other, small cones and leaves like *Larix* of Europe. *Juglans* sp., *porcina,*[2] nuts of them, trees from 20 to 40 feet high, make good fuel; the wood is not considered of any value. Two species (*alba* ?)[3] called Kœskatoma nuts. The nuts appear to be a very distinct variety, if not a species, shell thin, kernel large and well tasted. Mr. Lewis informs me that they are only in one other place, which is near Catskill, east of the Alleghany Mountains. Trees 60 feet high, wood hard, and makes excellent fences; much used in farmwork generally. *Juglans* var. (?) sp. (?). Trees very like the former, fruit smaller, longer, and acute at the point. Husk much larger than the former. *Castanea* sp., seems to be a large variety of the common—is it produced by the richness of the soil? After getting two plants of *Juglans alba* (?)[3] raised from nuts of the same trees of which the parcel of nuts is in this year's produce, some Indian corn, and a Honeysuckle, Mr. Lewis took me to the house of his friend and neighbour, James Thomson, Esq., to whom I had a letter of introduction from Mr. Clinton.

---

[1] *Q. acuminata*, see Sargent, Silva N. Am. viii. p. 55, *Q. Muehlenbergii*, Kew Hand-list of Trees and Shrubs, ed. 2, p. 697.

[2] *Carya porcina*, C.DC. in DC. Prod. xvi. ii. p. 144.

[3] *Carya alba*, C.DC. *loc. cit.*, p. 143.

Friday, October 17th.—When on my way up the country I was very kindly used by Mrs. Thomson and her son, who is much attached to mineralogy and painting. Mr. Thomson, Sen., was then at New York. Mr. Thomson and son who are now at Boston, took me from the inn and afforded me their house. In the afternoon General Lewis returned home.

18th.—Mr. Thomson's estate is about 600 English acres of rich soil in general, and about 200 acres of wood; his fields are all divided by walls four feet high. In clearing his ground he has left a few choice oaks, which form a pleasing prospect and give the place an English appearance. He has a fine garden and orchard; the garden has a dry light soil. Only peaches and small fruit; they were in a fine state of health and not affected with disease, as is the case in most places. Fine kitchen vegetables: carrots very fine; beets, parsnip, and celery of the first quality. An orchard very healthy, but in a young state. One fruit (an apple) Mr. Thomson observed in his wood self-sown, and after being grafted proved a fine fruit; it is not a large one, but fine quality, something like Lady Apple. I got two trees of it, one of Lady Apple, two *Aesculus* from Ohio, and another tree (?) from Ohio. In his woods west from the house grew *Lycopodium dendroideum* and *complanatum, Pteris atropurpurea*,[1] *Polypodium ilvense*[2] (?). I thought it to be *Cheilanthes vestita*; I rather take it to be *Polypodium*. On this spot I observed *Gerardia flava* in flower, but on examining it more closely it proved to be *G. Pedicularia*; of the former I took plants, and seeds of the latter. Of *Pyrola maculata*[3] and *P. umbellata*[4] I also took plants. In the afternoon I packed my plants, got on board the *Chancellor Livingstone* steamboat (which is celebrated for magnitude and elegance), and after a passage of nine hours landed at Newport at 5 o'clock on Sunday morning.

New York, Sunday, October 19th.—After getting to my lodging with all that I could carry of my little gleanings, I waited on Dr. Hosack at 8 o'clock in the morning. After breakfast I called on Mr. Hogg, arranged matters, &c., as to unpacking, went to dinner, and to church in the afternoon.

20th.—Employed getting the packages from the wharf to Mr. Hogg's, and unpacking the plants.

Tuesday, 21st.—In the morning went to Flushing and made arrangements with Mr. Prince as to taking up the trees; returned in the evening.

Wednesday, October 22nd.—Employed all day at Mr. Hogg's among the plants, securing, &c. I feel very sorry to find the Rose from Amherstburg *rubifolia*.[5] However, it is some consolation to see *Lonicera* in good order, with the majority of everything.

Thursday, October 23rd.—To-day employed giving fresh paper to seeds and specimens. Wrote a letter to Joseph Sabine, Esq. Called on Mr. Thorburn for the purpose of getting ready his parcel.

Friday, October 24th.—To-day was devoted with Mr. Floy, who has all

[1] *Pellaea atropurpurea*, Christensen, Ind. Fil. p. 478.
[2] *Woodsia ilvensis*, Christensen, *loc. cit.*, p. 656.
[3] *Chimaphila maculata*, A. Gray, Syn. Fl. N. Am. ii. i. p. 45.
[4] *Chimaphila umbellata*, A. Gray, *loc. cit.*, p. 45.
[5] *Rosa setigera* var. *tomentosa*, S. Wats. Bibl. Ind. N. Am. Bot. p. 313. Cf. Sept. 16th.

along manifested an unremitting desire to be useful to the Society. In the evening I attended a meeting of the Horticultural Society of New York. I cannot refrain from mentioning the great exertions which most of the efficient members have made in communicating anything worthy of notice. I feel glad to see it in such a state of perfection. A great many prominent inhabitants of New York have become members: De Witt Clinton, Dr. Hosack, General Lewis, the Mayor of the city, &c. Being the first effort to establish a society in America, they labour under many great disadvantages in having no other establishments to co-operate with them in their laudable exertion. The president, Martin Hoffman, Esq., is a very worthy respectable gentleman. His knowledge of gardening is but limited, but he takes a very spirited interest in promoting the science. Mr. Hogg, F. M. Floy, and Mr. Wilson are the chief of its practical members. Presented for their inspection some fine beet of very superior quality, indeed as fine as I ever saw; carrots, very fine; broccoli, very good; some fine specimens of Spanish tobacco; the gentleman kindly offered the Society some seeds before sailing. I was much pleased with the form of the proceedings. Some bunches of grapes of European origin were presented, they were considered good for America, but in my opinion were very inferior to what are to be seen in England.

Saturday, October 25th.—Wrote to W. Coxe, Esq., and to Mr. Dick, Philadelphia. Putting to rights specimens and seeds in the forenoon. At Mr. Floy's taking up plants, &c., till midday, when I went to Flatbush on Long Island, and returned late at night.

Sunday, 26th.—The fore part of the day I rectified some of the lists preparatory to going to Philadelphia. To-day was cold and had much the appearance of winter. Mr. Clinton was to dine at Dr. Hosack's; I was invited after dinner. I walked out to Mr. Hogg's and got to town after projecting a trip to Jersey in the morning.

Monday and Tuesday, October 27th and 28th.—Meeting Mr. Hogg, according to our last night's proposal, at half-past 5 o'clock in the morning, we crossed the Hudson River to New Jersey for obtaining *Sarracenia purpurea,* which was our chief object. The morning was fine and inviting, but before we had got to the desired spot, the rain fell in such torrents that we were urged to take shelter in the first place; towards midday on Tuesday we got out, and on looking on the face of the country beheld it deluged. Calculating that *Sarracenia* was inaccessible for the present, we bent our course to an adjoining wood where we were amply repaid. *Neottia repens* [1] in great profusion; cedar swamp, or rather anemones in the swamp; soil moderately dry, composed altogether of decayed leaves and branches of cedars. By setting our feet on the ground we soon sank, the soil being very soft. I took a good quantity of plants. In the same place *Vaccinium hispidulum* [2] and soil not differing from *Neottia.* Gathered seeds of *Rhododendron maximum* which is immensely large, fully 17 or 20 feet high; *Kalmia latifolia,* also vigorous; *Rhus Vernix* and a climbing species, with some species of *Smilax,* were twined round the trees. On a rising piece

---

[1] *Goodyera pubescens,* Ind. Kew. fasc. iii. p. 304.
[2] *Chiogenes hispidula,* A. Gray, Syn. Fl. N. Am. ii. I. p. 26.

of ground on dry soil grew *Gerardia quercifolia* : what I took from Canada of this having failed, and recollecting the difficulty of raising it from seed, without hesitation or reserve I secured many sets of seeds and plants. We got back at 6 o'clock in the evening, congratulating ourselves on having been more fortunate than we expected.

Wednesday, October 29th.—Waited on Mr. Kennedy and obtained some money of him, then set out for Philadelphia. Got to [New] Brunswick at 4 o'clock in the afternoon and then by stage to Trenton, which I reached at 9 at night.

Thursday, October 30th.—Left Trenton at 5 o'clock in the morning, and got to Burlington at 9. Went to Mr. Smith and proceeded to take up the trees. In the evening I called on W. Coxe, Esq., whom I am sorry to say I found very sick, so ill that he was by his medical attendant considered in danger. I am sorry ; I am afraid he will not be able to get out before I leave.

October 31st.—Finished taking up, and saw the trees packed by 11 o'clock. Went in search of *Epigaea repens* which failed with me in the summer, the soil light sandy loam. On the same place grew *Helonias* sp. ? *Gentiana* sp., Willow-leaved oaks (trees tall but not thick), on the same spot ; *Quercus obtusiloba* [1] beautiful foliage and fine clusters of fruit ; *Quercus triloba* [2] but sorry that no fruit could be found on them. These I carried to the tavern and packed, and at 6 o'clock in the evening went in steamboat to Philadelphia, which place I gained by 9 at night. The night was cold and rainy, with slight frost in the morning.

Philadelphia, Saturday, November 1st.—I waited on Mr. William Dick, janitor of the University of Pennsylvania, from whom I had already received sufficient testimony of his ability and will to make himself useful to me. I had the pleasure of meeting here Mr. Nuttall, whom I found very communicative. We looked round Mr. Dick's garden. At midday, with Mr. Dick, I set out for Messrs. Landreth's, west of Philadelphia, to whom I am also much indebted for very polite attention. (There is a great similarity of character between Messrs. Loddiges and Landreth.) I am again pleased to see *Maclura aurantiaca*. This night's frost had made them drop their leaves, and the tender shoots were injured a little. The Barberry is what Mr. Nuttall describes as *canadensis*, var., and he tells me that some alteration is necessary in that tribe of plants, for no species of *Berberis* are indigenous to America except *Aquifolium* and *nervosa* and this one. Mr. Landreth's one is, according to Mr. Nuttall, a new species and a good one (See Nuttall's ' Genera '). Mr. Landreth obtained it of his friend, Mr. John Fraser, when on one of his last routes to the western country. The particular place of habitat he does not know, as no memoranda were received from Mr. Fraser, but he thinks the mountains of Carolina. Mr. Fraser did not send any of these to England. *Gentiana* sp. Nuttall, on the Missouri; leaves ovate, nervous, and shining; flower large, of a greenish and yellow colour. Only one plant of it. *Rhododendron arboreum*, several species of *Erica, Mesembryanthemum, Stapelia*

---

[1] *Q. stellata*, Kew Hand-list of Trees and Shrubs, p. 709.
[2] *Q. cuneata*, Kew *loc. cit.*, p. 685.

from England—on the whole this is the finest collection of exotics in America. I called on Maccheus Collins, Esq., in the evening, a gentleman of high standing as a mineralogist in that quarter.

Sunday, November 2nd.—To-day I went in quest of *Nelumbium* (or *Cyamus*) *luteum*, which grows near Philadelphia, and which I saw in flower when here in August. I made an exertion to procure roots, but I am very sorry to say could not effect this as the roots run to an immense depth. However, I am so far pleased at having ten or eleven seeds of it, which I trust will grow. Mr. Bartram tried repeatedly to transplant it, without success. Hamilton of Woodlands, whose domains lie on the banks of the Schuylkill, some years before his death took roots from the fields south of Philadelphia, and planted them in the mud of the Schuylkill when the tide ebbed and flowed—they grew and flowered well; this is the only instance which has been successfully tried. Mr. Nuttall says on the Mississippi at St. Louis it grows much stronger and produces knobs or tubers at the root, which are greedily sought after by the natives. They are washed, boiled, and then beat up with butter or grease. He also informs me that when in that quarter he frequently had a dish of them and thinks them good. I tried for the tubers here, but the immense depth which they run in mud rendered it impossible. In the afternoon I called on Mr. Dick and from his seed cabinet had a portion of everything which I conceived might be useful or interesting to the Society. This gentleman would make a very useful Corresponding member. He has the friendship of all American travellers. Major Long has contributed to him largely in 1822 and is now just returning from an interesting spot; his expectations I hope will not be disappointed.

Philadelphia, Monday, November 3rd.—In company with Mr. Nuttall I set out this morning to the residence of the late Mr. Bartram; his niece is a considerable botanist and draws well. Mr. Carr, to whom she is married, has but a moderate share of knowledge; this deficiency, however, is made up by his pleasing manner. In front of the house stands a very large cypress, 90 feet high and 23 round, planted by the first John Bartram; his son William (the late) held the tree while his father put the earth round; it is eighty-five years old. At the foot of this a small pond in which many little and valuable treasures were; but since the death of the worthy protector, have been suffered to remain in a deplorable state. I made inquiry about *Quercus heterophylla*, as it would be an acquisition, but found it had been cut down by a servant of the person on whose estate it grew—cut down by mistake. Mr. Bartram was not reconciled about it so long as he lived. *Quercus lyrata* from Georgia and *Quercus macrocarpa* from the Alleghany do very well. I now feel happy: think some of my specimens from Canada and Michigan are fine. I am sure of *Q. macrocarpa*. On the margin of the pond *Andromeda arborea*,[1] fully 40 or 45 feet high, 19 inches round.[2] In summer it has been clothed with flowers and has now a great abundance of seed: obtained a paper of seed. Soil light, and it seems to like damp,

---

[1] *Oxydendrum arboreum*, A. Gray, Syn. Fl. N. Am. ii. I. p. 33.
[2] Sargent ' Silva N. Am:' v. p. 135 says: ' Occasionally 50 or 60 feet in height, with a tall straight trunk 12 to 20 inches in diameter.'—ED.

at any rate moisture. Mr. Bartram tried for upwards of forty years to raise it from seed, and always was unsuccessful till this season, when he had the gratification of transplanting an abundance of them in small boxes two days before he died. I have got a small box containing about twelve plants on condition that half should be given to Mr. George Loddiges. Some fine specimens of *Magnolia*—species *auriculata*,[1] *cordata*, *macrophylla*, *tripetala*, *grandiflora* but quite in miniature to the fine specimens in England. I have not seen any like those in Chelsea Botanical Gardens or at Kew. Two Roses from Missouri, only a plant of each—I wish they were propagated : perfectly hardy. *Sagittaria sagittifolia flore pleno*, I recollect being told of it by Mr. Loddiges when I was at Hackney a few days before leaving London. I could not get any of the tubers of the root : otherwise useless. On our way through the wood got three species of Oaks on rocky soil. Mr. Nuttall showed me *Asplenium rhizophyllum*[2] on a rock on our way home, four miles from Philadelphia. Called on Messrs. Landreth : obtained the trees with two species of *Phlox* from Georgia (Mr. Fraser), also roses Champneya, Cherokee, and two others. Got two fruits of *Maclura* and I shall put them with the other one : I shall pack in charcoal. Mr. Landreth expresses himself happy to do anything in his power for the Society. He is an acquaintance of Mr. Richard Williams, Turnham Green, to whom I am cordially to remember him. Got to town by 10 o'clock at night : put away seeds, &c.

Tuesday, November 4th and 5th.—In the morning went to Mr. Dick's and got up and packed and sent to the wharf. Mr. Dick has several species of *Cactus* from the Rocky Mountains, but they were too small to bear a voyage across the Atlantic. All the seeds of them were sown, I am sorry to know that I cannot get any of *Dionaea muscipula*, Mr. Dick's friend having died. Left Philadelphia at 12 o'clock and reached Burlington at 4 in the afternoon. Went to William Coxe, Esq., whom I found still very ill—but considerably better, since I saw him before. North of his house got two species of *Euphorbia*, those which I took in summer having failed. *Rhexia virginica*, seeds. *Rubus cuneifolius*, a handsome plant, fruit large and has a fine flavour, produces fruit in great abundance ; I am of opinion that it may prove good when cultivated. One other species, larger leaves and serrated, fruit red, hard, and dry even when ripe ; a stronger growing plant than *cuneifolius*. A third species, small, creeping : I did not see any fruit on it. All these were on dry light loam, and poor Mr. Coxe observes that they bear better fruit in that soil than in rich. *Monarda punctata* on dry light sandy soil : this, although a common species, is a fine one : got seeds. On sandy soil *Quercus ferruginea*,[3] small tree about 30 feet high, a singular and beautiful specimen. Obtained of Mr. Coxe one plant of Pennsylvanian rose, double ; originated in that State, which is all the history I could get. Two species of *Phlox* from the Southern States : one is the largest flowered specimen which I have seen, of a deep rose colour, about 2½ feet

---

[1] *Magnolia Fraseri*, S. Wats. Bibl. Ind. N. Am. Bot. p. 29.
[2] *Camptosorus rhizophyllus*, Christensen, Ind. Fil. p. 166.
[3] *Q. marylandica*, Sargent, Silva N. Am. viii. p. 161.

high (a plant and seeds); seeds only on the lateral shoot. The other species is fine but by no means equal to the former. The soil was light and dry. In the centre walk of the garden a row of the one on one side and of the other on the other side had a fine effect. Two specimens of Indian Corn. I regret much to mention my disappointment of the apples. He was not able to leave his room, indeed he was so ill before I left that to appearance he could not live long. I obtained all that was in the house, consisting of about eight or nine varieties and only two or three of each, with two bottles of cider seven years old—one made from Wine-sop, one from Virginian crab-apple: this is a present for Thos. A. Knight and Jos. Sabine Esqs., from Will. Coxe. I got also a few seeds of ornamental plants. I received from him and family all the attention they could give under the present circumstances. I had not time to call on Mr. Smith, and as I saw all the trees sent off before going to Philadelphia, left Burlington at 2 o'clock in the afternoon by steamboat for Bordentown, and then by stage to South Amboy. The night was exceedingly dark and rainy; the coach broke down at 11 o'clock at night, and having to stop two hours in repairing, got to South Amboy at 4 o'clock on Wednesday morning. This is a specimen of great speed—twenty-nine miles in thirteen hours.

Thursday, November 6th.—Left Amboy at 5 o'clock in the morning and got to New York at 11. Employed getting the trees to Mr. Hogg's and ordered some boxes for packing them in.

November 7th.—Finished taking up Mr. Floy's trees.

November 8th.—Wrote to Joseph Sabine, Esq., and at packing in the afternoon.

Sunday, November 9th.—As usual on Sundays took breakfast at Dr. Hosack's; packed up some seeds, &c., in the forenoon, and went to Mr. Hogg's in the evening.

Monday and Tuesday, November 10th and 11th.—Being unsuccessful with *Sarracenia* by the great rain a fortnight since, in company with Mr. Hogg we again crossed the River Hudson. We hired a car, and after driving and walking alternately, reached the spot. The swamp, as mentioned before, is large. We made an attempt to get over from the north side, but were obliged to return. Mr. Hogg led the way, when suddenly he went down to the middle in mud. On the south side we were more successful; after some difficulty and all besmeared with filth, reached them, having to carry them two miles through the swamp. Darkness put a stop to our pursuit before we could get enough plants. In the evening we sat by the fireside talking of our day's proceedings. We have now set out again. By 10 o'clock on Tuesday we had an abundance of plants and proceeded after something else. On a spot west of Hoboken swamp we found *Lycopodium dendroideum*; the soil was dry sandy peat. Intermingled with it, *Adiantum pedatum*; on the left, in the hollow where the soil was a little damp, another species of *Lycopodium*, a creeping one; a few ferns in wet places on the outskirts of the woods, but being killed by the frost cannot say what they are. In the bustle of our proceedings lost the whip, for which the man extorted two dollars in consequence of it being a favourite whip. On the rocks of the Hudson, seven miles from New York on the west side of the river, gathered *Asplenium ebeneum* [1] in abundance, two species of *Aster*, and one of *Solidago* : they were still in flower—*Aster*, both small flowers, blue : *Solidago*, small flower, yellow. *Neottia pubescens* [2] rather scarce in this place : I wish I could find some more of it. Brown loam on the north aspects of rocks. I must not forget *Pothos foetidus* [3] and the difficulty in taking it up, and *Orontium aquaticum*. *Orontium* seems to have few localities. Reached home : stand now dry and comfortable, a state we required after such undertaking.

12th.—Received a letter dated 5 Oct. from Joseph Sabine, Esq., accompanied by additional lists ; without delay wrote to Messrs. Landreth and Smith ; called on Mr. Floy ; Mr. Hogg then went to Flushing and returned in the evening. I must observe that Mr. Prince Jun., did not receive us with kindness but the reverse. If I can but obtain all the plants for the Society, I shall leave him for some other person.

Thursday, November 13th.—Wrote to Mr. Dick at Philadelphia a letter of thanks. Packed Mr. Floy's box ; and took up plants and packed Mr. Wilson's.

November 14th.—In the morning went to the vegetable market. Celery is fine, just like such as is to be seen in England, but raised at much less expense. Broccoli is inferior to European : I am told by many, cauliflower nor broccoli are not so good as in Britain. Carrots and beet are very fine ; they do not seem to possess many varieties. Turnips are very small and not of first quality. Potato is the worst vegetable in the market. In Canada I tasted some very excellent onions, here they are fine ; leeks also good. I cannot make any contrast between this and Covent Garden, as I never had many opportunities of seeing it, and I question much if those more capable of judging would have any reluctance in pronouncing equal in proportion to the town to Covent Garden. Fruit is brought in abundance ; certainly fine indeed, but to speak candidly I think much exaggerated, and on asking any person if they are good, they invariably say they are not so fine as last year, and I think had I been here then they would not have been so fine as the year before. Nuts from Europe, and indigenous to America are in quantity, with pineapples and coconuts from West Indies. I am a little disappointed as to price of vegetables, by no means so cheap as I was led to believe. Packing in the after part of the day.

November 15th.—Packing at Mr. Hogg's.

16th.—Wrote to Joseph Sabine, Esq., in the morning. Among specimens in the forenoon. Dinner at Dr. Hosack's. At Mr. Hogg's in the afternoon.

Monday, November 17th.—At Flushing and got part of the trees up. Could not get finished as Mr. Prince has not all the plants which are ardently needed. Many times he expressed himself pleased with the Society and is to put up a collection of fruits, which will be an acquisition,

[1] *Asplenium platyneuron*, Christensen, Ind. Fil. p. 126.
[2] *Goodyera pubescens*, Ind. Kew. fasc. iii. p. 304.
[3] *Symplocarpus foetidus*, Engler, in DC. Monog. Phan. ii. p. 212.

especially as I have almost failed with Coxe.   I am pleased to get *Berberis Aquifolium.*

From 18*th to* 22*nd.*—Employed packing plants, &c.   Got part of the boxes on board.

Sunday, November 23*rd.*—Wrote to Joseph Sabine, Esq., and finished putting up my dry plants.   At church in the afternoon.

Monday, 24*th.*—I am happy to hear from Messrs. Landreth and Smith as to the plants.   Mr. Landreth has again given another testimony of his kindness; he is certainly an attentive and punctual man.   Answered his letter.   I cannot but say that I am once more pleased.   I got part of the packages on board.

26*th.*—Received the trees from Philadelphia and Burlington, got them from the wharf, ordered boxes for them.   I now feel a little at ease.

Thursday, 27*th.*—Got remainder of trees, &c., packed and taken down to the vessel, and box from Baltimore.

29*th.*—Received boxes from Flushing, and transmitted on board ship; packing, and at Mr. Hogg's.

30*th.*—Accompanied Mr. Hogg to some of his friends in the country.

December 1*st,* 2*nd,* 3*rd,* 4*th.*—Finished packing and everything on board on the 5th.   I went to Flushing: our words were not of the most amicable tenor, and I am sorry to say that I must leave America without having good feeling towards every person; and the conduct towards the Society of Mr. Prince is by no means laudable and towards myself most infamous.   Mr. Kennedy (Mr. Maitland I never saw, he lives at Norfolk) has through the whole of my stay here been very kind, and I have experienced every facility in his power.   I need not say anything of Mr. Hogg, (his ability and willingness are well known to the Society) Mr. Floy, and Mr. Wilson, and in fact every person except Prince: this gentleman is now such as becomes the Society to withhold the correspondence.   The feelings of Dr. Hosack are already known.   I cannot but say, at all times and in all cases, I found his very philanthropic mind unchanged, and his advice as to my own comfort.   My feelings will never I hope, in any instance, speak but with gratefulness of Governor Clinton.   I can say nothing more than that Dr. Hosack is a Clinton, and Clinton a Hosack.

Monday, December 8*th.*—Paid Prince's bill, and settled account with Mr. Kennedy.

9*th.*—Expecting to sail next morning I had my Pigeons, Ducks and Quail on board, and it being my last day I spent it with Dr. Hosack.   At his house was a large party of friends who kept mirth up till a late hour in the evening.   In the course of the day General Lewis called, having come to town for the season; he has kindly invited me to visit him should I not sail.

10*th.*—Waited on the wharf till 12 o'clock noon.   I had the mortification to return.   According to promise went to General Lewis and had all the kindness renewed which I experienced at his house in the country.   At 7 o'clock in the evening took leave of Dr. Hosack, and Mr. Hogg saw me in the small boat, the ship having hauled out before dark.   Ship *Nimrod.*

12*th.*—Went under sail at daylight, having a keen frost and pleasant fresh breeze : sent a letter on shore to Dr. H[osack] and D. C[linton] by the pilot, who left us at midday. Our view of Long Island and New Jersey closed with the night and we beheld the land of Columbia no more.

13*th*–31*st.*—In the morning the wind from the north-east produced a heavy sea. In Lat. 39, long. 72, the water in the gulf of Florida stood at 83° Fahrenheit. My ducks were very sick for two days and ate nothing, the pigeons and quail continued well. On the morning of the 14th a favourable wind sprang up which continued to the end of the month with scarcely any alteration. Our passage was such as did not admit of an observance of the holiday [Christmas Day, &c.].

1824.

January 1*st*–3*rd.*—On the morning of the New Year, I had the gratification of seeing the rocky shores of Cornwall and with a continued steady wind came to an anchorage off Dover on Saturday morning.

4*th*–10*th.*—Gained the Downs on Sunday evening, where we lay becalmed for two days. Entered the river on Wednesday evening, and had the pleasure of arriving safe at London on Friday morning having had a highly interesting journey.

[*A number of side-notes, chiefly from Pursh or Michaux, have been added to this MS. in Douglas's handwriting. These have been incorporated in the text.*—ED.]

SOME ACCOUNT OF THE AMERICAN OAKS, PARTICULARLY OF SUCH SPECIES AS WERE MET WITH DURING A JOURNEY FROM * TO * , IN THE YEAR 1823 BY DAVID DOUGLAS, IN A LETTER TO THE SECRETARY, JOSEPH SABINE, F.R.S.

Sir,—Among the more important instructions with which I was charged by the Horticultural Society during my late mission to the State of New York, I was commissioned to employ that time which could not be devoted to the procuring of living plants for the garden, to obtaining whatever specimens might be in my power for the Herbarium. The only opportunity which I enjoyed of doing this was during a journey to * ; but the hasty manner in which I was obliged to travel, and the lateness of the season, prevented my forming an extensive Hortus Siccus. The oaks, from their number and beauty and the acknowledged utility of their wood, particularly attracted my attention, and of the thirty-four species enumerated by Pursh as natives of the vast continent of North America, I was so fortunate as to meet with no less than nineteen. Some account of these I have drawn up, adding to the specific characters, and a few synonyms, such information as I was able to collect in the country, together with some observations which I have extracted from the beautiful work of Michaux, a botanist whose long residence in the United States gave him such facilities in ascertaining the nature and uses of the forest trees as it has scarcely fallen to the lot of any other naturalist to possess. Such as these my imperfect notices are, I take the liberty of submitting them to your attention, and have the honour to be,

Your very obedient and obliged
humble Servant,
D. DOUGLAS.

In the arrangement of the following species of oak, I have thought it best to follow the divisions adopted by Michaux, *first* according to the annual or biennial fructifications (characters which he has fully explained in his admirable ' North American Sylva,' published at Philadelphia 1817), and secondly as to the modes of divisions of the leaves.

* The blanks occur in the MS.—ED.

A. *Annual Fructification.*

× *Leaves lobed.*

1. QUERCUS ALBA. *White Oak.*

Q. alba. Foliis oblongo-ovatis profunde pinnatifidis subtus glaucis basi attenuatis laciniis oblongis obtusis plerumque integerrimis, cupula hemisphaerica squamoso-tuberculata, glande ovata.

Q. *alba*, Linn. Sp. Pl. p. 1414, Willd. Sp. Pl. v. 4, p. 448 ; Michx. f. N. Am. Sylva (ed. Philad.) v. 1, p. 11, t. 1 ; Pursh, Fl. Am. Sept. v. 2, p. 633.

Q. alba, Willd. Sp. Pl. 4, p. 448; Kalm. It. 2, p. 357 ; Du Roi, Harbk. Baumz. 2, p. 270, t. 5, f. 5 ; Wangenh. Nordam. Holz. p. 12, t. 3, f. 6 ; Michx. Hist. Chênes Am. n. 4, t. 5, n. 1 ; Michx. f. Hist. Arb. Am. 2, p. 13, t. 1 ; Pursh, Fl. Am. Sept. 2 p. 633 ; Nutt. Gen. 2, p. 215; *alba virginiana*, Catesb. Carol. 1, t. 21 ; Sm. in Rees' Encyc. 29, part 1st, n. 69 ; Ait. Hort. Kew. ed. 2, 5, p. 293; β *alba repanda*, Michx. Hist. Chênes Am. n. 4, t. 5, n. 2 ; Pursh, Fl. Am. Sept. 2, p. 633.
Q. *alba*. On the fertile banks of Red River, situated in the 49° north latitude, this species attains no inconsiderable size—at least sufficiently large for domestic purposes. It is greatly used by the inhabitants of the colony formed on the banks of that stream. Forty miles to the south of the settlement, near Pembina, it is still more plentiful and increases the further we go to the south.   On the northern shores and islands of Lake Winnipeg in the 50°, on poor bleak soils of the shore, and on the Granite Islands where the soil is very thin, it dwindles to a small scrubby bush 6 to 10 feet high bearing abundance of acorns ; this is the most northern limits of this species.

A species remarkable for its deeply pinnatifid leaves and their nearly equal, and mostly entire, obtuse segments, the upper ones only being irregularly lobed, one of the lateral ones rarely so.   They taper gradually at the base, are of a bright, deep, and shining green above, glaucous and glabrous beneath, in all my specimens with ripe fruit ; sometimes, however, in young individuals especially, the leaves are pubescent beneath, and occasionally, according to Willdenow, green on both sides.   The acorn is large, handsome, ovate, almost an inch long ; the cup hemispherical, flat at the base, inserted singly or two together on a short stalk, and thickly clothed with tuberculiform scales.

It is one of the largest, most abundant, and useful species in the middle States, found upon every variety of soil.   Towards the middle and Western extremity of the State of New York in the Genessee Tract, where the soil is generally a dry rich black or brownish loam, it attains an immense size. On the banks of the River Detroit from Amherstburg to the junction of the Thames with the St. Clair in Upper Canada, and on the opposite banks, in the Michigan Territory, on a deep alluvial rich black soil, these trees frequently measure from 20 to 25 feet in circumference at 8 feet from the ground, and are from 80 to 100 feet high.   This species certainly appears to prefer a temperate, or rather a cold climate, for the individuals which I saw of it in Canada on the Niagara, and in the northern part of New York, were much larger than any which I observed in the States of Pennsylvania and Delaware, which were the most southern districts I visited.   Pursh, who mentions that it is found in fertile forests from New England to Carolina, says it grows to an immense size in the Middle States ; my observations make those produced in the more northern districts larger.

The wood is chiefly used for articles of husbandry and fencing. The Canadian cottages, commonly called log houses, are mostly built of it. In the navy yards of New York and Philadelphia it is much employed. I am informed by James Taylor of Lloyd's, who at present is engaged in superintending the building of vessels on an improved principle at Montreal and at St. John's, New Brunswick, that he finds abundantly in the neighbourhood of the last-mentioned place a sort of oak which proves to be much more durable than the common white oak to which it bears a slight resemblance. Mr. Taylor expects that it will become an article of great interest in shipbuilding. This specimen agrees perfectly with my *Quercus olivaeformis*. Valuable as is this species of oak to the Americans, Michaux does not agree with those writers who recommend its introduction into Europe, and its being employed in preference to European oak. He inclines to the opinion that our oaks are of a much superior quality, and rather advises that *they* should be naturalised on the American continent. In autumn the foliage changes, Michaux tells us, to a bright violet colour, which gives this tree a singular appearance, and renders it, from that point of view alone, a desirable acquisition to our woods and shrubberies.

Q. alba β repanda. Foliis levissime lobatis utrinque viridibus. Michx. Hist. Chênes Am. n. 4, t. 5, f. 2 ; Pursh, Fl. Am. Sept. 2, p. 633.

Found also, according to Pursh, in Pennsylvania and Carolina.

2. QUERCUS OLIVAEFORMIS. *Mossy-cup Oak.*

Q. olivaeformis. Foliis oblongis profunde inaequaliter sinuato-pinnatifidis, subtus glaucis subpubescentibus, fructu elliptico, cupula craterata squamosa margine fimbriata.

Q. olivaeformis, Michx. N. Am. Sylv. v. 1, p. 33, t. 3 ; Pursh, Fl. Am. Sept. v. 2, p. 632.

Q. olivaeformis, Michx. f. Hist. Arb. Am. 2, p. 32, t. 2 ; Pursh, Fl. Am. Sept. 2, p. 632 ; Nutt. Gen. 2, p. 215 ; Sm. in Rees' Encyc. 29, part 1st, n. 81.

Abundant at Palatine, forty miles west from Albany, State of New York, Amherstburg, and Sandwich, Upper Canada. Grows to a large tree from 50 to 70 feet high and proportionably thick. Branches small, drooping ; leaves long and deeply lobed ; acorns invariably in clusters at the point of the shoots. Michaux's figure of this species is by no means good, and I am much inclined to think it has been taken from an imperfect specimen. The acorn is much larger, the cup more fringed, but the seed is not so much enveloped in it. It is known in Canada and New York by the name of white oak. The wood resembles that of *Q. alba* in colour, but it is of a closer texture.

Michaux gives the banks of the Hudson river and the western parts of New York as its stations. Pursh found it in Pennsylvania and Virginia, on iron-ore hills.[1]

[1] The following is in Douglas's handwriting (Ed.) :—

EXTRACT OF A LETTER FROM MR. FLOY.

' *Quercus elongata* (*olivaeformis* of Michaux, the younger). I very much doubt if either Michaux or Pursh has seen this oak. I have never seen but one tree of it (and it was the most beautiful of oaks I ever saw) ; from this tree I obtained a few acorns and specimens in 1806. I gave Michaux the specimen from which his figure was

In all likelihood this is the tree which Mr. Taylor of St. John's prizes so highly for shipbuilding and considers superior to any except *Q. virens*.

3. QUERCUS MACROCARPA.   *Over-cup White Oak.*

Q. macrocarpa.   Foliis obovatis profunde lyratim sinuato-lobatis lobis superne dilatatis obtusis, subtus pubescentibus, fructu late ovato (maximo), cupula craterata squamosa margine fimbriata.

Q. macrocarpa, Michx. f. N. Am. Sylv. p. 35, t. 4; Pursh, Fl. Am. Sept. v. 2, p. 632; Willd. Sp. Pl. v. 4, p. 453.

Q. macrocarpa, Willd. Sp. Pl. 4, p. 453; Michx. Hist. Chênes Am. n. 2, t. 2, 3; Michx. f. Hist. Arb. Am. 2, p. 34, t. 3; Pursh, Fl. Am. Sept. 2, p. 632; Nutt. Gen. 2, p. 215; Sm. in Rees' Encyc. 29, part 1st, n. 80.

A species undoubtedly, in its foliage as well as its fruit, nearly allied to the preceding. The leaves are much shorter, however, and broader, of a deeper green hue, with the margins far less deeply cut, the lobes shorter, broader, and far more dilated upwards. The acorns are among the largest of the genus, and much broader in proportion to their length than those of *Q. olivaeformis*, but the cup is similar in its structure, downy and scaly, the upper scales lengthened into long hair-like points, which form a dense fringe around the margin. Plentiful on the shores of Lake Erie, near Amherstburg, and on the eastern shore of Lake St. Clair, Upper Canada, in dry sand and gravel, from 45 to 70 feet in height and from 8 to 10 feet in circumference.

Pursh gives its habitat as Kentucky, Tennessee, Illinois, Mississippi, and Missouri, within the mountains, on dry slate or limestone hills.

Said by Pursh to attain a large size on the Alleghany, its wood is of an excellent quality and is much used. The clusters of its large acorns with their mossy fringed cups, and the beauty of its deeply lobed leaves, give to this tree a singularly graceful aspect.

Q. macrocarpa β pendula ramis pendulis.

The variety is from 20 to 40 feet high and 2 or 3 feet in diameter, its branches drooping, leaves similar to those of *Q. macrocarpa*, acorns also large, but constantly borne in pairs at the extremity of the shoots, which is not the case in the general state of the tree.

Q. sp.[1]—From 70 to 100 feet high and 10 to 14 in circumference, leaves deeply lobed, pubescent underneath, and constantly smaller than

made: I considered it at that time a variety of *macrocarpa*, and two other kinds, I have given Mr. Douglas these specimens; and I am happy to see by his specimens he has discovered two or three kinds more. I doubt not but they are *species*, they are all of the Mossy Cup kinds, they seem to be intermediate from the *bicolor* of Pursh, to the *macrocarpa* of Michaux. It is certainly very singular these oaks should have been overlooked: Michaux never saw the *olivaeformis* on the Hudson river; and I think it very probable that Pursh saw some of the intermediate kinds in Pennsylvania but not the *olivaeformis*. I am inclined to believe the plant I now send you is the only one in England, perhaps in Europe. I have one large tree; I made several grafts from it, but this is the only one that grew. I shall try it next spring.'

Added to this, in Sir W. J. Hooker's handwriting, is the following (Ed.):—

' I do not understand this note. *Q. elongata* is the *Quercus falcata* of Michaux [see Number 26], *not Q. olivaeformis*. It had better perhaps be inserted, in the form of a NOTE to *Q. olivaeformis* (Mossy-cup Oak) and the name *olivaeformis* substituted for *Q. elongata.*'

[1] ' There is no specimen corresponding to this in Douglas's "Herbarium" sent to me.—H.' [This note was probably made by Sir William Jackson Hooker.—ED.]

Q. macrocarpa, to which it comes near; acorns oblong, larger than *Q. olivaeformis* and smaller than *Q. macrocarpa,* entirely enveloped in the cup, which is scaly, having a very fine filamentous fringe. The wood resembles *Q. alba.* More abundant on the east banks of Lake St. Clair than any other species of the genus.

4. QUERCUS OBTUSILOBA. *Upland White Oak* or *Post Oak.*

Q. obtusiloba. Foliis late obovatis sinuatis lobis obtusis superioribus dilatatis bilobis, fructibus subglomeratis (mediocribus), cupula hemisphaerica squamosa, glande ovata.

Q. obtusiloba, Michx. f. N. Am. Sylv. v. 1, p. 38, t. 5; Pursh, Fl. Am. Sept. v. 2, p. 632. *Q. stellata,* Willd. Sp. Pl. v. 4, p. 452.

Q. obtusiloba, Michx. Fl. Bor. Am. 2, p. 194; Michx. Hist. Chênes Am. n. 1, t. 1; Michx. f. Hist. Arb. Am. 2, p. 36, t. 4; Sm. in Rees' Encyc. 29, part 1st, n. 78; Pursh, Fl. Am. Sept. 2, p. 632; Nutt. Gen. 2, p. 215. *Q. stellata,* Willd. Sp. Pl. 4, p. 452; Ait. Hort. Kew. ed. 2, 5, p. 294. *Q. alba foliis ad modum anglicanae,* Clay Fl. Virg. p. 146, n. 467.

Usually 30 or 50 feet high and 2 or 3 feet in diameter, spreading and throwing out branches to the bottom of the trunk. Plentiful near Philadelphia; Burlington, New Jersey; and on the east banks of the Hudson, ten miles from New York. The wood is said to be exceedingly hard, and is extensively used in the navy yards and for other purposes.

Pursh says 50 or 60 feet high. It grows, according to Pursh, from Canada to Florida in most forests.

5. QUERCUS LYRATA. *Over-cup Oak.*

Q. lyrata. Foliis oblongis lyrato-sinuatis glabris, subtus albidis, lobis superioribus dilatatis, cupula muricato-squamosa subglobosa depressa, glande omnino tegente.

Q. lyrata, Walt. Fl. Car. p. 235; Willd. Sp. Pl. v. 4, p. 453; Michx. f. N. Am. Sylv. v. 1, p. 43, t. 6; Pursh, Fl. Am. Sept. v. 2, p. 632.

Q. lyrata, Willd. Sp. Pl. 4, p. 453; Walt. Fl. Car. p. 235; Michx. Hist. Chênes Am. n. 3, t. 4; Michx. f. Hist. Arb. Am. 2, p. 42, t. 5; Pursh, Fl. Am. Sept. 2, p. 632; Nutt. Gen. 2, p. 215; Sm. in Rees' Encyc. 29, part 1st, n. 79; Ait. Hort. Kew. ed. 2, 5, p. 295.

I had my specimen from the garden of the late Mr. Wm. Bartram, near Philadelphia.

Of all the oaks that grow in moist situations this is by Michaux considered the best, and it is held in considerable esteem for the uses to which its wood may be applied. It has the further advantage of attaining to very large dimensions, a height of about 80 feet, with a stem of 8, 10, or even 12 feet in circumference. Pursh indeed observes that its height is only from 8 to 15 feet.

Called also, according to Pursh, Swamp-post Oak and Water-white Oak. The acorns being entirely concealed by the cups, gives its other name.

It grows, according to Pursh, in swamps from Carolina to Florida and on the Mississippi.

×× *Leaves broadly and coarsely toothed (not lobed).*

6. QUERCUS BICOLOR. *Swamp White Oak.*

Q. bicolor. Foliis oblongis obovatis grosse inaequaliter sinuato-dentatis basi integerrimis, subtus albo-tomentosis, fructu longe pedunculato, cupula hemisphaerica squamosa, glande ovata.

Q. bicolor, Willd. Sp. Pl. v. 4, p. 440; Pursh, Fl. Am. Sept. v. 2, p. 633.

Q. Prinus discolor, Michx. f. N. Am. Sylv. v. 1, p. 47, t. 7.

Q. bicolor, Willd. Sp. Pl. 4, p. 440; Pursh, Fl. Am. Sept. 2, p. 633; Nutt. Gen. 2, p. 215; Sm. in Rees' Encyc. 5, part 1st, n. 50. *Q. Prinus tomentosa*, Michx. Hist. Chênes Am. n. 5, t. 9, f. 2; *Q. Prinus discolor*, Michx. f. Hist. Arb. Am. 2, p. 46, t. 6.

This oak probably combines beauty and utility to as great a degree as any in America. The stems are unusually tall and straight, having a few branches at the bottom; the leaves of a dark green on the upper side and clothed with a thick white pubescence underneath; smaller in every respect than *Q. Prinus*. The acorns, like those of the latter species, are sweet and eagerly sought after by pigs, squirrels, and pigeons. The wood, like most other species, is used for domestic purposes. It grows on the high ground of the Niagara, at the Falls, and on the banks of the Hudson, near Rhinebeck, in the State of New York.

Found by Pursh, from Pennsylvania to Carolina, growing to a large size in low wet woods.

Michaux observes that the wood is of a fine and pretty close grain, with the pores almost wholly obliterated; and that it is known to possess the qualities of strength and great elasticity, and especially that of easily splitting in a straight direction.

7. QUERCUS PRINUS. *Swamp Chestnut Oak.*

Q. Prinus. Foliis obovatis acutis subtus laeviter pubescentibus subaequaliter grosse sinuato-dentatis, cupula hemisphaerica squamosa basi subattenuata, glandula ovata.

Q. Prinus, Linn. Sp. Pl. p. 1413; Willd. Sp. Pl. v. 4, p. 439; Pursh, Fl. Am. Sept. v. 2, p. 633.

Q. Prinus palustris, Michx. f. N. Am. Sylv. v. 1, p. 52, t. 8.

Q. Prinus, Willd. Sp. Pl. 4, p. 439; Sm. in Rees' Encyc. 29, part 1st, n. 47; Ait. Hort. Kew. ed. 2, 5, p. 290; Pursh, Fl. Am. Sept. 2, p. 633; Nutt. Gen. 2, p. 215.
Q. Prinus palustris, Michx. Hist. Chênes Am. 5, t. 6; Michx. Fl. Bor. Am. 2, p. 196; Michx. f. Hist. Arb. Am. 2, p. 51, t. 7.
Q. Castaneae foliis procera arbor virginiana, Pluk. Almag. 309; Pluk. Phytog. t. 54, f. 3; Catesb. Carol. 1, t. 18.

A large tree of quick growth and very ornamental, of which the acorns vegetate immediately upon their falling from the tree. It is by far the largest of this section, and during the autumn presents many beautiful tints. I could not learn if the wood was considered good, but Michaux speaks of it as only a secondary quality, though much used for cartwrights' work and such purposes as require strength and durability, its numerous pores rendering it almost useless for making casks to hold spirits. The negroes weave baskets and brooms of it, because it splits well and in a straight direction, and can be divided into very thin slips. Hewn into rails for enclosing cultivated fields it lasts twelve or fifteen years, or a third longer than those which are made of the *Willow Oak* or *Water Oak*. For grandeur and majesty Michaux places this in the first rank of American forest-trees.

Called also Chestnut White Oak, and, to the South, according to Pursh, White Oak.

The acorns are large and sweet, like chestnuts, whence its name.

Pursh says it is useful, and found in low shady woods and on the banks of rivers from Pennsylvania to Florida.

8. QUERCUS MONTANA. *Rock Chestnut Oak.*

Q. montana. Foliis obovatis acutis subtus albo-tomentosis grosse dentatis dentibus subaequalibus dilatatis apice callosis, cupula hemisphaerica squamosa, glande ovata.

Q. montana, Willd. Sp. Pl. v. 4, p. 440; Pursh, Fl. Am. Sept. v. 2, p. 634.

Q. Prinus monticola, Michx. f. N. Am. Sylv. v. 1, p. 56, t. 9.

Q. montana, Willd. Sp. Pl. 4, p. 440; Smith in Abbot Insect. 2, p. 163, t. 82; Rees' Encyc. 25, part 1st, n. 49; Pursh, Fl. Am. Sept. 2, p. 634; Nutt. Gen. 2, p. 216. *Q. Prinus monticola*, Michx. Fl. Bor. Am. 2, p. 196; Michx. Hist. Chênes Am. n. 5, t. 7; Michx. f. Hist. Arb. Am. 2, p. 55, t. 8.

On both sides of the Niagara, from the falls to Queenston, in dry rocky situations, growing to a height of between 20 to 40 feet. The wood is very hard and much valued, and is useful for almost every purpose; the bark is largely employed in tanning.

Found by Pursh, from New England to Carolina, in rocky situations on the mountains, and he says it rises to about 60 feet.

Q. Castanea and *Q. Chincapin* are generally found with it.

Although I have followed Willdenow and Pursh in adopting the reference to Michaux for this and the preceding species, yet I think that if the two plants be really distinct the reference should be reversed. In regard to the leaves, there seems to be little or no difference between the two, but the cup of the acorn in *Q. Prinus monticola* is expressly described by Michaux as turbinate, implying a degree of attenuation at the base, (and as such he figures it), whereas in his *Q. Prinus palustris* he calls the cupule *craterata* and represents it as such, or hemispherical.

9. QUERCUS CASTANEA. *Yellow Oak.*

Q. Castanea. Foliis oblongis acuminatis subtus tomentosis albidis grosse subaequaliter dentatis, cupula hemisphaerica squamosa, glande ovata.

Q. Castanea, Willd. Sp. Pl. v. 4, p. 441; Pursh, Fl. Am. Sept. v. 2, p. 634.

Q. Prinus acuminata, Michx. f. N. Am. Sylv. v. 1, p. 61, t. 10.

Q. Castanea, Willd. Sp. Pl. 4, p. 441; Muhl. in Ges. Naturf. Fr. Neue Schr. 3, p. 396; Pursh, Fl. Am. Sept. 2, p. 634; Sm. in Rees' Encyc. 29, part 1st, n. 51; Nutt. Gen. 2, p. 214. *Q. Prinus acuminata*, Michx. Fl. Bor. Am. 2, p. 196; Michx. Hist. Chênes Am. n. 5, t. 8; Michx. f. Hist. Arb. Am. 2, p. 61, t. 9.

A large and handsome tree, which is from 30 to 60 feet high but does not become so thick as *Q. Prinus* or *Q. bicolor*. Like these species, with which it often grows, its wood is of extensive utility. All the species in this section retain their verdure much longer than those with lobed leaves. It is found on the high grounds of the Niagara and, Pursh says, in the Alleghany mountains and on the banks of the Delaware. The acorns are eatable.

10. QUERCUS PRINOIDES. *Small Chestnut Oak.*

Q. prinoides. Foliis obovato-lanceolatis acutis grosse sinuato-dentatis, subtus glaucis sub-tomentosis, cupula hemisphaerica squamosa, glande ovata.

Q. prinoides, Willd. Sp. Pl. v. 4, p. 440.

Q. Chinquapin, Pursh, Fl. Am. Sept. v. 2, p. 634.

Q. Prinus Chincapin, Michx. f. N. Am. Sylv. v. 1, p. 64, t. 11.

Q. Chinquapin, Pursh, Fl. Am. Sept. 2, p. 634; Sm. in Rees' Encyc. 29, part 1st, n. 48; Nutt. Gen. 2, p. 215. *Q. prinoides,* Willd. Sp. Pl. 4, p. 440. *Q. Prinus Chinquapin,* Michx. Hist. Chênes Am. n. 5, t. 9, f. 1; Michx. f. Hist. Arb. Am. 2, p. 64, t. 10.

A small bush or tree, seldom exceeding 8 or 10 feet in height. The acorns are small but very numerous, and at a short distance resemble common hazel-nuts. Although of humble growth, this species forms an interesting link with those of larger dimensions. In the same situation was found a variety with a deeper cup and longer seed; the leaves were also more elongated and deeply dentate, and placed upon longer footstalks.

According to Pursh it is known as the Chinquapin or Dwarf Chestnut Oak, does not exceed 3 or 4 feet in height, and is highly ornamental when in full bloom. It grows on dry mountain lands, Pennsylvania to Carolina.

All the species of this section are so closely allied one to another that they might, perhaps, be more naturally considered, as they are by Michaux, as only varieties of the same species, differing in appearance and in the quality of their timber, according to the soil and exposure which they inhabit.

B.—*Fructification Biennial (leaves mucronated).*

× *Leaves entire.*

11. QUERCUS PHELLOS. *Willow Oak.*

Q. Phellos. Foliis lineari-lanceolatis integerrimis glabris apice sebaceo-acuminatis (junioribus dentatis lobatisve), cupula brevi squamulosa, glande subrotunda.

Q. Phellos, Linn. Sp. Pl. p. 1412; Michx. f. N. Am. Sylv. v. 1, p. 78, t. 14; Pursh, Fl. Am. Sept. v. 2, p. 625.

Q. Phellos, Willd. Sp. Pl. 4, p. 423, Willow-leaved Oak Catesb. Car. 1, t. 16. *Quercus an potens Ilex Marilandica folio longo angusto Salicis,* Raii Hist. v. 3; Dendr. p. 8; Michx. f. Hist. Arb. Am. 2, p. 74, t. 12; Pluk. Amalth. t. 441, fig. 7; Abbot Insect. 2, t. 91; Ait. Hort. Kew. ed. 2, 5, p. 287; Smith in Rees' Encyc. 29, part 1; Pursh, Fl. Am. Sept. 2, p. 625; Nutt. Gen. 2, p. 214.

Q. Phellos humilis, Pursh, l.c. 2, p. 625. *Q. Phellos humilior salicis folio breviore,* Catesb. Car. 1, t. 22.

Q. Phellos sylvatica, Michx. Hist. Chênes Am. n. 7, t. 12; Wangenh. Nordam. Holz. t. 5, f. 11.

Phellos. Miller's Dict. no. 12.

Abundant near Philadelphia and Burlington, New Jersey: a tree of very large dimensions. My specimens were gathered too late in the season and in too imperfect a state to enable me to make any remarks on the species. Michaux considers this as among the least valuable of the oaks, neither deserving of encouragement in the United States nor worthy of being cultivated for its utility in our country. Found by Pursh in low swampy forests, near the sea coast from New Jersey to Florida, growing about 50 or 60 feet high.

Q. Phellos β humilis foliis brevioribus, Catesb. Car. 1, t. 22 ; Wangenh. Nordam. Holz. t. 5, f. 12 ; Pursh, Fl. Am. Sept. 2, p. 625.

This variety has shorter leaves and is of low straggling growth.

12. QUERCUS MARITIMA. *Marine Oak.*

Q. maritima. Foliis perennantibus coriaceis lanceolatis integerrimis glabris basi attenuatis, apice acutis mucronatis, cupula scutellata, glande subrotunda.—Pursh.

Q. maritima, Willd. Sp. Pl. v. 4, p. 424 ; Pursh, Fl. Am. Sept. v. 2, p. 625.

Q. Phellos maritima, Michx. Hist. Chênes Am. t. 13, f. 1.

Q. maritima, Willd. Sp. Pl. 4, 424.

Michaux in his 'North American Sylva' does not appear to have noticed this oak, unless he has included it in his account of the Willow Oak. Elliot observes that it comes very near to the *Q. virens*, and Pursh that it is a low shrubby species from 3 to 8 feet high, and found upon the sea coast of Virginia and Carolina.

13. QUERCUS SERICEA. *Running Oak.*

Q. sericea. Foliis deciduis lanceolato-oblongis integerrimis subundatis basi attenuatis obtusis, apice dilatatis acutis subtus sericeis, cupula scutellata, glande subglobosa.—Pursh.

Q. sericea, Willd. Sp. Pl. v. 4, p. 424 ; Pursh, Fl. Am. Sept. v. 2, p. 626.

Q. pumila, Michx. f. N. Am. Sylv. v. 1, p. 88, t. 17.

Q. sericea, Willd. Sp. Pl. 4, p. 424; Abbot Insect. 2, t. 51; Pursh, Fl. Am. Sept. 2, 626. Q. (pumila) foliis lanceolatis integerrimis glabris subtus glaucis, Walt. Car. 234. Q. pumila, Michx. f. Hist. Arb. Am. 2, p. 84, t. 15.

This is found exclusively in the maritime parts of the two Carolinas, Georgia, and the Floridas, and is the smallest of all the genus, seldom, according to Michaux, more than 20 inches high and two lines in diameter. Pursh says 2 feet high.

14. QUERCUS MYRTIFOLIA.

Q. myrtifolia. Foliis perennantibus coriaceis oblongis integerrimis glabris utrinque acutis supra nitidis margine revolutis.—Pursh.

Q. myrtifolia, Willd. Sp. Pl. v. 4, p. 424.

Q. myrtifolia, Willd. Sp. Pl. 4, p. 424; Pursh, Fl. Am. Sept. 2, p. 626; Nutt. Gen. 2, p. 214 ; Sm. in Rees' Encyc. 29, part 1.

Leaves about an inch long, like those of a large myrtle. Fruit unknown.

15. QUERCUS VIRENS. *Live Oak.*

Q. virens. Foliis perennantibus coriaceis oblongo-ellipticis integerrimis

*margine revolutis basi obtusis apice acutis muticis subtus stellatim pubes-
centibus, fructibus pedunculatis, cupula turbinata, squamis abbreviatis,
glande oblonga.*—Pursh.

Q. virens, Michx. f. N. Am. Sylv. v. 1, p. 67, t. 12 ; Willd. Sp. Pl. 4,
p. 425 ; Pursh, Fl. Am. Sept. v. 2, p. 626.

Q. Phellos, β Linn. Sp. Pl. p. 1412.

Q. sempervirens, Walt. Fl. Car. p. 234.

Q. virens, Willd. Sp. Pl. 4, p. 425 ; Michx. f. Hist. Arb. Am. 2, p. 67, t. 11 ; Pursh,
Fl. Am. Sept. 2, p. 626 ; Nutt. Gen. 2, p. 214 ; Ait. Hort. Kew. ed. 2, 5, p. 287 ;
Sm. in Rees' Encyc. 29, part 1.
Q. sempervirens, Walt. Fl. Car. p. 234.
Q. sempervirens, foliis oblongis non sinuatis, Catesb. Car. Tab. 17.

Found exclusively in the maritime parts of the Southern States of the
two Floridas, Lower Louisiana, and Virginia. It makes no part of the
forest, even at the distance of fifteen or twenty miles from the sea.

Pursh says near the sea coast, Virginia to Florida and Mississippi,
it grows to the height of 40 or 50 feet, spreading its branches wide
when not enclosed by other trees.

This is one of the most valuable of the timbers employed in the
United States, particularly for shipbuilding, but on account of its extreme
heaviness the workmen are obliged to insert in the upper frames of the
vessel pieces of red cedar, of the same diameter, the timber of which
being extremely light, nevertheless stands equally well the vicissitudes of
moisture and drought. In the south of the United States the wood is
employed by wheelwrights for the felloes and hubs of the larger carts, as
well as the teeth of mill-wheels and screw-presses. But such has been the
consumption of late years, and no attempt made for its reproduction, that
Michaux asserts that in a few years it will cease to exist in the United
States or be only found, like the *Quercus Ilex*, in the south of France, Spain,
or Italy, where it used to abound as a timber, in the form of a shrub or
low bush, merely useful as fuel. Yet the *Q. virens* is that which,
perhaps, will best bear the comparison, in point of the excellency of its
timber, with our European oak.

16. QUERCUS CINEREA. *Grey Oak, Upland Willow Oak.*

*Q. cinerea. Foliis perennantibus coriaceis lanceolato-oblongis integerrimis
margine revolutis basi attenuatis apice obtusiusculis mucronatis subtus
stellatim tomentosis, fructibus sessilibus, cupula scutellata, glande subglobosa.*—Pursh.

Q. cinerea, Michx. f. N. Am. Sylv. v. 1, p. 85, t. 16 ; Willd. Sp. Pl. v. 4,
p. 425 ; Pursh, Fl. Am. Sept. v. 2, p. 626.

Q. humilis, Walt. Fl. Carol. p. 234, nana foliis dentatis. *Q. nana*,
Willd. Sp. Pl. 4, p. 443.

Q. cinerea Willd. Sp. Pl. 4, 425 ; Michx. Hist. Chênes Am. t. 14 ; Michx. f. Hist.
Arb. Am. 2, p. 81, t. 14 ; Pursh, Fl. Am. Sept. 2, p. 626 ; Nutt. Gen. 2, p. 214 ;
Sm. in Rees' Encyc. 29, part 1st, no. 6 ; Ait. Hort. Kew. ed. 2, 5, p. 288 ;
Q. Phellos, β *sericea*, Ait. Hort. Kew. ed. 1, 3, p. 354.
Q. humilis, foliis lanceolatis integerrimis seta terminatis subtus tomentosis, Walt.
Fl. Car. 234.

This again is a maritime oak found in little groups in the midst of the barrens, where they reach to a height of 3 or 4 feet, with leaves one inch long, deeply toothed at the margin. This latter is the *Q. nana* of Willdenow, which Michaux has ascertained to be only a variety of *Q. cinerea*. This is on the authority of the younger Michaux, but Pursh doubts it.

Pursh places it in dry, barren soil and pine forests, Virginia to Georgia; very variable in size, sometimes 4 feet, at others 20 feet high.

This is useless as timber; the bark produces a fine yellow like that of the Black Oak.

17. Quercus imbricaria. *Shingle Oak.*

Q. imbricaria. Foliis deciduis oblongis utrinque acutis mucronatis integerrimis nitidis subtus pubescentibus, cupula scutellata, squamis lato-ovatis, glande subglobosa.—Pursh.

Q. imbricaria, Michx. f. N. Am. Sylv. v. 1, p. 82, t. 15; Willd. Sp. Pl. v. 4, p. 428; Pursh, Fl. Am. Sept. v. 2, p. 627.

Q. imbricaria, Willd. Sp. Pl. 4, p. 428; Michx. f. Hist. Arb. Am. 2, p. 78, t. 13; Michx. Hist. Chênes Am. no. 9, t. 15 and 16; Pursh, Fl. Am. Sept. 2, p. 627; Nutt. Gen. 2, p. 214; Ait. Hort. Kew. ed. 2, 5, p. 288; Sm. in Rees' Encyc. 29, part 1st, no. 15.

This species seems to be quite an occidental one, principally abounding to the westward of the Alleghany Mountains, growing on the banks of rivers, where it is known by the names of the Jack Oak, Black Jack Oak, and Laurel Oak; the French call it *chênes à lattes*.

The tree grows to a height of 40 or 50 feet, with a diameter of 12 to 15 inches.[1] The timber is heavy and hard, and although of very inferior quality is nevertheless used by the French of the Illinois (for want of better material perhaps) for making shingles.

18. Quercus laurifolia. *Laurel Oak, Swamp Willow Oak.*

Q. laurifolia. Foliis deciduis obovato-lanceolatis acutis basi attenuatis integerrimis utrinque glabris, cupula scutellata, squamis lanceolatis, glande subovata.—Pursh.

Q. laurifolia, Michx. Hist. Chênes Am. n. 10, t. 17; Willd. Sp. Pl. v. 4, p. 427; Pursh, Fl. Am. Sept. v. 2, p. 627.

Q. laurifolia, Willd. Sp. Pl. 4, p. 427; Michx. Hist. Chênes Am. no. 10, t. 17; Ait. Hort. Kew. ed. 2, 5, p. 288; Sm. in Rees' Encyc. 29. part 1st, no. 14; Pursh, Fl. Am. Sept. 2, p. 627; Nutt. Gen. 2, p. 214.

β *Q. foliis apice obtusis*, Michx. Hist. Chênes Am. t. 18; Pursh, Fl. Am. Sept. 2, p. 627.

This is spoken of by the elder Michaux as producing a valuable wood, almost preferable to that of *Q. virens*. The younger Michaux does not notice it, and Pursh suspects he considered it to be a variety of *Q. imbricaria*. Nuttall says it is a very doubtful species. Pursh refers to Michx. Hist. Chênes Am. t. 18, as his var. β ' *Q. foliis apice obtusis.*' It grows to about 50 or 60 feet high.

---

[1] Sargent says rarely exceeding 3 feet in diameter.—Syl. N. Am. viii. 175.—Ed.

× × *Leaves sinuate lobate or toothed.*

19. QUERCUS AGRIFOLIA.

Q. agrifolia. Foliis subrotundo-ovatis subcordatis utrinque glabris remote spinoso-dentatis, cupula hemisphaerica, squamis laxis, glande ovata acuta.—Pursh.

Q. agrifolia, Née in Anal. Cienc. Nat. 3, p. 271; Willd. Sp. Pl. v. 4, p. 431; Pursh, Fl. Am. Sept. v. 2, p. 627.

Q. agrifolia, Willd. Sp. Pl. 4, p. 431; Née in Anal. Cienc. Nat. 3, p. 271, an Ilex folio agrifolii americana, forte agria vel aquifolia glandifera ? Pluk. Phytogr. t. 196, f. 3; Sm. in Rees' Encyc. 29, part 1st, no. 29; Pursh, Fl. Am. Sept. 2, p. 627; Nutt. Gen. 2, p. 214.

Found upon the north-west coast of America, about Nootka Sound, but of the value of its timber nothing is known.

20. QUERCUS HETEROPHYLLA. *Bartram's Oak.*

Q. heterophylla. Foliis longe petiolatis ovato-lanceolatis oblongisve integris vel inaequaliter grandidentatis, cupula hemisphaerica, glande subglobosa.

Q. heterophylla, Michx. f. N. Am. Sylv. v. 1, p. 90, t. 18; Pursh, Fl. Am. Sept. v. 2, p. 627.

Q. heterophylla, Michx. f. Hist. Arb. Am. 2, p. 87, t. 16; Pursh, Fl. Am. Sept. 2, p. 627; Nutt. Gen. 2, p. 214.

All that is known of this oak is from a single tree upon the banks of the Schuylkill, in a field belonging to Mr. Bartram's farm, near Philadelphia; Pursh says on the banks of the Delaware. Michaux considers it to be a good species; Pursh supposes it may be a hybrid, but he does not say between what species; and Nuttall asks if it may be an anomalous variety of *Q. coccinea.*

21. QUERCUS AQUATICA. *Water Oak.*

Q. aquatica. Foliis obovato-cuneiformibus glabris integerrimis apice obsolete trilobis, lobo intermedio majore, cupula hemisphaerica, glande subglobosa.—Pursh.

Q. aquatica, Michx. f. N. Am. Sylv. v. 1, p. 92, t. 19; Willd. Sp. Pl. v. 4, p. 441; Pursh, Fl. Am. Sept. v. 2, p. 628.

Q. nigra, Linn. Sp. Pl. p. 1413; *Q. uliginosa*, Wangenh. Nordam. Holz. p. 80, t. 6, f. 18.

Q. aquatica, Willd. Sp. Pl. 4, p. 441; Walt. Fl. Car. 234; Q. folio non serrato in summitate quasi triangulari, Catesb. Car. 1, p. 20. t. 20; Michx. f. Hist. Arb. Am. 2, p. 89, t. 17; Clay, Fl. Virg. p. 149; Sm. in Rees' Encyc. 29, part 1, no. 52; Pursh, Fl. Am. Sept. 2, p. 628; Nutt. Gen. 2, p. 214.

Q. nana, Willd. Sp. Pl. 4, p. 443; *Q. aquatica* γ *elongata*, Ait. Hort. Kew. ed. 2, 5, p. 290. *Q. nana*, Abbot Insect. 2, t. 59. *Q. aquatica varietas*, Mich. Hist. Chênes Am. t. 20, f. 2.

In swamps and around ponds from Maryland to Florida, rising to a height of 40 to 45 feet, with a diameter in the stem of from 12 to 18 inches.[1] The leaves are extremely variable and are either deciduous or remain upon the tree for two or three years, according to the severity of the winters of the climate in which they are found. Its timber seems to be of very inferior quality.

[1] Sargent 'with a trunk from 2 to 3½ feet in diameter.'—Syl. N. Am. viii. 165.—ED.

22. QUERCUS HEMISPHAERICA.

Q. hemisphaerica. Foliis perennantibus oblongo-lanceolatis indivisis trilobis sinuatisque, lobis mucronatis, utrinque glabris.—Pursh.

Q. hemisphaerica, Michx. Hist. Chênes Am. t. 20, f. 2; Willd. Sp. Pl. v. 4, p. 443; Bartr. Itin. p. 320; Michx. Hist. Chênes Am. t. 20, f. 2; Pursh, Fl. Am. Sept. v. 2, p. 628.

Q. hemisphaerica, Willd. Sp. Pl. 4, p. 443; Pursh, Fl. Am. Sept. 2, p. 628; Nutt. Gen. 2, p. 214; Sm. in Rees' Encyc. 29, part 1, no. 55.

Inhabits Georgia and Florida, but seems now to be generally considered a var. of *Q. aquatica*.

23. QUERCUS TRILOBA. *Downy Black Oak.*

Q. triloba. Foliis oblongis cuneiformibus basi acutis, lobis aequalibus mucronatis, intermedio longiore, subtus tomentosis, cupula scutellata, glande depresso-globosa.—Pursh.

Q. triloba, Michx. Hist. Chênes Am. 14, t. 26. Willd. Sp. Pl. v. 4, p. 443; Pursh, Fl. Am. Sept. v. 2, p. 628. *Q. cuneata*, Wangenh. Nordam. Holz. p. 78, t. 5, f. 14.

Q. triloba, Willd. Sp. Pl. 4, p. 443; Ait. Hort. Kew. ed. 2, 5, p. 291; Abbot Insect. 1, p. 99, t. 50; Michx. Hist. Chênes Am. n. 14, t. 26; Pursh, Fl. Am. Sept. 2, p. 628.

Nuttall makes this the var. *β* of *Q. falcata* (Spanish Oak). The younger Michaux does not notice it. The elder says it forms a tree from 20 to 40 feet high, is of rapid growth, and extremely well suited for enclosing lands, which Pursh copies and adds, in pine-barrens, near the sea coast, New Jersey to Georgia.

24. QUERCUS NIGRA. *Barren Oak* or *Black Jack Oak.*

Q. nigra. Foliis cuneatis basi subcordatis apice dilatatis abruptis breviter lobatis, lobis mucronatis subtus ferrugineis, cupula hemisphaerica squamis membranaceis, glande breviter ovata.

Q. nigra, Linn. Sp. Pl. 1413; Pursh, Fl. Am. Sept. v. 2, p. 629.
Q. ferruginea, Michx. f. N. Am. Sylv. v. 1, p. 95, t. 20.

Q. nigra, Willd. Sp. Pl. 4, p. 442; Sm. in Rees' Encyc. 29, part 1, no. 53; Sm. in Abbot Insect. t. 58; Ait. Hort. Kew. ed. 2, 5, p. 291; Pursh, Fl. Am. Sept. 2, p. 629; Nutt. Gen. 2, p. 214. *Q. ferruginea*, Michx. f. Hist. Chênes Am. n. 12, t. 22, 23; Michx. f. Hist. Arb. Am. 2, p. 92, t. 18; Mill Dict. no. 10; Quercus (forte) Marilandica folio trifido ad sassafras accedente, Raii, Catesb. Car. 19, t. 19.

From 20 to 30 feet high, trunk small and scrubby, principally used for fuel, for which purpose, according to Michaux, it is held next in estimation to the Hickory; seldom more than 2½ feet in diameter. Said to be greatly used and very durable. Abundant in dry barren soil, from Burlington to South Amboy, New Jersey.

Pursh places it in barren sandy or gravelly woods, New Jersey to Florida, and says it bears abundantly a fine mast for hogs.

25. QUERCUS ILICIFOLIA. *Bear Oak*, called also *Black Scrub Oak* and *Dwarf Red Oak.*

Q. ilicifolia. Foliis obovato-cuneiformibus 3–5-lobis, lobis acutis mucro-

nulatis subtus tomentosis, cupula hemisphaerica squamis membranaceis, glande ovata globosa.

Q. *ilicifolia,* Wangenh. Nordam. Holz. p. 79, t. 6, f. 17 ; Willd. Sp. Pl. v. 4, p. 447.

Q. *Banisteri,* Michx. f. N. Am. Sylv. v. 1, p. 99, t. 21 ; Pursh, Fl. Am. Sept. v. 2, p. 631.

Q. *Banisteri,* Michx. Fl. Bor. Am. 2, p. 199 ; Michx. Hist. Chênes Am. n. 15, t. 27 ; Michx. f. Hist. Arb. Am. 2, p. 96, t. 19 ; Sm. in Abbot Insect. 2, t. 79 ; Pursh, Fl. Am. Sept. 2, p. 631 ; Nutt. Gen. 2, p. 215.
Q. *ilicifolia,* Willd. Sp. Pl. 4, p. 447 ; Sm. in Rees' Encyc. 29, part 1st, n. 66 ; Ait. Hort. Kew. ed. 2, 5, p. 292.

Near Flushing, Long Island, on dry poor soil, and in rocky situations of the Hudson, at Rhinebeck, in the State of New York. This very dwarf kind of oak is occasionally employed about New York in forming fences, and Michaux strongly recommends it to the owners of great estates in Europe (and it is sufficiently hardy to bear our climate) to form copses of it for the protection of game. The acorns are described as being produced so abundantly upon the bushes of this oak that they sometimes cover the branches, and the bears, deer, and hogs feed on them whilst growing on the plants.

Pursh places it in dry barren fields and on the mountains, New Jersey to Virginia. He calls it a shrub, about 4 to 6 feet high, covering large tracts which are called Oak-barrens.

26. QUERCUS ELONGATA. *Spanish Oak.*

Q. *elongata. Foliis sinuatis lobis tribus vel pluribus subfalcatis aristato-acuminatis terminale elongato subtus tomentosis, cupula hemisphaerica squamis membranaceis, glande subglobosa.*

Q. *elongata,* Willd. Sp. Pl. v. 4, p. 444.

Q. *falcata,* Michx. f. N. Am. Sylv. v. 1, p. 106, t. 23 ; Pursh, Fl. Am. Sept. v. 2, p. 631.

Q. *falcata,* Michx. Fl. Bor. Am. 2, p. 199 ; Michx. Hist. Chênes Am. n. 16, t. 28 ; Michx. f. Hist. Arb. Am. 2, p. 104, t. 21 ; Pursh, Fl. Am. Sept. 2, p. 631 ; Nutt. Gen. 2, p. 214. Q. *elongata,* Willd. Sp. Pl. 4, p. 444 ; Sm. in Rees' Encyc. 29, part 1st, n. 57 ; Ait. Hort. Kew. ed. 2, 5, p. 291. Q. *discolor,* Ait. Hort. Kew. 3, p. 358 [now Q. *velutina*].

A very large and lofty tree, frequently 80 or 90 feet in height, of which I could not obtain any perfect specimen. Plentiful at Burlington, New Jersey, and on the islands in the River Delaware, 26 miles from Philadelphia. In the Southern States it is called the Red Oak, and Pursh places it in sandy soil, near the sea coast, New Jersey to Georgia ; the wood, according to Michaux, is employed for making the staves of barrels, but only such as contain salt provisions or dry wares. It is never used in shipbuilding. Cartwrights make the felloes for the wheels of heavy carriages of it, because it is less apt to split than the white and other oaks. To compensate for the inferiority of its timber in some measure, its bark is preferable to that of the other oaks for tanning coarse leather, and it sells at a high price.

The leaves of the younger individuals are almost invariably simply trilobed; those which have larger, more numerous, and falcate segments belong to the older trees.

27. QUERCUS TINCTORIA. *Black Oak.*

Q. tinctoria. Foliis late obovatis profunde sinuosis lobis oblongis inaequaliter setaceo-denticulatis subtus pubescentibus vel glabris, cupula hemisphaerica squamis majusculis submembranaceis, glande ovata rotundata.

Q. tinctoria, Michx. f. N. Am. Sylv. v. 1, p. 112, t. 24; Willd. Sp. Pl. v. 4, p. 444; Pursh, Fl. Am. Sept. v. 2, p. 629.

Q. discolor, Willd. Sp. Pl. v. 4, p. 444; Pursh, Fl. Am. Sept. v. 2, p. 629.

Q. tinctoria, Willd. Sp. Pl. 4, p. 444. *Q. discolor*, Willd. in Ges. Naturf. Fr. Neue Schr. 3, p. 399. *Q. tinctoria angulosa*, Michx. Hist. Chênes Am. 13, t. 24; Pursh, Fl. Am. Sept. 2, p. 629; Nutt. Gen. 2, p. 215; Sm. in Rees' Encyc. 29, part 1st, no. 58. *Q. velutina*, Lam. Encyc. i. p. 721.

Q. discolor, Willd. Sp. Pl. 4, p. 444. *Q. tinctoria sinuosa*, Michx. Hist. Chênes Am. n. 13, t. 25; Pluk. Phytog. t. 54, f. 5; Sm. in Abbot Insect. 2, p. 111 and 56; Sm. in Rees' Encyc. 29, part 1st, no. 59; Pursh, Fl. Am. Sept. 2, p. 629.

A large tree, full 90 feet in height and from 4 to 6 feet in diameter. The wood is considered good, and the bark is more used for tanning than that of any other species. On October 18 it continued quite green, though very keen frosts had prevailed for several nights. The acorns when broken are of a bright yellow colour within. It inhabits Rhinebeck and Albany, in the State of New York. Pursh says in all large woods, particularly in the mountainous parts, New England to Georgia. The wood, we are told, is employed at Philadelphia as a substitute for white oak in house-building. Its bark yields abundantly the tanning principle, and would be of great value but that it gives out a yellow colour which requires to be removed by a particular process. The colour is contained in the cellular part of the bark, according to Michaux, and produces the famous Quincitron, so much employed in France as well as this country. The discovery of it was made in 1784, by Dr. Bancroft, since which time it has been generally used by dyers, especially for colouring wool, silk, and paper hangings for rooms. Michaux's figure above quoted admirably represents the specimen I gathered. The *Q. tinctoria* of Willd. is described by him as having slightly sinuated leaves, for which reason I have quoted it with a doubt. His *Q. discolor* quite accords with the *Q. tinctoria* of Michaux, and he observes of it that during autumn its foliage is quite free from pubescence, which is the case with my plants.

× × × *Leaves deeply toothed and lobed.*

28. QUERCUS COCCINEA. *Scarlet Oak.*

Q. coccinea. Foliis late ovatis profunde sinuatis glabris lobis inaequaliter dentatis setaceo-acuminatis, cupula hemisphaerica basi attenuatis squamis majusculis submembranaceis, glande ovata rotundata.

Q. coccinea, Michx. f. N. Am. Sylv. v. 1, p. 115, t. 25; Willd. Sp. Pl. v. 4, p. 445; Pursh, Fl. Am. Sept. v. 2, p. 630.

Q. coccinea, Wangenh. Nordam. Holz. p. 44, t. 4, f. 9; Willd. Sp. Pl. 4, p. 445; Michx. Hist. Chênes Am. n. 18, t. 31, 32; Michx. f. Hist. Arb. Am. 2, p. 116, t. 23; Ait. Hort. Kew. ed. 2, 5, p. 292. *Quercus rubra β coccinea*, Ait. Hort. Kew. 3, p. 357, Sm. in Rees' Encyc. 29, part 1st, no. 61; Pursh, Fl. Am. Sept. 2, p. 630; Nutt. Gen. 2, p. 214.

Plentiful in the middle States, upon every kind of soil. Between Amherstburg and Sandwich, Upper Canada, it constitutes with *Q. rubra* and *Q. alba* the largest trees in the forest. Its leaves fade early in the season, and, as is justly observed by Pursh, who places it in fertile woods (New England to Georgia), it gives the woods a very picturesque appearance. The wood is of an inferior quality. The cup of the acorn of this oak, besides being attenuated at the base, is incurved at the margin, and Michaux further informs us that though this species is certainly allied in the structure of its leaves and acorns to the black oak (*Q. tinctoria*) yet it may always be known from it by the acorns of the scarlet oak being white within, whilst those of the black oak are yellowish. The timber is reddish and of coarse texture, and being liable to decay soon, it is used as little as possible for housebuilding and wrights' work. Staves are made from it, which are sent to the West India Islands. The bark is thick and much employed in tanning. Sir J. E. Smith, in Rees' Cyclopedia, observes that this was among the first American trees that were brought to Europe, it having been cultivated before the end of the seventeenth century by Bishop Compton, as well as by the first Earl of Portland. A plantation was made of these trees from acorns sent by the elder Michaux, and exists at Lambouillet in France, the trees of which in thirty years' time had attained the height of about 45 feet.

29. QUERCUS BOREALIS.   *Grey Oak.*

Q. borealis. Foliis sinuatis glabris, sinubus subacutis, cupula subscutellata, glande turgide ovata.—Michx.

Q. borealis, Michx. f. N. Am. Sylv. v. 1, p. 119, t. 26.

Q. ambigua, Michx. f. Hist. Arb. Am. 2, p. 120, t. 24 ; Pursh, Fl. Am. Sept. v. 2, p. 630.

Q. ambigua, Michx. f. Hist. Arb. Am. 2, p. 120, t. 24 ; Pursh, Fl. Am. Sept. 2, p. 630 ; Nutt. Gen. 2, p. 215. Exclud. Sm. in Rees' Encyc. 29, part 1st.

Q. ambigua is found in great abundance in the northern parts of California from the 43° of North latitude southwards. As far as my observation went it is not found immediately on the sea coast, but on the undulating ground of the interior in a thick stratum of alluvial soil on a substratum of clay. It forms belts and groups over the face of the country, and is comparatively of diminutive growth to the timber found on the Atlantic side of America; few trees will give a plank exceeding a foot broad and 12 or 20 long. The northern limit of this species is near the sources of River Multnomah.

My specimens of this oak perfectly accord with the beautiful figure of it in Michaux, yet I must confess there is some truth in the observation of Pursh that it seems intermediate between *Q. coccinea* and *Q. rubra* ; its fruit however differs from that of either of these species, being smaller and more clustered. It forms a fine large tree with large and very variable foliage, of a vivid green above and brownish hue underneath. I am not aware of the wood having any peculiar quality, and Michaux reckons it among the least valuable of the American oaks. Michaux first named this species *ambigua* to indicate its alliance with *Q. coccinea* and *Q. rubra* ; but afterwards, on discovering that this appellation was already applied to an oak of New Spain, by M. De Humboldt, he altered it to *Q. borealis*, since it extends further North than any other.

It grows on the high grounds of the Hudson, near Rhinebeck; near the Falls of Niagara, on the Canadian side; and at Sandwich, Upper Canada. Michaux adds on Hudson's Bay and in Nova Scotia.

30. QUERCUS RUBRA. *Red Oak.*

Q. rubra. *Foliis longe petiolatis glabris obtuse sinuatis, lobis acute dentatis setaceo-mucronatis, cupula hemisphaerica squamosa basi plana, glande ovata elliptica.*

Q. rubra, Linn. Sp. Pl. p. 1413; Michx. f. N. Am. Sylv. v. 1, p. 125, t. 28; Willd. Sp. Pl. v. 4, p. 445; Pursh, Fl. Am. Sept. v. 2, p. 630.

Q. rubra, Willd. Sp. Pl. 4, p. 445; Michx. Hist. Chênes Am. n. 20, t. 35, 36; Michx. f. Hist. Arb. Am. 2, p. 126, t. 26; Pluk. Phytog. t. 54, fig. 4; Ait. Hort. Kew. ed. 2, 5, p. 292; Sm. in Rees' Encyc. 29, part 1st, n. 60; Pursh, Fl. Am. Sept. 2, p. 630; Nutt. Gen. 2, p. 214.
Q. rubra is found forming belts, clumps, and open woods in the northern parts of California in conjunction with *Q. Garryana.* The same observation made on *Q. ambigua* is in every way applicable to this species—namely, confined to the interior of the country, few in number, and of diminutive growth to that seen on the Atlantic side of the continent in similar parallels of latitude; the northern range of this species is 44°, but it does not become in any wise plentiful until 40°, which may be said to be the southern limits of the pine tribe.

One of the most magnificent of trees, which grows upon all soils in abundance, attaining and even exceeding the height of 80 or 100 feet. The diversified form of its foliage, and its large acorns placed in shallow cups, must always render this oak an interesting and worthy object of cultivation, as far as regards appearance; its timber is, however, of very inferior quality, scarcely fit for any purpose but the making of staves for such barrels as are only destined to contain molasses or dry goods.

Pursh places it in fertile forests, Canada to Pennsylvania, and in the Alleghany mountains, and says that the name of Red Oak exclusively belongs to this, though in parts of America it is applied to *Q. tinctoria, coccinea, falcata, palustris, Phellos, ambigua,* and *aquatica.*

31. QUERCUS PALUSTRIS. *Pin Oak;* also *Swamp Spanish Oak.*

Q. palustris. *Foliis longe petiolatis profunde sinuatis glabris, axillis venarum subtus villosis, lobis divaricatis dentatis acutis setaceo-mucronatis, cupula plano-hemisphaerica squamosa, glande brevi-ovata.*

Q. palustris, Michx. f. N. Am. Sylv. v. 1, p. 121, t. 27; Willd. Sp. Pl. v. 4, p. 446; Pursh, Fl. Am. Sept. v. 2, p. 631.

Q. palustris, Willd. Sp. Pl. 4, p. 446; Michx. Hist. Chênes Am. n. 19, t. 33, 34; Michx. f. Hist. Arb. Am. 2, p. 123, t. 25; Du Roi, Harbk. Baumz. 2, p. 268, t. 5, f. 4; Wangenh. Nordam. Holz. p. 76, t. 5, f. 10; Sm. in Rees' Encyc. 29, part 1st, n. 63; Ait. Hort. Kew. ed. 2, 5, p. 292; Pursh, Fl. Am. Sept. 2, p. 631; Nutt. Gen. 2, p. 214.

It inhabits poor and sandy or peat soils, wherever the ground is rather wet; it reaches from 40 to 70 feet in height. The leaves are small and finely divided, the acorns are small also, and produced in great abundance. Although not so fine a tree as the preceding (*Q. rubra*) yet when growing among other species it produces a pleasing effect. It is of frequent occurrence. Pursh places it in low, swampy woods, New England to Pennsylvania.

The acorns of *Q. palustris* are smaller than those of any other family.

The present individual, as well as *Q. tinctoria, coccinea, elongata,* and *palustris,* are included under the denomination of Red Oaks in the United States, and it must be owned that they are very nearly allied species for which it is very difficult to find distingushing characters.

32. QUERCUS CATESBAEI. *Barren Scrub Oak.*

Q. Catesbaei. Foliis brevissime petiolatis basi cuneatis oblongis profunde sinuatis glabris, lobis tribus quinisve divaricatis dentatis acutis setaceo-mucronatis, cupula turbinata ampla, squamis obtusis marginalibus introflexis, glande subglobosa.—Pursh.

Q. Catesbaei, Michx. f. N. Am. Sylv. v. 2, p. 103, t. 22 ; Willd. Sp. Pl. v. 4, p. 446 ; Pursh, Fl. Am. Sept. v. 2, p. 630.

Q. Catesbaei, Willd. Sp. Pl. 4, p. 446 ; Michx. f. Hist. Chênes Am. n. 17, t. 29 ; Michx. f. Hist. Arb. Am. 2, p. 101, t. 20 ; Catesb. Car. i. t. 23 ; Sm. in Rees' Encyc. 29, part 1st, no. 62 ; Pursh, Fl. Am. Sept. 2, p. 630 ; Nutt. Gen. 2, p. 214.

Found in the lower parts of the two Carolinas and in Georgia, growing in a wretched meagre soil. The wood serves only for fuel. Pursh says in pine-barrens, Carolina to Georgia, not above 15 feet high ; abundant.

33. QUERCUS GARRYANA.[1]

Q. Garryana. Foliis petiolatis deciduis oblongis superne latioribus subtus pubescentibus, sinubus acutioribus, angulis obtusis, calycibus fructus sessilibus subtus planis.

Flower in May ; acorns ripe in September of the second year.

Leaves deciduous, alternate, with short footstalks ; oblong, dilated upwards ; regularly sinuated, with obtuse, rounded, pointless, entire lobes, varying in length from 3 to 5 inches, and from 2½ to 3 inches broad ; of a strong coriaceous texture, mossy-green above, with a white stellate pubescence underneath, which is very conspicuous in the young half-expanded leaf. *Nerves* or ribs strong and prominent, covered with minute linear, blunt, brown, chaffy scales. *Stipules* deciduous, linear-spatulate, densely hirsute, half an inch in length.

Male flowers in pendulous, dense, hairy, yellow spikes, 1½ inch in length, springing from the buds below the leaves. *Segments of the calyx,* usually fine, fringed with long, white, silky hairs. *Stamens* generally seven, with a dotted central disc. *Calyx of the female,* coriaceous, scaly, downy, globose. *Style,* cylindrical, stiff, short. *Stigmas,* five, purple. *Germ,* quite sessile, ovate. *Acorns,* sessile, in pairs, usually one abortive, nearly 1 inch long, about the size of a large filbert, with a shallow flat cup, and having an even edge and linear obtuse pointed scales, soft and silky on the outside.

This is a handsome straight tree of considerable dimensions, 40 to 100 feet high, varying in thickness from 18 inches to 3 feet in diameter, widespreading and much divided by its branches at the top. Few trees have trunks exceeding 30 or 35 feet of clean wood or undivided by branches, much the greater number about 25. There is a great peculiarity in the bark of this species, different from any other with which I am acquainted, which is in the close resemblance it has to the bark of *Fraxinus*—namely, being divided by regular oblique shallow reticulated fissures, the

[1] *Q. Garryana,* Dougl. ex Hook. Fl. Bor. Am. ii. p. 159.

surface presenting an appearance of network of a white-grey colour. The bark of the young trees is smooth and that of the twigs somewhat downy.

For various domestic purposes the wood of the tree will be of great advantage, more especially in shipbuilding. Its hard, tough texture and durability, qualities of rare occurrence in the same species of oaks yet found in America, gives the present a decided superiority over every other native of that country; indeed, not one for general purposes can exceed it. It has been submitted to the usual tests and has been found in every point equal in strength and capable of receiving a polish in the same degree as *Q. Robur* of English botany, to which in appearance and in the quality of the wood it is so nearly akin.

The wood in a green state splits freely and very regularly, and makes excellent barrel-staves vastly superior to that (*Q. rubra*) generally used in the Canadas and in the United States of America, but the seasoned wood is very hard to work. Among the American species of oaks it comes between the White (*Q. alba*) and Iron Oak (*Q. stellata*, Willd.; *Q. obtusiloba*, Michx.). From the former of these species it is readily distinguished by the leaves, by the smaller fruit, and by the cup not being tuberculated; and from the latter it is equally readily separated by the leaves being less deeply and openly lobed, more strongly nerved, more of a coriaceous texture, sessile and larger fruit, and in having a more shallow and flatter cup.

The wood is much harder, tougher, and more durable than the former, and is not quite so hard as the latter, and from the greater toughness of *Q. Garryana* is less liable to splinter, and I believe I am not far wrong in saying it is for all purposes greatly superior to either.

Common in alluvial deposits, on a substratum of clay, on the low banks of the Columbia, but never at any time exceeding two hundred miles from the sea. Plentiful on the north banks of that stream sixty miles from the ocean, and from that circumstance named by Capt. Vancouver " *Oak Point,*" 1792.

It does not form thick woods as is the case with the Pine tribe, but is interspersed over the country in an open manner, forming belts or clumps along the tributaries of the larger streams, on which conveniently it could be floated down. Common from the 40° to the 50° North lat.

I have great pleasure in dedicating this species to N. Garry, Esq., Deputy Governor of the Hudson's Bay Company, as a sincere though simple token of regard.

⁎ The description of this last species seems to have been added by Douglas at a later date, after his journey to North-west America.—ED.

The following pages 51–76 consist of a condensed account in Douglas'
own handwriting of the journeys which are afterwards expanded and re-
counted in detail.   It has been thought better to reproduce it exactly
as it stands, as it is in many respects easier reading than the expanded
account, and also because it contains several not uninteresting little
additions.—ED.

A Sketch of a Journey to the North-Western Parts of the Continent of North America during the Years 1824, 1825, 1826, and 1827. By David Douglas, F.L.S.

WHILE so much geographical information has of late years been added to the general stock of knowledge, and so many distinguished individuals have assiduously devoted their talents to the investigation of the northern parts of this continent, the Horticultural Society being desirous of making known to the British gardens the vegetable treasures of those widely extended and highly diversified countries, resolved on sending a person qualified in the modes of collecting and preserving botanical subjects and of transmitting seeds to England. I had the pleasure of being the person selected, having previously extensively travelled in the eastern parts of the same continent for the same purpose. Before entering on this brief statement I must beg leave to return my grateful thanks to John Henry Pelly, Esq., Governor, and Nicholas Garry, Esq., Deputy-Governor of the Honourable the Hudson's Bay Company, for the kind assistance I on all occasions experienced, for much valuable information, both before and after my arrival in England; to whose enlightened zeal for all that tends to promote scientific objects they have ever taken a deep and warm interest, and I am truly happy of this opportunity of testifying my thanks. To the residents, partners, and agents I am also deeply indebted, and I sincerely thank them individually and collectively.

1824. July 25.

I embarked on board the Hudson's Bay Company's brig *William and Ann*, Captain Henry Hanwell, destined for the entrance of the River Columbia. To eke the monotony attending long voyages, I held myself fortunate in having a companion in John Scouler, M.D., skilled in several, and devotedly attached to all, branches of natural history, who undertook the voyage in the capacity of surgeon to the ship that he might have an opportunity of prosecuting his favourite pursuit. A few days' favourable weather carried us clear of the shores of England, and on August 9 we passed the high grounds of the Island of Porto Santo in the morning. About 4 P.M. of the same day we had the pleasure to anchor in the Bay of Funchal. So far as the experience of two days went on this highly delightful island I was amazingly gratified : visited the summit of one of the highest mountains in company with my companion the doctor; collected for our respective herbaria several interesting though not new plants; visited the vineyards in the neighbourhood of the town, hospital, church, and other establishments, when we resumed our voyage in the evening of the 12th towards Rio de Janeiro. From August 12th to 24th, as we approached the Equator, the temperature increased—the maximum 76°, minimum 59°; the greatest height was 84° in the shade on the 21st at 3 P.M.

The mornings were particularly pleasant and fine. Near to the Cape Verde Islands *Exocoetus volans* were skipping from wave to wave and sometimes fluttering on board, our vessel being low in the water, more especially in the evenings ; and the screaming noise of *Phaethon aethereus*, with the never-absent *Procellaria pelagica*, were the only alleviation from sky and water. Ten degrees on each side of the Equator the weather was very variable : calms, thunder and lightning, and sudden gusts of wind, which made this part of our voyage somewhat tedious.

On September 3rd the south-east trade sprang up, which carried us to within sight of Cape Frio on the 26th. Towards evening the ship was surrounded by a vast variety of sea-birds—*Procellaria*—and for the first time I saw *Diomedea exulans*. The magnificent prospect to the harbour of Rio is well known. One feature in Brazilian scenery strikes the European eye—that is, the leading one, the palms being larger on the summit of the highest hills. During my stay I had the pleasure to become known to William Harrison, Esq., residing at Botafogo, through whose exertions many beautiful plants have been introduced to England, and who bestows great pains in procuring subjects in other departments of natural history illustrative of Brazil. In company with this gentleman and Mr. Henry Harrison, a relation of his, who afforded me great assistance and showed me every attention which a good [1] can suggest, I made a short journey to Tijuca, where I was extremely gratified with the rich luxuriance of the forest (though seen to disadvantage, being too early), and was particularly delighted with the varied and endless forms of *Orchideae*.

Mr. Harrison cultivated with great success about seventy species of this family of plants by being simply nailed to his garden wall, having the benefit of the bark or wood on which they grew still given them for support. He had also an aviary containing several rare as well as beautiful native birds. I became known also to the late John Dickson, Esq., surgeon in the Royal Navy, who was never more happy than when he had an opportunity of doing an act of kindness. I left on October 15th this charming place with much regret, more especially having scarcely been able to put myself in possession of any dried plants from the earliness of the season and from the continued rain. For a few days, until we got clear of the land, the weather was changeable, accompanied by rains in the evenings.

At 4 A.M. on Tuesday the 19th a fine breeze sprang up and we bore away for the south, gradually leaving the warm weather. Off the Plate, in lat. 37° S., long. 37° W., immense shoals of *Fucus pyriformis* passed the ship, some of which measured 60 feet in length, and a stem at the thickest part 3 inches in diameter. On the root was a variety of *Asteria, Beroë* and other *Molluscae*. In this parallel *Procellaria capensis*

---

[1] Blank left in MS.—ED.

and *P. fuliginosa* began to be numerous, many of which I captured with a small hook and line. In passing between the mainland and the Falkland Islands (November 5th) an indescribable piercing chilliness told us we were drawing near to the dreary inhospitable region of Cape Horn, which in a few days longer we fully shared.

While within the parallels of 50° and 59° S. lat. I captured in all sixty-nine birds of *Diomedea*, consisting of *D. exulans*, *D. fuliginosa*, and *D. chlororhynchos*; the last, though a smaller bird than the first, reigns as lord over the others, and all give way to him on his approach. It is stated by most authors that these birds are more easily taken during calm weather. I have invariably found the reverse : it was only during the driving gusts of a storm that I could secure them, and on such occasions they fight voraciously about the bait, the hook often being received into the stomach. The appearance of these birds is majestic and grand. The largest of which I ever measured was 12 feet 4 inches from tip to tip between the wings, and 4 feet from the point of the beak to the point of the tail. As regards flight, all that can be said of the one species seems to be applicable to all. When sitting on the water the wings are gently raised like the swan ; when feeding, they are raised somewhat higher and curved, continually in tremulous motion, like the hawk tribe ; and when rising from the water to soar in the atmosphere they partly walk the water, tipping the surface with the points of the wings for the distance of several hundred yards ere they can raise themselves sufficiently high to soar, which they do with the greatest gracefulness, with scarcely any apparent motion of the wings. In illustration of this fact, they cannot rise from the deck of a vessel. They are bold and savage in disposition ; at least, they display these qualities on being made captives. Of *Larus* and *Procellaria* I captured vaster numbers by the same means—a hook baited with the fat of pork. In these latitudes a white-striped porpoise was observed, smaller in size but equally rapid in his motion with the well-known speed of the common one. Till we passed the parallel of 50° S. lat. on the Pacific side we were subject to boisterous weather, high seas, hail, rain, and thick fogs. On December 14th the Island of Mas-a-Fuera was distinctly seen at the distance of seven leagues, and appeared like a dark bare rock. We passed sufficiently close to see that it was far from being fertile, though a little verdure could be seen in the valleys, some stunted trees on the hills, and a few goats browsing on the rocky clefts.

A high surf breaking on the beach prevented a boat from being sent on shore ; we consequently bore away for the Island of Juan Fernandez. The wind having failed us, we did not reach it till two days afterwards. This classic island, which might be properly termed the Madeira of the South, is very mountainous and volcanic; the hills beautifully clothed with verdure to their summits, which, save in very clear weather, are enveloped in the clouds ; the red burnt-like rocky soil well contrasted with the rich varied deep green of its beautiful vegetation. As our boat approached the shore in Cumberland Bay we were not a little surprised to see smoke issuing from a small straw-thatched hut, a small neat coppered schooner

riding at anchor in the bay.   But our astonishment was still more increased
when, on the eve of landing, a person sprang from the thicket behind,
saluted us in English and directed our boat to a sheltered creek, carried us
to a convenient place to water, and for wild fruits to the neglected gardens
of the Spanish settlements.   This person's name was William Clark,
native of Whitechapel, London, an English sailor who had been five
years on the coast of Chile and had been some months on this island
with five Spaniards, procuring seals, goats, skins, and making junk
from the wild cattle.   He commanded the little bark (25 tons burden).
His intelligence was considerably beyond his sphere, his mode of life
altogether worthy of being recorded.   The others I did not see, for
they were on the opposite side of the island procuring furs ;   while
Mr. Clark's province was to look to his vessel, which he appeared quite
equal to, and seemed to be happy though far from being comfortable.

No pen can correctly depict the rural enchanting appearance of this
island and the numerous rills descending through the valleys shaded
by rich luxuriant verdure terminating in the dark recesses of rocky dells,
while the feathery fronds of *Lomaria, Aspidium*, and *Polypodium*—several
species of which are new and truly princely—form a denseness to the
forests.   On the hills were several species of *Escallonia, Berberis, Lobelia,
Hordeum*, and *Avena*.   During my short stay I secured seventy distinct
and exceedingly interesting plants for my herbarium.

The birds were few as to new species, and not beautiful.   I killed
one of *Strix* and several of a dark species of *Columba*, which were
exceedingly abundant.

In this bay, about 200 yards from the tide mark, is seen, overgrown
by nettles, thistles, and a curious shrubby *Campanula*, a circular oven
built of London fire-brick, about 7 feet diameter, marked on the top 1741.
This oven did not appear to have been recently used, as this species of
bird had converted it into a cote, out of which I took some eggs, but no
young.   This oven probably was built by Anson.   Twenty-six pieces of
cannon lay on the beach, and the ruin of an old church is to be seen.
Over the door is the following : ' La casa de Dios puerta del cielo y so
colocoesta a 24 Septiembre 1811 '[1] (The house of God consecrated
24 September 1811).   In the old gardens we found some pears, apples,
prunes, and figs.   There were also some vines, but these were not ripe ;
also one kind of strawberry.

I sowed a quantity of garden seeds here and at Cruz Bay.   The doctor
and myself made some little additions to our new friend's stock of clothing,
for which he was thankful.   He presented us with a goat, which we
took on board alive.   We left Clark standing on a large stone on the
beach wishing well to the ship, but scarcely had we reached her when
a strong easterly wind compelled us to stand out to sea, and then we were
obliged to leave this island and poor Clark.   Our course was then directed
towards the Galapagos Islands, which we pleasantly gained on the
Sunday, January 9th, 1825.

[1] We cannot vouch for the correctness of the Spanish, but it is given exactly as it
is written in the MS.—ED.

1825.

We passed along the eastern side of Chatham Island, and on to James Island, thirty-seven miles further to the west. The whole of these islands are mountainous and volcanic, with numerous vestiges of volcanoes to be seen, and in many places covered with lava. The verdure is scanty compared with tropical countries, arising no doubt from the uniformity of soil, more especially from the parched ground and nearly total want of springs.

The birds are exceedingly numerous, and so little acquainted with man's devices that the smaller ones we killed with sticks. A species of Rock Cod was taken in great abundance near to the steep rocks on the north-east side of the bay, without any bait, and so voracious were the sharks that they kept biting the points of the oars as they left the water. The woods teemed with land tortoise of prodigious size. The sandy part of the shore was slated [1] with turtle; we took in one day thirteen in the water while they were performing their sacrifices to Venus. They are powerful.

Of the many plants and choice collection of birds I had prepared with no ordinary care, I lost, I may say, all, in consequence of the almost incessant rains for twelve days after leaving the island, not being sufficiently dry to pack away. Nothing did I regret so much as a new species of *Lacerta*, 20 to 30 inches long, of a dark orange colour, a rough warty skin, and which made good soup.

We then proceeded towards the Columbia. A few degrees north of the Equator we were becalmed for nearly ten days, but we found a powerful current which carried us 10° to the westward from forty-five to forty-six miles a day. The remainder of our voyage within the tropics was attended with lightning in the evenings, variable winds, and frequent rains.

In the latitude of 34° N., long. 130° W., I caught an undescribed species of *Diomedea*, akin to *D. fuliginosa*, but a less powerful bird. *D. exulans* found in the higher latitudes in the Pacific is a much smaller bird than that found in the Southern hemisphere, and will on comparison prove a distinct bird. On February 12th we were in the latitude of the river, long. 136° W., but the weather was so terribly boisterous, with such a dreadfully heavy sea, that we were obliged to lay to, day after day, till April 1st. During this period we endured a thousandfold more bad weather than on all our voyage. At the beginning of the month the weather became a little inviting, and we drew in to the land. On Sunday the 3rd, at daybreak, Cape Disappointment was seen at twenty-eight miles distant. With a fine breeze we came close in with the shore, when a violent westerly blast obliged us to again run to sea. No channel whatever could be seen, only the water breaking furiously the whole breadth of its bar. The following Thursday (7th) the wind again moderated, and we made a second attempt. At 6 A.M. we were only forty miles distant when all sail was set, and, the breeze improving as the day advanced, we had the pleasure of entering in perfect safety this dangerous place,

[1] Probably meaning 'as close together as the slates on a roof.'—Ed.

and anchored in Baker's Bay, having had an interesting though in some points a comparatively tedious voyage.

The rain on Friday fell in torrents without interruption. On Saturday the 9th, the doctor and I went on shore in Baker's Bay, where immediately on stepping out of the boat we found *Rubus spectabilis* and *Gaultheria Shallon* growing close to the water's edge. We picked many plants only known to us as in the herbaria, others only by name. On returning to the ship we found that a Canadian had been sent from the establishment, accompanied by some natives, with fresh provisions for us. Several canoes of Indians visited the ship and behaved civilly, bringing dried salmon, fresh sturgeon, and dried berries of various kinds, which they readily gave in exchange for trinkets, molasses, and bread. On Monday the 11th the vessel drew up to opposite the Fort and anchored at 'Point Ellis' on the north side of the river. On Tuesday the 12th we were received by the late Mr. Alexander McKenzie, who showed us all attention in his power. The country along the coast is mountainous and woody, consisting principally of pines of various species, for a more particular description of which the reader is referred to the account of this family from the pen of Joseph Sabine, Esq.

Cape Disappointment, on the northern point of the river, is a remarkable promontory considerably elevated above the level of the sea, and forms a good landmark for ships entering the river. This noble stream is 4½ miles in breadth at its confluence with the sea, widening into deep bays ten to fourteen miles above with a deep rapid current. On every occasion we continued our little excursions when the weather permitted, when our attention was continually arrested by some new or strikingly grand objects. Nothing gave us more sincere pleasure than to find abundance of *Hookera lucens* [sic] in the damp shady forests, and *Menziesia ferruginea*. All my papers and trunks were sent on shore on the 16th, and on the 19th I embarked in a small boat with Dr. John McLoughlin, chief factor, who received me with every demonstration of good feeling and gave me all kindly offices which he had in his power to bestow.

Here I may be permitted to tender my warmest thanks to my friend the captain for his kindness during my voyage. The following night, at ten, we arrived at Fort Vancouver, ninety miles from the sea, a few miles below Point Vancouver, the spot where the officers of that expedition terminated the survey of the river in 1792. The scenery from this place is sublime—high well-wooded hills, mountains covered with perpetual snow, extensive natural meadows and plains of deep fertile alluvial deposit covered with a rich sward of grass and a profusion of flowering plants. The most remarkable mountains are Mounts Hood, St. Helens, Vancouver, and Jefferson, which are at all seasons covered with snow as low down as the summit of the hills by which they are surrounded. From this period till May 10th my labour in the neighbourhood of this place was well rewarded by *Ribes sanguineum, Berberis Aquifolium, B. glumacea, (B. nervosa,* Pursh), *Acer macrophyllum, Scilla esculenta,*[1] *Pyrola*

[1] *Camassia esculenta,* Hook. Fl. Bor. Am. ii. p. 186.

aphylla, Caprifolium ciliosum,[1] and a multitude of other plants. I cannot pass over the grandeur of *Lupinus polyphyllus* covering immense tracts of low land on the banks of streams, with here and there a white variety. This beautiful plant attains the height of 6 to 8 feet where partly overflowed by water. My friend Dr. Scouler having visited me for a few days, I returned with him to the sea in order to have the residue of my articles brought up, with among other curious plants a noble species of *Arbutus, A. procera.*[2] We had abundance of salmon brought to us by the native tribes, which was purchased cheap and which we found excellent. I returned to Fort Vancouver at the end of the month, having increased my collection by seventy-five species of plants, a few birds and insects, and four quadrupeds.

I resumed in the vicinity of Fort Vancouver, procuring seeds of early-flowering plants and adding others to my collection till June 20th, when I availed myself of the despatch of the boats to the inland establishments, of going with them as far as it appeared to me to be advisable. The current at this season of the year being exceedingly powerful from the melting of the snow in the mountains, the boats made but little way, so that in the open parts, unobstructed by timber or rocks, I botanised along the banks. The second day we passed the Grand Rapids, forty-six miles above the Fort. The scenery at this place is romantic and wild, with high mountains on each side clothed with timber of immense size. The rapid is formed by the river passing through a narrow channel 170 yards wide. The channel is rocky, obstructed by large stones and small islands, with a descent of 147 feet, being about two miles long. Whole petrified trees, both of pine and *Acer macrophyllum*, are visible close to the edge of the water.

This being the season of salmon-fishing, I had opportunities of seeing prodigious numbers taken, simply with a small hoop or scoop net fastened to the end of a pole. The salmon is excellent in quality, averaging 15 lb. in weight. The seine is resorted to as a means of taking salmon in the still parts of the stream with great success; spindle-formed pieces of the wood of *Thuya plicata*, which is very buoyant, attached to the net by the smaller end, act as corks and oblong stones as lead, which both serve the purpose well. The rope of the net is made from the bark of a species of *Salix*, some of *Thuya*; the cord of *Apocynum piscatorium*,[3] a gigantic species peculiar to that country, which affords a great quantity of flax. From the Grand Rapids to the Great Falls (70 miles) the banks are steep, rocky, and in many places rugged. The hills gradually diminish in elevation, and are thinly clothed with stunted timber, the shrubs only a few feet high. We are no longer fanned by the huge pine, *Thuya* or *Acer*, or regaled by *Populus tremuloides* for ever quivering in the breeze. As far as the eye can stretch is one dreary waste of barren soil thinly clothed with herbage. In such places are found the beautiful *Clarkia pulchella, Calochortus*

[1] *Lonicera ciliosa,* A. Gray, Syn. Fl. N. Am. i. II. p. 16.
[2] *Arbutus Menziesii,* A. Gray, *loc. cit.,* ii. I. p. 27.
[3] *Apocynum cannabinum* var. *glaberrimum,* A. DC. in DC. Prod. viii. p. 439.

macrocarpus, Lupinus aridus, L. leucophyllus, Brodiaea, &c. The present bed of the river at the Falls is 600 feet below the former bed of the river of decomposed granite. I could not this season go higher than a few miles above the Falls, but was amply repaid by *Purshia tridentata, Bartonia albicaulis,*[1] *Collomia grandiflora,* and several Pentstemons, and by seeds of many desirable plants. In this journey many fine plants were secured. Early in the morning of July 19th I descended the river in an Indian canoe for the purpose of prosecuting my researches on the coast, which was in a great measure frustrated by the tribe among whom I lived going to war with the natives to the northward, the direction I intended to follow. During my stay several persons were killed and some wounded in a quarrel. The principal chief in the village, Cockqua, behaved to me with great fidelity, and had built for me a small cabin in his own lodge; but by reason of the immense number of fleas and the great inconvenience suffered thereby, I preferred to put up at my own camp, a few yards from the village, on the shore of the river. He was so deeply interested in my safety that he watched himself the whole night on which he expected the war party.

In the evening about three hundred men in their war garments danced the war dance and sung several death songs, which to me alone imparted an indescribable sensation. The following day seventeen canoes of warriors, nearly four hundred men, made their appearance, when, after several harangues, hostilities were for the present suspended.

A sturgeon was caught by one of my chief's fishermen, which measured 12 feet 9 inches from the snout to the point of the tail, and 7 feet round at the thickest part. The weight might on a moderate calculation exceed 500 lb. Among the plants I found on this occasion were *Lupinus littoralis, Carex Menziesii*[2] [*sic*], *Juncus Menziesii,*[3] *J. globosus,*[4] *Vaccinium ovatum, V. parvifolium* and *ovalifolium,* and seeds of the beautiful *Spiraea ariaefolia,*[5] *Gaultheria, Ribes sanguineum, Berberis,* and of several other valuable plants. Before taking leave of my Indian friend I purchased from his people several articles of wearing apparel, gaming articles, and things used in domestic economy, for which I gave trinkets and tobacco.

I arrived again at Fort Vancouver on August 5th, and was employed until the 18th drying the plants collected, making short journeys in quest of seeds and other plants, in the course of which time my labours were greatly retarded by rainy weather. On the 19th I left again for the purpose of ascending the River Multnomah, one of the southern tributaries of the Columbia. This is a fine stream with very fertile banks. Thirty-six miles from the Columbia are falls of 43 feet perpendicular pitch across the whole breadth of the river, in one sheet at this season of the year, but during the spring and autumn divided into three channels. There is but little current thus far, the water being gorged back by the waters of the Columbia. Making the portage over the falls was no small undertaking. I killed

---

[1] *Mentzelia albicaulis,* S. Wats. Bibl. Ind. N. Am. Bot. p. 389.
[2] err. for *C. Menziesiana = C. macrocephala* Boott, Illustr. Carex, i. p. 27.
[3] *Juncus falcatus,* Ind. Kew. fasc. ii. p. 1258.
[4] See Hook. Fl. Bot. Am. i. p. 190 as a syn. of *Juncus polycephalus.*
[5] *S. discolor* var. *ariaefolia,* S. Wats. Bibl. Ind. N. Am. Bot. p. 321.

several of *Cervus leucurus*, or Long White-tailed Deer, as well as some of the Black-tailed Deer.

Two days more took me to the village of the Calapooie nation, a peaceful, good-disposed people, twenty-four miles above the falls, where my camp was formed for several days. A hunting party left me here, they going over the ridge of mountains to the westward. Near to my camp was a saline spring to which the deer and a beautiful ringed species of *Columba* resorted in the mornings for the purpose of licking and picking the saline particles, which afforded me great amusement. In the extensive plains bounded on the east by the mountainous woody parts of the coast and a ridge of high mountains on the west, on the banks of the River Sandiam, one of the rapid branches of the Mult-nomah, was abundance of *Eschscholzia californica*, *Iris tenax*, *Nicotiana multivalvis*,[1] two new species of *Trichostema*, and many other delightful plants. Of animals I procured some curious species of *Myoxus*, *Mus*, *Arctomys*, a new species of *Canis* of singular habits, and a new genus of animals ² of Richardson ('Fauna Am.,' page ²). In the tobacco pouches of the Indians I found the seeds of a remarkably large pine which they eat as nuts, and from whom I learned it existed in the mountains to the south. No time was lost in ascertaining the existence of this truly grand tree, which I named *Pinus Lambertiana*, but no perfect seeds could I find. I returned again to my rendezvous, Fort Vancouver, richly laden with many treasures.

A few days was devoted to the putting in order of my last collection and drying the seeds collected, when, without loss of time, on September 5th, having engaged a chief as my guide and accompanied by one Canadian, I left on a journey to the Grand Rapids. Two days were spent ascending the stream, though I was favoured with a sail wind. I pitched my camp close to Chumtalia's (my guide's) house, taking the precaution to have the ground well drenched with water to prevent the annoyance of fleas, which did not entirely banish those pestiferous insects.

On Saturday morning, when Chumtalia learned that it was my wish and intention to visit the summit of the mountains on the north side of the river, he became forthwith sick, and soon found an excuse for exempting himself from this undertaking. He, however, sent a younger brother instead, accompanied by two young men of the village. The Canadian remained at the tent to take care of my boots, &c., injunctions being given to Chumtalia to supply him with salmon and to see that nothing came over him [*sic*]. To encourage my guides I was under the neces-sity of giving them the whole of the provisions the first day, excepting four small biscuits and a little tea and sugar. At our first encampment, about two-thirds up the mountain, we left our blankets, intending, after having visited the top, to return there and sleep. But our path being so dreadfully fatiguing, climbing over the shelving detached rocks and fallen timber, that night had closed in on us ere we had reached the summit; I killed a young or half-grown brown eagle, on which we

---

[1] *N. quadrivalvis* var. *multivalvis*, A. Gray, Syn. Fl. N. Am. ii. I. p. 243.
[2] Blank left in MS.—ED.

fared, and with a little tea made in an open kettle and drunk out of bark dishes, and a good fire, passed a tolerable night without any bedding. Previous to lying down I took the precaution to dry all my clothing, which was soaked with perspiration from the violent exertion of the day. The following day, in the dusk of the evening, I reached my camp faint and weak, but was pleased to learn that all things were going on smoothly. My feet suffered so much on the last three days' journey, that I was totally unable to think of continuing my journeys until refreshed. I therefore amused myself fishing and shooting seals (*Phoca vitulina*) which were sporting in vast numbers in the rapid, where salmon were particularly abundant. In two days I got my guide Chumtalia to attend me to the mountain on the south side of the river, which he did most willingly. I found this of easier ascent, and reached the top after a laborious march of fifteen hours. I had the good fortune to find two new species of pines, *Pinus nobilis*[1] and *P. amabilis*,[2] two of the noblest species of the tribe; *Helonias tenax*,[3] a new species of *Rhododendron*, and a scarce species of *Pterospora* ; also some interesting species of *Ribes*. In the rocky places of the mountain *Arbutus tomentosa* [4] was not a stranger. I procured seeds of several species of *Pentstemon*.

On the morning of the 13th I embarked in my canoe, and shortly after midday I once more reached Fort Vancouver, where I had the pleasure to find Dr. Scouler returned from his northern voyage, and was delighted to hear of his success. We sat and talked over our several journeys, unconscious of time, until the sun from behind the majestic hills warned us that a new day had come. We therefore retired for a few hours. The remainder of this month was devoted to packing up my gleanings of dried plants, consisting of sixteen large bundles of American and eight from other places, a large chest of seeds, one of birds and quadrupeds, and one of various articles of dress &c. A portion of each of the varieties of seeds was reserved for the purpose of sending across the continent in the ensuing spring.

In consequence of a slight wound I was unable to continue my labour for some weeks, which greatly distressed me at this important season of the year—October 22nd to November 15th. Having learned that the ship was still detained by contrary wind, and finding myself somewhat better, I determined on visiting Whitbey Harbour, situated in 48° N. lat., in search of seaweeds and other plants. I left the establishment in company with the late Mr. Alexander McKenzie and a small canoe and four Indians. A boisterous westerly wind obliged us to camp at the junction of the River Multnomah, having made only twenty miles progress. The following day we proceeded, and at eight at night reached the village of Oak Point, where I had a letter given to me from my friend Dr. Scouler, who informed me they would probably be in the river for some days longer. We lost no time, but proceeded to McKenzie's encampment, four miles below the village, where we supped and resumed our route.

[1] *Abies nobilis*, Mast. in Journ. R. Hort. Soc. xiv. p. 193.
[2] *Abies amabilis*, Mast. *loc. cit.*, p. 189.
[3] *Xerophyllum asphodeloides*, var. Baker, in Journ. Linn. Soc. xvii. p. 467.
[4] *Arctostaphylos tomentosa*, A. Gray, Syn. Fl. N. Am. ii. I. p. 28.

At four in the morning of the 24th, with the setting in of the tide, a violent westerly wind obliged us to keep alongshore on the north side of the river, and being unable to double Tongue Point, we hauled the canoe over that neck of land. The wind moderated about nine and veered round to the north-east, when the ship in the bay weighed anchor and put to sea. We thus lost her only by an hour. We slept at the village of Com Comly, the chief of the Chenook, on the north side of the river. This old man sent his canoe and twelve Indians to ferry us across the river to Baker's Bay on the south side, which they performed with great skill, though we had the misfortune to be taken in a violent storm while in the centre of the channel. We reached the shore in perfect safety, but with the loss of a few pounds of flour and a little tea—all our provisions with the exception of a few ounces of chocolate which I had in my pocket— by the sea breaking over us, which put us to some inconvenience. We camped for the night near Knight's River in Baker's Bay.

The following day we made a short portage over the bite or neck of Cape Disappointment to a small lake, out of which flowed a narrow stream to the bay northward of the cape, which we descended, and put on shore at dusk on Cape Foulweather. The rain fell in torrents without intermission throughout the day. We sent the canoe to the Columbia from this place in the evening, the Indians being anxious to return hastily as we had not the means of feeding them. The wind about midnight increased to a hurricane with sleet and hail, and twice were we obliged to shift our camp, the sea rising so unusually high. We had no protection save what a few pine branches and our wet blankets afforded, and no food. Long ere daylight we were ready to leave Cape Foulweather, which name it truly deserves, and we walked along the sandy beach sixteen miles to Whitbey Harbour, where we found the village deserted, our prospect not in the least bettered. We remained here several days, faring scantily on roots of *Sagittaria sagittifolia* and *Lupinus littoralis*, called in the Chenook tongue *Somŭchtan,* and from continual exposure to the cold and rain and the want of proper sustenance I became greatly reduced. So soon as the weather permitted us to make a large fire to attract the notice of our guide's friends residing on the opposite side of the bay, they sent a canoe for us, with whom we stayed some days, sharing of what they had to spare.

We ascended this stream, the River Cheecheeler, sixty miles in an Indian canoe, when from the deluges of rain and the advanced season our zeal cooled and we gave up our voyage which was to have gone to its source. We therefore paid our last guide, 'The Beard,' his fees for his attendance and loan of his canoe, and took in our employ two fresh guides to conduct us overland from this stream to the River Cow-a-lidsk, a stream of considerable size which empties itself into the Columbia forty miles from the ocean. This distance, though not more than forty miles, took two days. The low places on the plain were lakes, the rivulets had overflowed their banks, and the difficulty of ascending and descending the low woody hills was increased thereby. It rained both days. We used all the berries I had collected on this journey, and Mr. McKenzie suffered some inconvenience from having eaten a few roots of a species of *Narthecium.*

Fortunately we found at the Cow-a-lidsk a small boat which Schach-anaway, the chief, had borrowed from the establishment a few days before. He gave us a few roots, some dried salmon, and a goose (*Anser canadensis*). The following day we descended the river to the Columbia, and on November 15th landed at the establishment. Seeds of *Helonias tenax*,[1] *Rubus spectabilis*, and a few others were the only things I saved on this unfortunate journey ; all marine birds and seaweeds I was obliged to leave. I had suffered so much from my last trip that little hope was left me for being able to do much good for this season, at least in botany.

1826.

When opportunity favoured I collected woods, and gathered *Musci*, &c., and from this time to March 20th I formed a tolerable collection of preserved animals and birds, but this desirable object was frequently interrupted by heavy rains. Among the birds and animals deserve to be mentioned *Tetrao Sabini*, *T. Richardsonii*, *Sarcoramphos californica*, *Corvus Stelleri*, an endless variety of *Anas*, several species of *Canis*, *Cervus*, *Mus*, and *Myoxus*. The winter was uncommonly mild, scarcely any frost, and no snow until February 20th.

March 20th.—Having resolved to devote a season in the interior parts of the country skirting the Rocky Mountains, Dr. McLoughlin, who was unremitting with his kind attentions, allowed me to embark in the spring boat for the interior with two reams of paper, which was an enormous indulgence. Rather than go unprovided in this respect I curtailed the small supply of clothing. We reached the falls on the 24th. From this point to Wallawallah, the first inland post, the country is hilly, destitute of timber, the soil sandy and barren, the banks of the river rocky. I walked along the banks of the river save when the boat was under sail, and, though early in the season, added some plants to my collection—the beautiful *Lilium pudicum*,[2] Pursh, two new genera of *Cruciferae*, and *Ribes cereum* ; but what gave me most pleasure was finding a new species of *Wulfenia*. The banks of the river as high up as the junction of Spokane River are steep, bold and rocky, the river rapid and difficult to ascend. This part of the country is barren, sandy soils and very parched. A few deer, *C[ervus] leucurus* (*C. macrotis*, Say), bears, wolves, foxes, and badgers are occasionally seen among the brushwood, which consists principally of *Purshia tridentata* and *Artemisia arborea* [*sic*]. The principal of the feathered tribe are *Tetrao urophasianus* and *T. urophasianellus*, at this season of the year celebrating their nuptials on the gravelly shores of the streams. In several contracted parts of the river, where the breadth does not exceed 200 yards, the water during the melting of the snow rises to the amazing height of 43 feet perpendicular above its ordinary level. On April 11th we arrived at the junction of Spokane River, where Mr. John W. Dease had fixed his camp. I took an opportunity of sending letters

---

[1] *Xerophyllum asphodeloides*, var. Baker, in Journ. Linn. Soc. xvii. p. 467.
[2] *Fritillaria pudica*, Baker, loc. cit., xiv. p. 267.

to England across the continent to Hudson's Bay, and then accompanied Mr. Dease to Kettle Falls on the Columbia, where a new establishment was about to be formed by him. Of all the places I have seen this is by far the finest : high rugged mountains, fertile valleys, and this immense body of water dashed over a pitch 24 feet perpendicular, the country abounding with game. Many fine plants were collected here not before in my collection ; and *Tetrao Richardsonii* and *T. urophasianellus* were so plentiful that they formed a principal part of food.

Mr. Dease kindly gave me the use of three of his best horses and engaged two hunters to attend me to Spokane, distant about seventy miles in a northerly direction from this spot, which was his former residence. On May 2nd I set out for that purpose and travelled slowly, finding the country interesting, and arrived late on the third day. Mr. Jacques Finlay was here, and obligingly undertook to repair the lock of my gun, and on this occasion I felt happy in having it in my power to give him some assistance in provisions. For several days he had had nothing excepting a sort of cake made of *Lichen jubatum*, Linn., and a few roots of *Scilla esculenta* [1] and of *Lewisia rediviva*. I spent a few days here and returned to Kettle Falls on Sunday the 14th. On this journey I found *Ribes viscosissimum, R. petiolare*,[2] *R. tenuiflorum*,[3] *Astragalus glareosus*, a new species of *Pinus, P. ponderosa*, and two new species of *Viscum*—one, the larger species, on the pine mentioned, the smaller on *P. Banksiana*, which is not rare, though a smaller tree than that found on the east side of the Rocky Mountains. One large bear, *Ursus horribilis*, was killed by young Finlay ; it was too bulky to be preserved.

I continued in the neighbourhood of this place till June 5th, when I availed myself of the opportunity of descending in one of the boats with Mr. William Kittson to Wallawallah, on the plains, where I should remain for four weeks in order to secure the herbage of these regions. I had already three large bundles of select plants and upwards of forty varieties of seeds, among which were *Chelone Scouleri*,[4] *Claytonia lanceolata*,[5] *Erythronium maximum*,[6] and *Rubus nutkanus*; of birds a very interesting species of *Scolopax*.

At 6 P.M. on the 8th our descent had averaged 115 miles a day, and the high water was not then at full height. On my descent I found *Pentstemon deustus* on the narrows above Okanagan, *Malva Munroi* [sic][7] on the same place, *Eriogonum sphaerocephalum* on the Stony Islands, *Abronia vespertina*[8] and *Phacelia ramosissima* near the Priests Rapid ; and on the clay hills above Okanagan and near the Big or Great Bend, *Pentstemon speciosus*.[9]

[1] *Camassia esculenta*, Hook. Fl. Bor. Am. ii. p. 186.
[2] *R. hudsonianum*, var. S. Wats. Bibl. Ind. N. Am. Bot. p. 334.
[3] *R. aureum* var. *tenuiflorum*, S. Wats. *loc. cit.*, p. 332.
[4] =*Pentstemon Scouleri*, see Bot. Reg. t. 1277 = *P. Menziesii* var. *Scouleri*, A. Gray, Syn. Fl. N. Am. ii. I. p. 260.
[5] (?) *C. caroliniana* var. *sessilifolia*, S. Wats. Bibl. Ind. N. Am. Bot. p. 117.
[6] *E. grandiflorum* var. *gigantum*, Baker, in Journ. Linn. Soc. xiv. p. 298.
[7] *Malvastrum Munroanum*, S. Wats. Bibl. Ind. N. Am. Bot. p. 138.
[8] *A. mellifera*, Choisy, in DC. Prod. xiii. II. p. 435.
[9] *P. glaber*, A. Gray, Syn. Fl. N. Am. ii. I. p. 262.

Saturday, June 17th.—To a ridge of high snowy mountains about a hundred miles distant south-east, having taken measures to have horses and guides provided some days before. I left about 2 P.M. in an easterly direction along the southern banks of the River Wallawallah, which stream I followed to the base of the Blue Mountains.

On Tuesday the 20th, with a view of crossing this ridge to the opposite side, I started early in the morning and stayed near a small well to breakfast about midday, having passed over the first snow-capped hump, made a short stay and proceeded very slowly (for the horses were weak) till 4 P.M., when I was reluctantly obliged to stop, seeing no hope of the horses being able to cross the snow, which was very deep. I camped on its verge, left the two guides at the camp, and proceeded on foot to the summit, which I gained at six o'clock. The lower part was difficult to ascend, the snow being soft, but after passing over this part of it, a crust being on that above, I gained it without any difficulty. Scarcely had half an hour gone when a dark cloud passed over me and a dreadful thunderstorm commenced, with lightning in massy sheets, mixed with forked flashes and hail and large pieces of ice, and the thunder resounding through the deep valleys below. In the dying gusts of the storm one of the most sublime spectacles in Nature presented itself: the declining sun had just partially gilt the top of the snowy mountains, and below a magnificent rainbow was nearly a perfect circle. All tended to impress the mind with reverential awe. I might have crossed this ridge at a more southerly point, but the boy who served me as interpreter found little difficulty in preventing my guides from accompanying me.

I contented myself by botanising over the eastern declivities of the mountains for a few days, and returned to the Columbia on Sunday the 25th. All this time, toil, and some vexation, were not spent without being productive of some pleasure. In those untrodden regions on the verge of eternal snow were *Paeonia Brownii*, the first ever found in America ; and at a lower elevation the whole declivities of the mountains were covered with *Lupinus Sabinii* (whose beautiful golden blossoms gave a tint to the country that reminded me of *Spartium scoparium*),[1] *Trifolium megacephalum* (*Lupinaster macrocephalus*, Pursh), *Trifolium altissimum*, and several new species of *Phlox*.

Near to the junction of Lewis and Clarke's River with the Columbia I remained till July 8th, when tired of this barren country which scarcely afforded food—more especially this season, for, from the unusually high state of the water, no *barriers* could be made in the small stream to take salmon and no net could be used in the Columbia ; horse-flesh was therefore the principal food. I descended the river to the Gr[eat] Falls, where I had the good fortune to meet the Brigade of Boats on their way to the interior and letters for me from England. I returned and walked to Wallawallah a second time, on Saturday the 15th.

July 16th to August 5th.—In the hope of increasing my collection and knowledge of the country, I left this place, attached to a party of twenty-eight men commanded by Messrs. Archibald McDonald and John Work,

[1] *Cytisus scoparius*, Hook. f. Stud. Fl. Brit. Isl. ed. 3. p. 92.

to whom I was known and from whom I had many good offices, for the purpose of purchasing horses of several Indian tribes who had assembled at the forks of Lewis and Clarke's River for that purpose. This is a fine large stream flowing through the same sort of country as the interior parts of the Columbia. The weather was intensely warm, and we were obliged to travel by short stages in order to devote time to procure fish. These, however, were scarce, and we were under the necessity of destroying our horses as a means of support, and some of the men preferred dog's flesh. We arrived on the 25th, and the following day was devoted to dancing and racing, in testimony of their respect for me. From this point I again visited the Blue Mountains, the same ridge I had been on two weeks ago. The want of provisions obliged me to return sooner than I should have otherwise done. I brought with me *Pentstemon glandulosus, Ribes irriguum*,[1] and many fine seeds.

One of the principal chiefs whose nation was the most powerful, not fewer in all than 700 warriors in his band, got into a quarrel with Tosand, the interpreter, respecting his not translating faithfully to Mr. Work. Seeing our party was weak, he did not fail to take advantage of the circumstance by summoning his men to arms, in which the different tribes assembled joined. Our party being divided, fourteen on one side of the river and fifteen on the other, surrounded by a powerful armed party, our state was not an enviable one. The cool way in which we looked to these proceedings had a good effect on these persons. Rather than come to extremities their ill-temper was removed by presents. The next day was devoted to peacemaking, smoking, haranguing, and dancing. Though friendship was apparently restored, I could not trust myself from the camp, but continued to pick up what appeared curious in its immediate vicinity. On August 31st [*sic*] [2] Mr. Work, with six Canadian hunters and a band of 114 fine horses, left this place for Spokane overland to the north, whom I joined, intending, if possible, to spend a few days at the Kettle Falls. We passed over an undulating woodless country of good soil, but not well watered; we were obliged to cook from stagnant pools full of lizards, frogs, water snakes, *Lemna* and *Utricularia*. Thirty miles south of the Spokane River we came to a fine small lake, very deep, clear and cool water, in a pine wood, where we stayed a few hours, and arrived the following morning at Spokane. From this place to the Kettle Falls I went by the same route I followed in May last, and arrived on August 5th.

August 6th to 15th was devoted to collecting seeds of all the important plants seen here in the early part of summer, and making other additions. Learning from Dr. McLoughlin the ship would be despatched for England on the first of the month, I felt desirous of forwarding by her the whole of my collection. I consequently made arrangements through Mr. Dease to pass me on from one tribe to the other between this place and the sea. 'The Little Wolf,' chief of a tribe near Okanagan Lake, a useful fine fellow, was selected to go with me by land the 200 miles.

---

[1] *R. divaricatum* var. *irriguum*, S. Wats. Bibl. Ind. N. Am. Bot. p. 333.
[2] Probably a mistake of the pen and should have been July 31.—ED.

A party of Cootanie Indians arrived, with whom and the Little Wolf's people an old grievance existed, when war was instantly declared by the Wolf. This delayed me some days, waiting on him until peace was restored, when a feast was given which took several days. Mr. Dease in the meantime provided me with three fine horses and a good little man, Robado, belonging to the Spokane tribe, to conduct me to Okanagan. I passed over a vastly interesting country, the former channel of this princely stream, to which justice cannot here be done. I spent one day with Mr. McDonald, who hired for me a small canoe, in which I embarked with an old man and his son, a youth about fourteen years old. Five miles below, passing a rapid, I lost the whole of my provisions, the canoe having swamped. I had to pay all my guides' losses. I descended the whole chain of this river from the Kettle Falls to the sea, a distance of 800 miles, 600 of that in Indian canoes. In the narrow below the Great Falls, a dangerous rapid part of the river, my canoe was wrecked (Sunday, August 27th). Melancholy to relate, I lost the whole of my insects, a few seeds, and my pistols. The first was a heavy loss; this was replaced by Packenawaka,[1] the chief on the Falls, by a much smaller and inferior canoe, so small that with difficulty it held me and two guides. I procured one of my guides, 'Red Coat,' who was a valuable man, as I understood his language.

This part of the river has little current—that is, as far as the Grand Rapids. I resolved to drift day and night in order to secure the ship. All went well with me, but I was obliged to leave the canoe fifteen miles above the Grand Rapids and proceed by land; a strong easterly wind had raised such a swell that no progress could be made. I repaired to the home of my old guide Chumtalia, and got him to carry me to Fort Vancouver in one of his canoes, which he did with great speed. I arrived in safety at four in the morning of August 30th. I learned that all my articles had arrived safe, and the 31st was devoted to packing and writing home to England.

September 1st to 20th.—This time was spent making short journeys, and preparing for a journey to Northern California.

On Wednesday the 20th proceeded up the River Multnomah by the same conveyance as last year and arrived at the old encampment on the evening of the 22nd. Several days were spent collecting the horses and getting the baggage in travelling condition. The party consisted of thirty, commanded by Mr. A. R. McLeod.

On the 28th (Thursday) and the two following days we travelled in a south-west direction over a fertile woody country abounding with small streams, plains, and belts of large oaks; deer being scarce in consequence of the plains being burned by the Indians to compel these animals to seek food in certain parts more convenient for hunting. On the low hills on gravelly soil *Arbutus procera*[2] attains a much greater size than any on the Columbia; they are frequently 15 inches to 2 feet in diameter near the root, and 30 to 45 feet high.

---

[1] Spelled Pawquanawaka elsewhere.—ED.
[2] *A. Menziesii*, A. Gray, Syn. Fl. N. Am. ii. I. p. 27.

From the 2nd to the 9th of October the weather was remarkably fine, generally clear and dry, with cool dewy or foggy evenings. Our course was southerly, and we generally marched from ten to sixteen miles a day, obliged to go sometimes more, sometimes less, as we found suitable places for our horses and hunting. The country is highly diversified with hill and dale, and well watered by numerous small streams. Though late as the season is, I found many curious and some beautiful plants.

On Monday the 9th we crossed a low hill elevated about 2500 feet above its platform, covered with wood, principally pine. But what delighted me greatly was finding *Castanea chrysophylla*,[1] a princely tree 60 to 100 feet high, 3 to 5 feet in diameter, evergreen, the leaves having a dark rich glossiness on the upper surface and rich golden-yellow below. Nothing can exceed the magnificence of this tree, or the strikingly beautiful contrast formed with the sable glory of the shadowy pine among which it delights to grow. In thickets of *Pteris lanuginosa*,[2] excellent covert for deer, where we killed several very large males of *Cervus alces*, the meat of which, as might be expected at this season of the year, was but indifferent.

10th to 15th.—This morning we passed a hill of similar elevation and appearance to that passed yesterday. Several species of *Clethra* were gathered—one in particular, *C. grandis* [*sic*], was very fine—and many birds of *Sarcoramphos californica* and *Ortyx californica*, and two other species of great beauty were collected. This part of the time was rainy, ill-adapted for hunting. The last two days' march we descended the banks of Red Deer River, which empties itself into the River Arguilar or Umpqua, forty-three miles from the sea.

Monday the 16th was dull, foggy and raw in the morning. We passed through about two miles of open hilly ground, and then entered a thick wood where it became necessary to lop the branches off to allow the horses to pass; crossed three low ridges of hills, the highest about 2700 feet from its base, descending which, some of the horses were considerably injured and rendered incapable of proceeding for some time. In the deep dark valleys of this stream, Red Deer River, I had the pleasure to find *Laurus regia*,[3] a beautiful evergreen tree, a decoction made from the bark of which was used by the hunters as a beverage. So exceedingly powerful is the fragrant scent which it emits by the rustling of its leaves that it produces sneezing; the smell is precisely like that of the well-known *Myrtus Pimenta*. Shortly after midday we had the pleasure to arrive on the River Arguilar and camped close to the junction of Red Deer River. Arguilar River is here ninety yards broad, clear and rapid, on a bed of soft white sandstone cut and divided into narrow chinks and separate channels. My horses were of those broken-down, but my little articles did not suffer, and Mr. McLeod was kind enough to give me part of his bedding.

17th to 23rd.—Some days ago I had attempted to gain the higher

---

[1] *Castanopsis chrysophylla*, Sargent, Silva N. Am. ix. p. 3.
[2] *Pteridium aquilinum*, Christensen, Ind. Fil. p. 600.
[3] *Umbellularia californica*, Sargent, Silva N. Am. vii. p. 22.

parts of the river with an Indian boy of Baptist Mackay as my guide, but owing to an accident which befel me, and some little difficulties having arisen, I was obliged to return lamed and broken down. In the meantime we had gone down the southern bank of the river within twenty-three miles of the ocean, near to a small rapid, the termination of the tide as well as the termination of the woody country. Our new Indians from the village a mile below brought us some very fine salmon trout of large size, 15 to 25 lb. in weight.

23rd to 31st.—The chief from the upper village, ' Centrenose ' by name, having come to pay us a visit and who seemed well disposed towards us, undertook to accompany Mr. McLeod and party along the sea-shore, while his eldest son would accompany me to the higher parts of the river, which was forthwith agreed to. I took only two horses, for my new guide had no experience with such animals, and with other things this was as much as I could well manage. I retraced my steps to the place whence I was obliged to return a few days since, and crossed the river at the house of my new guide's father, where I was civilly and kindly entertained with salmon trout, hazel nuts and nuts of *Laurus regia*[1] roasted in the embers. Among the riches of his father's home were fifteen wives, one of whom he was at pains to make me understand was his mother. He had himself been wedded only a few days, and had some reluctance of leaving his young bride. I camped four miles higher up the river, close to an Indian village, at dusk, when the kind inhabitants kindled my fire, some brought me nuts, another salmon trout, a third water from the river to drink.

Tuesday 24th, Wednesday 25th.—These were days of hard labour, wet and dreary, with hail and thunder and lightning during both nights. Having to pass along the brow of two high hills that were greatly obstructed by fallen timber, my poor horses were worn down to the greatest extremity.

I had left my new guide at the camp and proceeded in a south-east direction, and had only crossed a low hill when I came to abundance of *Pinus Lambertiana*. I put myself in possession of a great number of perfect cones, but circumstances obliged me to leave the ground hastily with only three—a party of eight Indians endeavoured to destroy me. I returned to the camp, got the horses saddled, and made a speedy retreat. To-day I killed two large bears, and one next morning. With hostile Indians and incessant rains, together with the lateness of the season, fatigued and broken down I could have but little zeal to continue my exertions; and after consideration what I ought to do, I resolved to return to the sea and there wait the arrival of Mr. McLeod from the south. Accordingly I did so.

November 1st.—On arriving at the old camp I found Michel Laframboise and Jean Baptist Mackay, who informed me the Indians were far from being friendly disposed; we were therefore under the necessity of keeping a vigilant watch. My new guides behaved in every way I wished, and I paid considerably beyond what I promised, which pleased them well.

[1] *Umbellularia californica*, Sargent, Silva N. Am. vii. p. 22.

Mr. McLeod returned on the 4th; he had passed over a very interesting country, but found the tribes very hostile. One of his party was killed, and an Indian woman (the wife of one of the hunters) with five children were carried off; what became of them we never learned.

I remained here a week, putting in order my collection. Mr. McLeod being about to send two men to the Columbia, I thought it better to return with them, particularly as I was not well equipped for spending all the winter in the woods; and another thing, I was fearful if I should lose the opportunity of crossing the Rocky Mountains the following spring.

We returned nearly by the same way we had come, in twelve days' hard labour, with great misery, hunger, rain, and cold; but what gave me most pain was the nearly total loss of my collections crossing the River Sandiam, one of the tributaries of the Multnomah.

On the 20th, at the Fort, I found letters for me from London, which were pleasant. I remained here till December 9th, when I undertook a voyage to the coast in hope of replacing some articles lost last winter. This was a still more unfortunate undertaking, for I had my canoe wrecked; and from the wet and cold returned home sick, having added nothing to my collection save one new species of *Ledum*. This winter was spent in the same way as the former.

1827.

On Thursday, March 6th, I again visited the sea, but by reason of the rain returned for the last time, having failed a third time. The remainder of my time on this coast was spent packing up my collection.

March 20th.—By the annual express and in company with Dr. McLoughlin, I left Fort Vancouver for England, where I spent, if not many comfortable days, many pleasant ones.[1] Though happy of the opportunity of returning to my native land, yet I confess I certainly left with regret a country so exceedingly interesting.

I walked the whole distance from this place to Fort Colville on the Kettle Falls, which occupied twenty-five days. Not a day passed but brought something new or interesting either in botany or zoology. The beautiful *Erythronium maximum*[2] and *Claytonia lanceolata*[3] were in full bloom among the snow.

April 18th.—Mr. Edward Ermatinger, seven men, and myself took our departure from this place for the Rocky Mountains early in the morning. Nothing of importance transpired. On Friday the 20th we entered the lower lake and used our sail. The wind being favourable we went on very prosperously, and reached the termination of the upper lake on Sunday the 22nd at four o'clock.

Twenty-eight miles above the upper lake the river takes a sudden bend and to all appearances tosses itself in the mountain. A scene of the most terrific grandeur presents itself. The river is confined to the breadth of

---

[1] The entries after this point appear to relate to his journey towards England by 'the annual express,' and the not comfortable days' probably refer to the journey across the continent.—ED.

[2] *E. grandiflorum* var. *giganteum*, Baker, in Journ. Linn. Soc. xiv. p. 298.

[3] *C. caroliniana* var. *sessilifolia*, S. Wats. Bibl. Ind. N. Am. Bot. p. 117.

35 yards, rapids, whirlpools, and eddies; on both sides mountains 6000 or 8000 feet from their base, with rugged perpendicular rocks to the bed of the river, covered with dead timber of enormous growth, the roots of which have been laid bare by the torrents and were hurled by the violence of the wind from the high precipices, bringing with them immense masses of granite attached to their roots, spreading devastation before them. Passing this place just as the sun was tipping the mountain tops, his feeble rays now and then seen through the shady pine imparted a melancholy sensation on beholding this picture of gloomy wildness.

On the 25th we passed the 'Narrows of Death,' a terrific place in the river, which takes its name from a melancholy story which I cannot here relate, where ten individuals endured almost unparalleled suffering and at last were all released by death but one.

At noon on April 27th we had the satisfaction of landing at the boat encampment at the base of the Rocky Mountains. How familiar soever high snowy mountains may have been to us, where in such a case we might be expected to lose that just notion of their immense altitude, yet on beholding the grand dividing ridge of the continent all that we have seen before disappears from the mind and is forgotten, by the height, the sharp and indescribably rugged peaks, the darkness of the rocks, the glacier and eternal snow. The principal branch of the Columbia is here 60 yards wide, the Canoe river 40 yards, the middle branch 30 yards, the one on whose banks we ascend.

Saturday, 28th.—Having the whole of my journals, a tin box of seeds, and a shirt or two tied up in a bundle, we commenced our march across the mountains in an easterly course, first entering a low swampy piece of ground about 3 miles long, knee deep of water and covered with rotten ice through which we sunk to the knees at every step. Crossed a deep muddy creek and entered a point of wood principally consisting of pine— P[*inus*] *balsamea*,[1] *P. nigra*,[2] *P. alba*,[3] and *P. Strobus*, and *Thuya plicata*. About eleven we entered the snow, which was 4 to 7 feet deep, moist and soft, and together with the fallen timber, made heavy walking on snow-shoes. Camped on the west side of the middle branch of the Columbia. Of animals we saw only two species of squirrels.

Sunday, 29th.—Minimum heat 23°, maximum 43°. After a sound, refreshing night's rest we started this morning at four, due east for six miles, in the course of which we made six traverses or fordings of the river, which was 2½ to 3 feet deep, clear, and with a powerful current. Though the breadth did not exceed 25 to 50 yards, the length of time in the water was considerable, for the feet cannot with safety be lifted from the bottom, but must be slided along—the moment the water gets under the sole, over goes the person. It is necessary in very powerful currents to pass in a body, the one supporting the other, in an oblique direction. This is a level valley 3 miles broad, dry at this season, but during the summer is an inland lake bounded by the mountains. Our

---

[1] *Abies balsamea*, Mast. in Journ. R. Hort. Soc. xiv. p. 189.
[2] *Picea nigra*, Mast. *loc. cit.*, p. 222.
[3] *Picea alba*, Mast. *loc. cit.*, p. 221.

course was then four miles north-east over the same sort of country, and in this short distance we made seven fordings more. We did not require snow-shoes, for there was a fine hard solid crust. On coming out of the water and trotting along on the hoarfrost we found it intensely cold, and all our clothing that was wet immediately became cased with ice, still withal no inconvenience whatever was sustained. About nine we entered a second point of the wood, where we had recourse to our snow-shoes. Towards noon the snow became so soft that we were obliged to camp for the day. Our progress was to-day fifteen miles.

Of plants *Aralia ferox* [*sic*],[1] *Dryas*, and *Betula* were the only specimens added to the catalogue. Of animals one large Wolverine came to our camp to steal, for which he was shot ; great numbers of *Anas canadensis*, and one female *Tetrao canadensis*.

Monday, 30*th*.—The minimum heat this morning at four was 22°, maximum at noon 43°, at an elevation of 700 feet above the level of the river. We resumed our route this morning on snow-shoes in the wood about three-quarters of a mile, when we entered a second valley, in every way similar to the one passed yesterday, following a north-easterly course. We rested a few minutes after having travelled 2¼ miles in this valley and made seven fordings of the same stream that we forded yesterday, for we were obliged to cross it, keeping a direct line from point to point.

Four miles and four more fordings over this stream took us to the termination of this platform or valley. Here the river divides into two branches, the larger one flowing from the north, the smaller from due east. We crossed at the angle between the two streams and commenced our ascent of the ' Big Hill.' The snow being so deep, exceeding 6 feet, the footpath markings on the trees were hidden, so that some difficulty was experienced in keeping the way ; the steep ascent, the deep gullies, and brushwood and fallen timber rendered it laborious. Camped two miles up the hill, having gained in all nine miles. The timber gradually becoming smaller, no new animals or plants were added to the list.

May 1*st*, *Tuesday*.—This morning the thermometer stood 2° below zero, and the maximum heat at noon was 44°. We continued our ascent, and at ten had the satisfaction to reach the summit, where we made a short stoppage to rest ourselves, and then descended the eastern side of the Big Hill to a small round open piece of ground, through which flowed the smaller or east feeder of the Columbia and the same stream we left yesterday at the western base of the Big Hill. On the right hand is a small point of low stunted wood of *Pinus nigra*,[2] *P. alba*[3] and *P. Banksiana*. Near this point we put up at midday. One fine male bird of *Tetrao Franklinii* was killed during the day, which I preserved with great care. After breakfast, about one o'clock, being well refreshed, I set out with the view of ascending what appeared to be the highest peak on the north

[1] *Panax horridum* is given instead of this, in the account of Douglas's Journey, printed in Hooker's Comp. Bot. Mag. ii. p. 135, which is referred to *Fatsia horrida*, Benth. and Hook. f. Gen. Pl. i. p. 939.

[2] *Picea nigra*, Mast. in Journ. R. Hort. Soc. xiv. p. 222.

[3] *Picea alba*, Mast. *loc. cit.*, p. 221.

or left-hand side. The height from its apparent base exceeds 6000 feet, 17,000 feet above the level of the sea.

After passing over the lower ridge of about 200 feet, by far the most difficult and fatiguing part, on snow-shoes, there was a crust on the snow, over which I walked with the greatest ease. A few mosses and lichens, *Andreae* and *Jungermanniae*, were seen. At the elevation of 4800 feet vegetation no longer exists—not so much as a lichen of any kind to be seen, 1200 feet of eternal ice. The view from the summit is of that cast too awful to afford pleasure—nothing as far as the eye can reach in every direction but mountains towering above each other, rugged beyond all description ; the dazzling reflection from the snow, the heavenly arena of the solid glacier, and the rainbow-like tints of its shattered fragments, together with the enormous icicles suspended from the perpendicular rocks ; the majestic but terrible avalanche hurtling down from the southerly exposed rocks producing a crash, and groans through the distant valleys, only equalled by an earthquake. Such gives us a sense of the stupendous and wondrous works of the Almighty. This peak, the highest yet known in the northern continent of America, I felt a sincere pleasure in naming MOUNT BROWN, in honour of R. Brown, Esq., the illustrious botanist, no less distinguished by the amiable qualities of his refined mind. A little to the south is one nearly of the same height, rising more into a sharp point, which I named MOUNT HOOKER, in honour of my early patron the enlightened and learned Professor of Botany in the University of Glasgow, Dr. Hooker, to whose kindness I, in a great measure, owe my success hitherto in life, and I feel exceedingly glad of an opportunity of recording a simple but sincere token of my kindest regard for him and respect for his profound talents. I was not on this mountain. *Menziesia, Andromeda hypnoides,*[1] *Gentiana, Lycopodium alpinum, Salix herbacea, Empetrum,* and *Juncus biglumis* and *triglumis* were among the last of Phanerogamous plants observed.

Wednesday, 2nd.—At three o'clock I felt the cold so much, the thermometer stood at only 2° below zero, that I was obliged to rise and enliven the fire and have myself comfortably warmed before starting. Through 300 yards of gradually rising open low pine wood we passed, and about the same distance of open ground took us to the basin of this mighty river, a circular small lake, 20 yards in diameter, in the centre of the valley, with *a small outlet at the west end—namely, the Columbia* ; and *a small outlet at the east end—namely, one of the branches of the Athabasca which must be considered one of the tributaries of the McKenzie River.* This is not the only fact of two opposite streams flowing from the same lake.

This, ' The Committee Punchbowl,' is considered the halfway house. We were glad the more laborious and arduous part of the journey was done. The little stream Athabasca, over which we conveniently stepped, soon assumed a considerable size, and was dashed over cascades and formed cauldrons of limestone and basalt seven miles below the pass ; like the tributaries of the Columbia on the west side, the Athabasca widens to a narrow lake and has a much greater descent than the Columbia.

[1] *Cassiope hypnoides,* A. Gray, Syn. Fl. N. Am. ii. I. p. 36.

At this point the snow was nearly gone, and the temperature greatly increased. Many of the mountains on the right are at all seasons caped with glacier. At ten we stayed to breakfast fifteen miles from the ridge, where we remained four hours. The thermometer this morning stood at 2° below zero, and at 2 P.M. at 57°. We found it dreadfully oppressive.

In starting in the afternoon a little before the party, I missed the path and went out of the way. As the sun was edging on the mountains, I descried east of me, about a mile behind a low knoll, a curling blue smoke issuing from among the trees, a sign which gave me infinite pleasure. I quickened my steps and in a few minutes came up to it, where I found Jacques Cardinal, who had come to the Moose encampment and brought with him eight horses to take us along. He gave me an excellent supper of the flesh of *Ovis montana* (Geof.) and regretted he had no spirit to offer me. After supper he jocularly said, pointing to the stream, ' This is my barrel and it is always running.' He also afforded me part of his hut.

Thursday, 3rd.—My companion, Mr. Ermatinger, and the party were brought up by Cardinal this morning ; they were distressed about me, not having made my appearance last night. We breakfasted, and proceeded on the banks of the river, I preferring walking on foot to riding on horseback. The road was still soft and heavy from the recent melting of the snow, and strewn with timber of small size. The difference of climate and of soil, and the amazing difference of the variety and size of vegetation, are truly astonishing ; one would suppose he was in another hemisphere, the change is so sudden and so great. We crossed the principal branch of the Athabasca, a stream 70 yards broad, where it is joined by the branch on the banks of which we descended. There it was our intention to put up for the night, but Cardinal found his horses so unexpectedly strong that the route was continued to the ' Rocky Mountain House,' where we should find canoes, and which we gained shortly after 6 P.M. Several partridges were killed ; but the only plant new to me was *Anemone Nuttalliana*,[1] which was in full flower. The scenery here is fine, with a small lake and open valley commanding a sublime prospect of the mountains. Our distance was to-day thirty-four miles.

Friday, 4th.—In the fine light birch canoes we embarked at daylight and went rapidly before the stream. The banks of the river are low and woody, some places narrow, others widening out to narrow lakes full of sand shoals. Stayed to breakfast on a small low island in the upper lake, where we had some mountain sheep's flesh given to us by Cardinal's hunter. Continued our route, and passed on our right a ridge of high rugged mountains, and five miles lower down on the left a ridge of lower elevation which is the termination of the dividing ridge on the east. Arrived at Jasper House at 2 P.M. The minimum heat of to-day was 29°, the maximum 61°.

Saturday, 5th.—This day admits of scarcely any variety. The river is 100 to 140 yards wide, shallow and rapid, with low gravelly woody

[1] *A. patens* var. *Nuttalliana*, S. Wats. Bibl. Ind. N. Am. Bot. p. 5.

banks of poplars and pines. This river abounds with wildfowl, and the Northern Diver charmed us with his deep mellow melancholy voice in the evenings. Gained ninety-three miles.

On the 6th we had only proceeded three miles when we were detained by the ice, and here we found Mr. George McDougall. Little progress was made this day on account of a portage—that is, the canoes were carried over a considerable distance to the main channel, which was clear, when we proceeded rapidly and arrived at Assiniboine, 184 miles from Jasper House. The next day we started for Lesser Slave Lake in hope of finding Mr. John Stuart, as this place did not afford much food.

On arriving here with Mr. Stuart, who treated me with the greatest kindness on every occasion, I travelled partly with the brigade and latterly alone with one guide, on foot, to Fort Edmonton on the Saskatchewan River, where I arrived on Monday the 21st. I found here my small chest of seeds well attended to, through the kindness of Mr. Rowand. On the plains I killed several curlews, and in the woods a number, both male and female, of *Tetrao phasianellus*, the Pin-tailed Grouse of Edwards, and abundance of *T. canadensis*.

May 21st to 31st.—Around Edmonton the country is woodless and uninteresting. Embarked in Mr. Stuart's boat, in company with others, to Fort Carlton House. Our mode of travelling gave me little time to botanise; the only times were the short stay made to breakfast, the dusk of the evening before camping, and the most when a delay was made for the purpose of hunting buffalo and red deer.

The scenery in this river in some places is varied and highly picturesque, particularly near the Red Deer and Eagle Hills. The soil is dry and light, but not unfertile; a rich herbage and belts and clumps of wood interspersed over it give it a romantic, beautiful appearance. Near to these parts many buffalo were killed, a few red deer and antelope of the plain. This animal has so much curiosity about him that he will approach within 100 yards, particularly if the hunter has any of his clothing red, which seems to attract them. The buffalo is easily approached by a skilful hunter and readily destroyed.

Among a variety of the plants not before in my herbarium were *Astragalus pectinatus*, *A. Drummondii*, *Phlox Hoodii*, *Thermopsis rhombifolia*, *Hedysarum Mackenzii*, *Astragalus succulentus*,[1] *A. caryocarpus*, and seven species of *Salix*. On one of the hunting excursions Mr. F. McDonald was dreadfully lacerated by a wounded buffalo bull in the back part of the left thigh, had some ribs broken, his left wrist dislocated, and was otherwise severely bruised. These animals have a disposition not to destroy life at once, but delight to torture. On first striking the object of their revenge, if he is stunned or feigns to be dead there is some chance of his escape; the animal in the meantime will lie down beside his victim, keeping a steadfast eye on him, and the moment there is the least motion up he gets and gives another blow. Poor Mr. McDonald was so placed for two and a half hours, bleeding and at the point of death, and that under cloud of night, which gave us scarcely any opportunity of rescuing, for the

[1] *Astragalus caryocarpus*, S. Wats. Bibl. Ind. N. Am. Bot. p. 192.

animal was within a few paces and we were fearful to fire lest a shot should take him. By the activity of Mr. Harriott and my assistance he was saved. I bound up his wounds and afforded all the assistance a small medicine chest and my slender knowledge could suggest. We passed hastily on in hope of finding Doctor Richardson, but on our arrival found the doctor had gone to Cumberland House. At Carlton House I had the pleasure to meet Mr. Drummond, of Captain Franklin's party, who spent the greater part of his time in the Rocky Mountains contiguous to the sources of the Rivers Athabasca and Columbia.

Mr. Drummond had a princely collection. I had intended to cross the plain from this place to Swan and Red Rivers, but from the hostile disposition of the Stone Indians deemed it unsafe. I descended to Cumberland House, and found there Dr. Richardson, who kindly showed me parts of the princely collection of natural history made during the expedition. This part of the country has been so well described in the former narrative of Captain (now Sir John) Franklin that it leaves me no room ; I shall therefore only notice my stages. After leaving Cumberland, two days took us to the Grand Rapids, the entrance of Lake Winnipeg, where we were detained by the ice ; a few hours more it became *rotten, sank*, and disappeared ; and we proceeded under sail to Norway House with an open sheet of water. The shores of the lake are clothed with diminutive trees, *Pinus alba,*[1] *nigra,*[2] *microcarpa,*[3] *Betula papyracea, nigra, Populus trepida,*[4] with sphagnous swamps of *Ledum, Kalmia,* and *Andromeda,* and near springs or pools strong herbage of *Carex.* On the 16th we arrived at Norway House, where I had letters from England. The following day George Simpson, Esq., the resident Governor of the Hudson's Bay Company, arrived, from whom I had great kindness. A few days were spent here, when Captain Sir John Franklin arrived, who politely offered me a passage in his canoe through the lake as far as the mouth of Winnipeg River on my way to Red River, which was gladly embraced. This was at this season of the greatest moment, for I gained twelve days on the ordinary time usually taken to perform the trip.

Captain Sir John Franklin left me for England on July 9th, and the following day I proceeded to the settlement on the Red River, where I arrived on the 12th. I took up my abode with Donald McKenzie, Esq., Governor of the Colony, an excellent and good man, who during the whole of my stay showed me great kindness and afforded me much assistance. With Mr. McKenzie I passed an agreeable time, for his knowledge of that country, particularly of that west of the Rocky Mountains, was great, where he had spent many years.

I had the pleasure to be made known to the Rev. David Jones and R. W. Cockran at the English Mission House, an excellent establishment which owes its merit to the unremitting care and zeal of these gentlemen, whose useful lives are devoted to the little flock over which they preside.

---

[1] *Picea alba,* Mast. in Journ. R. Hort. Soc. xiv. p. 221.
[2] *Picea nigra,* Mast. *loc. cit.,* p. 222.
[3] *Larix pendula,* Mast. *loc. cit.,* p. 218.
[4] *P. tremuloides,* Sargent, Silva N. Am. ix. p. 158.

I became acquainted also with the Rev. J. N. Provenchier, the Roman Catholic Bishop, a gentleman of liberal disposition and highly cultivated mind, who lives only to be useful and to do good. The soil—a deep alluvial stratum of brown loam on gravel and limestone—is exceedingly fertile, capable of bearing every kind of produce. The settlers live comfortably and are seemingly happy. The crops are liable to be attacked by grasshoppers, but the wheat is exempt from *smut* and *rust*. Cattle thrive, as well as pigs and horses; sheep had not then been introduced.

During a month's residence I formed a small herbarium of 288 species, many of which were new to me, and I felt truly happy at having devoted a little time to it; for several plants were added to the flora, and had I stayed with Mr. Drummond or Dr. Richardson on the Saskatchewan these would have been omitted. I left with Mr. Hamlyn, the surgeon to the colony, had a somewhat tedious passage through the lake, and had the pleasure to arrive at York Factory, Hudson's Bay, where I was kindly received by John George McTavish, Esq., Chief Factor, who had had the kindness to get made for me some clothing, my travelling stock being completely worn out.

Here my labours ended; and I may be allowed to state, when the natural difficulties of passing through a new country are taken into view, the disposition of the native tribes—in fact, the varied insufferable inconveniences that daily present themselves—I have great reason to look on myself as highly favoured. All that my feeble exertions may have done only stimulated us to future exertion. The whole of my botanical collection, save a few that came intimately within the Society's Minute, were, agreeably with my anxious wishes, given for publication in the forthcoming American Flora from the pen of Dr. Hooker.[1]

I sailed from Hudson's Bay on September 15th and arrived at Portsmouth on October 11th, having enjoyed a most gratifying trip.

D. D.

[1] The title-page of 'the forthcoming work referred to here is as follows :—

'Flora Boreali-Americana; or, the Botany of the Northern Parts of British America : compiled principally from the Plants collected by Dr. RICHARDSON and Mr. DRUMMOND on the late northern expeditions, under command of Captain Sir JOHN FRANKLIN, R.N., to which are added, (by permission of the Horticultural Society of London,) those of Mr. DOUGLAS, from North-West America; and of other Naturalists.' By WILLIAM JACKSON HOOKER, London [1829]—1840. 2 vols. [4to.   ED.

JOURNAL OF AN EXPEDITION TO NORTH-WEST AMERICA; BEING THE
   SECOND JOURNEY UNDERTAKEN BY DAVID DOUGLAS, ON BEHALF
   OF THE HORTICULTURAL SOCIETY.

1824.

Saturday, July 24th.—After several weeks' preparation for a voyage
to the Columbia river on the west coast of North America, on the afternoon
of Saturday parted with J. Sabine, Esq., and all other friends. In the
evening wrote a letter to my father, to Dr. Hooker, and Mr. Murray of
Glasgow.

July 25th.—Left London at half-past eight o'clock in the morning
from the Spread Eagle office, Piccadilly, by the Times coach in company
with my brother for Gravesend.

The morning was very pleasant, cloudy and calm. Passed some
fields of rye, cut down; wheat, oats, and barley nearly ready for the
sickle. At Gravesend I met Mr. John Scouler of Glasgow, who was
going on the same voyage to officiate in the capacity of surgeon. This
was to me news of the most welcome kind, being previously acquainted
with each other and on the strictest terms of friendship. At twelve
o'clock went on board the Hudson's Bay Company's ship *William and
Ann*, Captain Hanwell, bound for the Columbia river, north-west coast
of America, and came on shore again at two o'clock. At 4 o'clock in
the afternoon saw my brother in the steamboat for London, who was
affected at parting with me, and returned to the ship.

Monday, 26th.—In the morning, employed stowing away all my baggage
&c. Went under weigh at four o'clock p.m., having a fine breeze with
rain; thermometer 58°. We made only seven miles and then let down
the anchor at darkening.

Tuesday, 27th.—Cold with thick fog; passed the Nore at daylight;
at seven the ship struck on the " Shivering Sands " and beat about dreadfully for an hour. Fortunately the wind was moderate with little swell
at the time, otherwise our situation must have been perilous. On being
rescued from our unhappy situation, it afforded the captain and pilot,
as well as all on board, much pleasure to learn that the vessel had
sustained little or no injury.   I confess it gave me pleasure to be enabled
o proceed, as delays in such undertakings are by no means agreeable.
The pilot left us off Deal at six o'clock in the evening (thermometer 63°).
Did not write as the captain intended to put into Portsmouth, to await

the orders of the Company as to his proceeding to sea, not knowing how far he might be justified in his present circumstances : 14 inches of water in the hold.

Wednesday, 28*th*.—Very pleasant; thermometer at midday 73°. Passed Dover. Towards evening almost calm.

Thursday, 29*th*.—Passed Dungeness, having a very favourable breeze ; at noon a perfect calm off the Isle of Wight. The vessel being in much better condition than was anticipated, the captain abandoned the idea of putting in at Portsmouth and would go to sea at daylight. The vessel made only 2 inches of water during the last twenty-four hours. From the appearance of the sky it must have been a very warm day on shore. Thermometer at eight o'clock this morning 57°, at twelve 73°—at eight in the evening 64°.

Friday, 30*th*.—A light air of wind sprung up at midnight, and before daylight the Portland lights were observed. Middle of the day calm and warm (thermometer at 70°). At noon fine wind, which continued during the night.

Saturday, *July* 31*st*.—Morning cool, accompanied with rain ; midday warm and dry. Passed the Lizard at eleven o'clock. In the afternoon a strong wind from the south, with intervening showers of rain. At twelve o'clock the thermometer was 62° ; at 4 P.M., 57°. In the evening our delightful view of the rocky shores of Cornwall closed.

Sunday, *August* 1*st*.—During the night strong wind from the northeast with heavy showers of rain ; at eight this morning the sun broke through and I had an observation ; 8° west longitude. Towards afternoon wind moderate, and pleasant in the evening. Passed a schooner at four o'clock. During the whole of our progress down Channel only saw about fifty gulls and a few other sea-birds. Thermometer at five in the evening at 66°. Our scenery, sky and water, but in these a great variety is seen. At night when there is a gentle ruffling on the water the *Medusae*, *Physalae* and other zoophytes giving off their phosphoric or illuminating particles over a vast expanse of water produce a very fine effect.

Monday, *August* 2*nd*.—Very pleasant, favourable wind from the north, lighter at noon, freshened again in the evening. Lat. 47° north, Long. 11° west. Thermometer 63° in the water and 67° exposed to the air.

August 3*rd*.—Wind from the north, rather cold (thermometer 62° at twelve o'clock) ; sun visible. Lat. 45° 32' Long. 12° 43'. A year has elapsed to-day since I arrived at Staten Island, near New York. It was warm and pleasant, and afforded gratification after a tedious passage of fifty-seven days from Liverpool. Is there anything in the world more agreeable to the feelings of a prisoner than liberty ? Saw only two birds, resembling gulls ; they seemed shy.

August 4*th*.—Wind the same as yesterday, course south-west : seven miles, at an average, all day. Sun visible, Lat. 41° 38', Long. 14° 12' ; thermometer 68° at twelve o'clock, in the shade.

Thursday, *August* 5*th*.—Wind from the north, very pleasant ; warm during the middle of the day, with a fanning breeze, succeeded by a cool damp in the evening. Thermometer 74°.

Friday, August 6th.—Pleasant and dry; sun visible, Lat. 47°. Wind northerly. Thermometer at noon 68°, on the surface of the water 62°; winds usually moderate towards midday and freshen in the evening.

Saturday, 7th.—Pleasant, but cloudy in the morning. Sun visible at ten o'clock; an observation was had which gave 45° 41'. Cloudy in the afternoon; sensible of increasing heat in the evenings and during the night. At two passed a vessel going eastward, distance two leagues. The night was observed as is usual with sailors, succeeded by some songs. The good will shown was more to be admired than the melody.

Sunday, August 8th.—Pleasant fanning wind throughout the day and, as is usually the case, freshened towards dusk. Prayers read by the surgeon at the captain's request. At three o'clock the peaks of Porto Santo were observed and up to dusk became more and more visible. Sail was shortened to come up with the land at daybreak. In the morning shall be regaled with a fine view.

Monday, August 9th.—In the morning at daylight the Island of Porto Santo was perfectly visible, about four leagues distant. With a pleasant fanning breeze we passed the north-east point at seven o'clock.

On the east side the hills rise into high rugged barren peaks. Several rocks of considerable size, with numerous detached pieces from them, are seen above water two hundred to four hundred yards from the shore. The town (Porto Santo) is pleasantly situated on a gentle declivity, having on the left the high peaks mentioned and on the right some conical hills, but neither so high nor so rugged as those on the eastern side.

The houses are generally low, built of stone, and whitewashed. Each house has a large garden, or piece of ground under cultivation, attached to it, which gives it more the appearance of a village than a town. Vines do very well here, with most of the fruits cultivated at Madeira. On the western side high rocks rise out of the ocean, larger by far than those on the eastern side. There was a sudden rise of the thermometer. Yesterday in the shade it stood at 71°, and now at eight o'clock 79°. At eleven o'clock the high mountains of Madeira showed themselves with their tops enveloped in the clouds, objects which we hailed with pleasure. Reached Point Lorenza, the eastern extremity of the island; no cultivation is seen for several miles to the south-west. The rocks are low on the east, of a copper or blackish colour, and look much like volcanic remains.

On the left, at the distance of three leagues, are two islands called the Deserters, where the Portuguese transport their criminals. They appear almost barren.

Towards three o'clock the wind failed us and left us to contemplate several rich and romantic valleys near the side of the ocean from four to ten miles from Funchal, the capital of the island.

Scarcely any sea-birds were seen. Towards evening the wind freshened and at seven o'clock we anchored in the Bay of Funchal. We were visited by boats from the Customs and Board of Health. The latter made considerable noise as the captain had no bill of health from London. Fair words and a good deal of courtesy had to be used before matters could be adjusted. The number of souls on board, with all the usual formalities, were gone through, and on his taking leave he said that at daylight a white flag must be at the masthead, a sign of quarantine. My imprisonment on a former voyage came into my mind, attended with all its consequences; but in the meantime the health of all friends was drunk with much pleasure; and with these sensations on my mind which will ever afford satisfaction, I wished my fellow-voyagers "Good night," and went to sleep. Thermometer at twelve o'clock, in the shade, 80°. In the water, 76°.

Tuesday and Wednesday, August 10*th*, 11*th*.—Tuesday: In the morning as follows to Joseph Sabine, Esq., &c :—

Madeira, *August* 10*th*, 1824.

DEAR SIR,—It gives me much pleasure to inform you of my arrival here last night, having enjoyed a very pleasant passage from England. I have not yet been on shore, but will as soon as I get permission. The vessel will stay to-day and probably part of to-morrow, which will enable me to see a little of the island. I sincerely wish they would make a stay of a few days. The woods and valleys have a beautiful appearance from the water, and no doubt contain many interesting things. If I have the least time to spare I shall make a point of seeing Henry Veitch, Esq., who will probably facilitate my movements. Should the vessel touch at Rio (as I hope it may), or some other place equally good, and make a stay of ten or twelve days, I hope to collect some of the yet hidden treasures.

I feel extremely gratified by the kindness and comfort shown me by Captain Hanwell—he is a very attentive man. Anything that will come by the return of the vessel will be carefully looked to.

I do not think anything has been omitted that could add to my comfort. I regret that I could not see Mr. Turner before I left.—Dear Sir, I am,

Your obedient humble servant, D. DOUGLAS.

Afterwards wrote to my brother, took breakfast, and went on shore at eight o'clock without delay : my first inquiry was for the vegetable market, which is situated near the south-west side of the town. It is a square of fifty yards, enclosed on the south and west by sheds which are fitted as stalls, on the east by a shed or house serving as the butcher market, on the north by a high iron rail with one gate, which is the only entrance to it. There are four rows of houses or shops, five in each row, equally divided, in the centre of the square, built of wood, ten feet square with pavilion roofs. The whole is neatly paved with round stones kept very clean and has quite a genteel appearance. There is a daily market, continued throughout the day; officers are in attendance whose duty it is to see that business is conducted with propriety. Their services I valued much; particularly as it was evident, from the movements of merchants, they were not strangers to deception. There appeared to be a scarcity of vegetables. The following is a list of what I saw :—

Cabbage of inferior size, seeming a late variety ; had no opportunity of tasting it. The leaves of *Arum esculentum* [1] used as spinach. The roots,

---

[1] *Colocasia antiquorum*, var. *esculenta*, Engl. in DC. Monog. Phan. ii. p. 492.

called *yams*, used in lieu of potatos, and preferred by the inhabitants to them ; they are used in the same way as potatos are in England. This is a vegetable admirably calculated for taking to sea, being not so liable to grow in warm latitudes as potatos, and (according to experienced seamen) will keep much longer. Onions are large, of one sort, red and flat, but much milder than those in England. From the large quantity exposed for sale I judge it to be here a favourite vegetable. Potatos are neither large nor are they very good, mostly red with pink eyes. Pumpkins of several varieties, and cucumbers chiefly of one variety—short and prickly. Two varieties of cayenne pepper : one small longish fruit of a red colour ; one small round yellow ; the former is a native of the island. The common red tomatos, neither remarkable for size nor quality. Turnips, carrots, parsnips, cauliflowers, celery ; not even a single blade of parsley could be seen. In fruits they are richer and generally of better quality. The banana, *Musa sapientum*, is extensively cultivated and perfects its fruit in abundance. It is usually eaten without any preparation, but when fried in a little butter it tastes like a good pancake. Lemons larger and better than any we are ever accustomed to see ; I think them very fine. One variety of apple like Summer Redstreak ; which it probably is. It is large, round and redstreaked, dry and mealy, an insipid fruit. Three different pears ; one large, in form and colour like Jargonel, but entirely destitute of its flavour, mealy and tasteless ; a second of a much smaller size, yellow colour almost approaching to a sulphur-yellow, with a flat compressed head and gradually tapering towards the stalk (like the former a dry fruit, but as all which I saw appeared to be too ripe, probably they may be better in an earlier state) ; a third resembling what is called in Scotland 'Crawford' pear, both in size, form and flavour; as in all likelihood it is. This is by far the best fruit. Of grapes there was an abundant supply, of four or five varieties ; one like what is called *Black Cluster*, small close bunches, rarely branched, small globular berries, generally with only one or two perfect seeds, having short footstalks. One with large and much-branched bunches with large globular white berries. It seems to be a shy bearer, as few berries in comparison to the other varieties were on a bunch. It tastes similar to Muscat of Alexandria. A third with long bunches of a brownish-black or copper colour, of a rich fine flavour. It looks and tastes like *Grizzly Fontignan*. A fourth, the largest of any, much-branched, with large black round berries ; the flavour is not so fine as the former one.

Peaches of two varieties : one a large, long and pointed fruit towards the top, in form much like a lemon, with a yellow, thin and downy skin ; the pulp is also a bright yellow, rather, if anything, coarse flavour, and adheres to the stone, which is large in proportion to the fruit. The other, a smaller and very different fruit, nearly globular, thin, white, delicate downy skin, pulp white and of a delicious flavour ; it also adheres to the stone, which is small and flat.

Mr. Atkinson presented me with a supply of grapes of several varieties. I believe one is the White Muscat of Alexandria. I saw only a few which might be said to be weighty bunches, none with large berries. I am informed that the smaller bunches are brought to market and all the good fruit reserved for wine. Most people have fruit of their own, so that it is chiefly for vessels calling and for the destitute ; a cluster weighing about a pound may be had for three farthings.

The banana is sold according to the number of fingers on a bunch ; one with forty to fifty may be had for 1s. 6d. sterling, which will serve six or seven individuals for one meal. Pears are sold by number, thirty for $2\frac{1}{2}d$. Potatos are sold by weight and are dear in proportion to other things. Two varieties of figs, both of excellent quality : one long black, thin delicate skin and purplish pulp ; one green, small, short and flat fruit, if anything finer than the black one ; fifty of the green for $4\frac{1}{2}d$., and thirty of the black for the same price.

Soft fruits, figs, grapes, &c., are carried to market on the head in baskets, like strawberries in England. Pears, apples, lemons, &c., in bags made of hogskin, and some of hemp, on mules, and the poorer class carry them on their backs. Two plums : one small round green, in form like greengage, colour a brighter green than it, but the flavour far inferior ; the other a longish oval fruit, black-skinned and insipid taste, something like ' Orleans.'

Having satisfied myself at market, I made a journey north of the town, on the hill where most of the principal vineyards are.

In planting the vines no situation or aspect is studied ; they thrive in the valleys and deep ravines, on the little eminences and high grounds, and even on the top of old walls and on the roofs of old thatched cottages. The soil in general is a light brown like burned sand. I could not learn, what (or if any) manure is used. They are planted from 6 to 12 or 14 feet apart, and supported on horizontal rough railing of wood 4 or 5 feet from the ground, inclining sometimes to the north, sometimes to the south, but in all cases following the declination, which on a small space varies much.

In pruning very little old wood is left, leaving spurs of 2 or 4 inches on the principal shoots. From the rivulets, which are numerous, small chinks or channels are dug along the sides of the valleys for the purpose of conveying water to the plantations, which is let off when required—a cheap mode of watering and at the same time adds beauty to the place. A great portion of the clusters were daubed over with mud and some with lime, to prevent the attacks of mealy-bug and also probably to prevent the ravages of wasps and other insects ; the undersides of the leaves are almost covered with a species of white mealy-bug similar to that which infests pineapple plants in England. Old women and young girls are employed pinching off the leaves that shade the fruit. This is done by stooping under the trellis-work : a delightful occupation, screened from the influence of a scorching sun. The observation made to me in the market as to the best fruit being reserved for wine I found correct. On the whole, I confess to be somewhat disappointed about the size and quality—although certainly good, by no means what they are generally represented. Few clusters exceed $2\frac{1}{2}$ lb. or $3\frac{1}{2}$ lb. ; the greater part 1 lb. I learn the clusters are thinned early in the season when too numerous ; thinning berries of course is not practised. In passing along the lanes that lead through the vineyards I was invariably escorted by an elderly matron with her distaff, and a little girl or boy whose suspicious eyes indicated their profession. The banana has the best and most conspicuous place of the garden assigned it, at the end of houses or in small well-sheltered courts, to prevent its massive leaves from being destroyed by the winds. Figs thrive uncommonly well, particularly in low, moist situations on the margins of woods, being partially shaded.

Arum esculentum[1] and Yams are planted in low level beds on the margins of rivulets, so placed that water from the chink is let on the ground when required, in the same manner as for vines. It is studied to keep them almost in a state of saturation. The soil which they thrive best in is a rich, black, alluvial soil carried by the currents from the high grounds. Lemons are to be seen in great luxuriance in moist shady places in northern or eastern aspects. *Eugenia Jambos*, called 'Jambos,' matures its fruit in abundance with little attention. It is considered one of the finest fruits by the Portuguese. It thrives well in dry, light, sandy soil.

I saw the following native plants, if not introduced: *Dracaena Draco*, from 25 to 30 feet high and a proportionate thickness; a species of *Pinus* found on the mountains, from 35 to 40 feet high, with flat round tops and large round cones—seldom more than 18 inches to 2 feet in diameter. This with a species of Juniper, which attains a considerable size, form the only conspicuous objects on the mountains.

Castanea is the only wood of the forest that the inhabitants seem to pay attention to. All the valleys and the less fertile spots on the high grounds are planted with this tree. Most of the large trees have a large protuberance, occasioned by tying ligatures round them. I thought they were grafted, but on asking was informed that it is done by way of ornament; this is done 4 feet from the ground. The Common Myrtle, *Rosmarinus officinalis*, *Fuchsia coccinea*, and a species of *Jasminum* decorate old walls and hedges that surround the vineyards, forming a delightful fence.

Rosa sp. from 2 to 5 feet high, branches twiggy, thickly set with small hooked prickles; leaves smooth, nearly ovate, serrated; berries smooth, globose; not in flower, fruit not ripe. On the summit of the high hill north of the town. *Capsicum* sp., herbaceous; leaves alternate, lanceolate; flowers white, solitary; fruit globose, small, of a bright yellow colour and very acrid taste. This one is sold in market. *Crepis* sp.; leaves lanceolate, the cauline ones smaller and more pointed; stem spreading, having long linear bracteas. On the high grounds, *Holcus* sp., like *H. odoratus*[2] of Britain, abundant; *Aspidium* sp., tall, on moist rocks on the hills; *Thymus* sp., shrubby, on the high ground; this differs little from *Thymus vulgaris* except in smell. *Plantago* sp., resembles *P. lanceolata* of Britain, only being very woolly, a small plant scarcely exceeding 3 or 4 inches high. *Asplenium Adiantum nigrum* on moist rocks and old walls. *Adiantum Capillus Veneris* in the same situations. *Cineraria cruenta*[3]? in hedges and near streams. *Rubus* sp. *fruticosus*? *Psoralea* sp. *bituminosa*? on dry rocky situations, one of the finest plants on the island.

[1] *Colocasia antiquorum*, var. *esculenta*, Engl. in DC. Monog. Phan. ii. p. 492.
[2] *Hierochloe borealis*, Sowerby, Engl. Bot. ed. 3, xi. p. 16.
[3] *Senecio cruentus*, DC. Prod. vi. p. 410.

Hypericum sp., *Origanum* sp., *Veronica* sp., leaves ovate-lanceolate, serrate, flowers small, purplish-blue; on the margins of rivulets very plentiful. *Asplenium trichomanes*, one species of *Scirpus*, one of *Carex*, and one of *Poa* ; *Plectranthus* sp., shrubby, in dry places ; *Scutellaria* sp., one of *Lobelia* sp. annual ; flowers small, blue ; on the side of rills, and one of *Asclepias—Blechnum* sp. I called at the house of Henry Veitch, Esq., but he had some weeks since gone on a tour to Italy. In his garden, which is laid out and kept with considerable taste, there were many fine specimens of trees larger than I have seen anywhere, &c. Banana, *Eugenia Jambos*, lemons, oranges, grapes, peaches, &c., and in a high state of perfection. Groups of flowers of warmer regions. In the centre of the garden a fine specimen of *Artocarpus incisa* (Bread Fruit) in fine health, just coming into flower ; requires no protection ; fully 20 feet high and 10 inches in diameter.

Thermometer on board at ten o'clock this morning 80° ; at the same hour, 800 feet above the level of the sea, 80°.

On the summit of the highest peak at four o'clock afternoon 72°. The sun shone in full vigour ; the fatigue of descending after a laborious day's work made me enjoy a night's rest. Thermometer on the beach at noon 84° ; on the surface of the water in thirty fathoms 79°. Purchased in conjunction with Mr. Scouler $\frac{1}{16}$ of a pipe of wine (about 6 and $\frac{1}{2}$ dozen) for which we paid £7. Went under weigh at six o'clock in the evening of Wednesday with a pleasant fanning breeze from the south-east, being, for the short stay I made, much gratified.

Thursday, August 12*th.*—Towards midday the mountains of Madeira were out of sight and at noon the wind changed to north-east which carried us speedily on our voyage. Thermometer 80° and 78°.

Friday, 13*th.*—Some sea-birds hovering round the vessel, but very shy. More wind than yesterday, and being farther from the land the thermometer at noon stood at 78° and 74°.

Saturday, 14*th.*—Wind from the same direction as the last two days. The few specimens obtained at Madeira afford a fine amusement during some of the tedious hours ; put them in fresh paper. Thermometer 79° and 75°.

Sunday, 15*th.*—Wind north-east, morning cool and pleasant ; thermometer 70° at six o'clock this morning. Saw only two gulls and for the first time immense schools of flying-fish (*Exocoetus volans*) sporting round us. Thermometer at noon 78°, on the surface of the water 75°. As usual we had service performed.

Monday, August 16*th.*—Wind the same as yesterday ; sun obscured. Thermometer 76° and 71°.

Tuesday, August 17*th.*—Very high wind from north-east with intervening showers. Thermometer 77° and 75°.

Wednesday, August 18*th.*—A pleasant breeze during the whole day. Thermometer 76° and 72°.

Thursday, 19*th.*—A vessel bound eastward, but having a favourable wind we did not speak her. Thermometer 79° and 77°.

Friday, 20*th.*—Cool and very pleasant during the whole day. Our only visitants are porpoise and flying-fish. Thermometer 78° and 76°.

Saturday.—Close and warm. Thermometer 82° and 79°; heavy showers of rain in the afternoon and the evening.

Sunday, August 22nd.—Gentle winds in the morning from north-east, accompanied by light showers. A vessel bound southward. Sun visible at midday. Thermometer 82° in the shade; on the surface of the water 77°. Continued heavy rain from two o'clock afternoon until ten at night; intermediate showers during the night.

Monday, August 23rd.—Heavy showers of rain during night, with gusts of wind. Morning still and warm. Sun visible at eight o'clock; cloudy during the remainder of the day, with intervening showers. Evening cool and nearly calm; thermometer in the shade 84°, in the water 78°. Caught two specimens of seaweed: one a variety of *Fucus natans*, attached to it some small shellfish; one probably a species of *Confervae* in close thready tufts of a bright olive colour. Intermingled with the *Confervae* a small species of *Fucus* with circular branches and minute flattened bladders. On the same, two very minute insects of the same species having a beautiful shining azure colour. Sixty miles off Cape Verde Islands.

Tuesday, 24th.—Light showers of rain and gentle wind; cloudy during the greater part of the day. Thermometer at noon in the shade 84°, on the surface of the water 81°. Several flying-fish were washed or flew on board during the preceding night, but being much damaged were not worth preserving, and a very curious species of *Beroë* which retained its transparency for a long time in a bucket of water. At eight o'clock in the evening thermometer in cabin 86° with all the air that could be given.

Wednesday, August 25th.—Thermometer 87° in the shade, in the water 81°. Light winds from the west with showers.

Thursday, 26th.—Wind from the west, with very heavy rains. Thermometer at noon 86°, and 81° in the water; cool and pleasant in the evening.

Friday, August 27th.—High wind from the west with showers at two o'clock afternoon; moderate. Thermometer 81°, in the water 78°.

Saturday, August 28th.—Pleasant and cool in the morning, with a fine breeze. Thermometer 81° in the shade, on the surface of the water 79°. Several large birds in the evening hovered round the vessel; some were large, brown on the back, white on the belly and under the wings. Probably a species of gull.

Sunday, 29th.—Light wind from south-east. Thermometer 83°, in the shade 80°. Cool and pleasant in the evening.

Monday, 30th.—Wind from the north-west. Very warm during the middle of the day. Thermometer 86° in the shade, 84° on the surface of the water.

Tuesday, 31st.—Cloudy, with little wind from west. Thermometer 81° in the shade, 70° in the water; heavy rain during the evening.

Wednesday, September 1st.—Very pleasant and cool in the morning. Thermometer 82° in the shade, on the surface of the water 79°. Warm and close in the afternoon, with light showers. Spoke ship *Jane* of

Philadelphia, forty-seven days from that port ; bound for Valparaiso, we cherished the hope of keeping company with her, but during the night a violent storm came on and we saw no more of her.

Thursday, September 2nd.—Very light wind and continued heavy rain throughout the day ; from seven o'clock in the evening until ten at night it fell in torrents. Thermometer 82° and 78°.

Friday, September 3rd.—Cool and pleasant in the morning. Thermometer 84° and 79°. Heavy rain afternoon. Preserved three specimens of a curious zoophyte which surrounded the vessel last night in immense shoals, producing a beautiful illumination. On being put in a bucket of sea-water they retained the same power for a few minutes and then died.

Saturday, September 4th.—Damp and cool ; sun invisible at noon. Thermometer 81°, and 77° in the water. As usual, heavy rain in the evening.

Sunday, September 5th.—Light airs of wind from south-west. Visited by great flocks of a small sort of bird, about the size of the European lapwing ; the form of the wings and manner of flying also resembled it. Thermometer at noon in the shade 82°, on the surface of the water 79°. Exceedingly pleasant in the afternoon and evening.

Monday, September 6th.—Wind from the south. Thermometer 81°, and 77° in the water, one degree colder than yesterday, arising from the greater agitation of the water. Great numbers of petrels following the vessel. Lat. 4° north long.

Tuesday, September 7th.—Morning cool and pleasant ; fanning wind from the south. Thermometer 84° and 80°. Continued the same throughout the day.

Wednesday, September 8th.—Morning pleasant with light wind from the south. Thermometer 81°, and in the water 76° ; the difference to-day between the temperature of the air and water is much greater than a person would suppose.

Thursday, September 9th.—As is usual, the pleasantest, and consistently the coolest time of the day is from daylight until eight or nine o'clock in the morning. Wind from the south. Thermometer 79°, and in the water 76°. Very pleasant towards evening ; heavy dews during the night, and particularly so as the day has been warm, succeeded by a calm night. Saw no birds for some days past.

Friday, September 10th.—Wind light and cool from the south-west. Warm during the middle of the day. Thermometer 86° in the shade, in the water 79° ; the difference to-day greater than ever before, being 7° at eight o'clock in the evening. The god of the seas paid us a visit and informed us that he would hold a levée the following day.

Saturday, September 11th.—Heavy rain during the most of last night, with light wind from the south-east. At ten o'clock this morning Neptune, accompanied with his guard of honour, fulfilled his promise made last night, when all his unqualified sons had an interview with his Majesty. The day was passed with much pleasure.

Sunday, September 12th.—Fine, clear, and dry wind, south-N.W. Thermometer 82° and 80° ; cool and pleasant in the evening.

Monday, September 13*th.*—Wind south-east; thermometer 79° in the shade and 79° in the water.   This is the first time that the temperatures came to each other.   Several birds in the evening.

Tuesday,   September   14*th.*—Fanning   wind   from   south-east. Thermometer 79° in the shade, in the water 70°.

Wednesday, September 15*th.*—Wind and thermometer the same as yesterday.

Thursday, September 16*th.*—Morning pleasant; thermometer at five o'clock 71°; at noon, 81° in the shade and 80° in the water.   Very pleasant throughout the day.

Friday, September 17*th.*—Wind south-east, cool; thermometer 81° and 80°.   A small bird, not unlike a sand-snipe, sought refuge on the vessel; unfortunately I could not make him a prisoner.   A beautiful sky at sunset.

Saturday, September 18*th.*—A fine easterly wind.   Thermometer 83° in the shade, in the water 81°.   Visited by flocks of birds, some of which seemed land ones, and insects; welcome visitors, the sight of which tends to make people uneasy, particularly when they cannot be caught.

Sunday, September 19*th.*—Thermometer this morning at eight o'clock 78°; at noon, 83°, and 81° in the water.   Passed a schooner on the west, bound to the north.   Pleasant in the afternoon, with a very favourable breeze from the east.   Lat. 15° 30′ S.

Monday, September 20*th.*—Wind east, warm and clear.   Thermometer 82° in the shade, in the water 79°; pleasant in the evening.   Made some paper bags for seeds against reaching Rio Janeiro.   Lat. 17° 52′ S.

Tuesday, 21*st.*—High wind and rain from the south, producing a very heavy sea, which continued throughout the day.   Caught one large butterfly.   Several birds followed us all day.   Sun obscured. Thermometer 73° in air , in the water 74°; one degree warmer.

Wednesday, 22*nd.*—Wind from the south.   Thermometer at eight, morning, 70°; noon 75°, in the water 74°; at six o'clock in the evening 72°. Very pleasant but felt cold towards evening.   Killed a bird called by sailors Cape pigeon.   There being a great swell of sea a boat could not be lowered down.   Lat. 19° 6′ S.

Thursday, 23*rd.*—Strong wind from the south-east.   Thermometer 72° in the shade, 70° in.the water.   Number of birds increasing.

Friday, 24*th.*—Morning cold (68°) and wet, with strong north wind. At noon it cleared away for a short time, which gave an opportunity of finding our latitude.

In the afternoon the weather became more boisterous, and continued so during the night.   It was exceedingly dark.   The waves breaking over the vessel, very little sleep was had by any person.

Saturday, September 25*th.*—Morning cloudy; loud wind from the north, with heavy showers; the sea running very high.   Towards midday it cleared away; the wind became more moderate.   At two o'clock noon Cape Frio was seen, about eight leagues distant.   Great flock of birds with innumerable swarms of butterflies came to invite us to their coast. Thermometer 75°, and 72° in the water.   The evening cool but pleasant.

Heavy dew during the night; at four o'clock in the morning thermometer only 63°.

Sunday, September 26th.—Morning clear and pleasant, but rather cool. Had a fine view of the high mountains of Cape Frio and along the coast. Thermometer at noon 75°, and 72° in the water. Great numbers of sea-birds, some of which are very large. Rough in the afternoon with high sea. One large turtle (at least from 250 lb. to 300 lb. weight) passed the ship yesterday; one to-day rather smaller. Our attempts to take them were ineffectual. The vessel on tack all day, wind being foul.

Monday, September 27th.—Morning pleasant, with light airs of wind from the north-west. Thermometer at midday 76°, and 73° in the water. Calm during the night.

Tuesday, September 28th.—Morning pleasant and warm, with light variable wind from the west and south. Entered the mouth of the river at noon and came to anchor at two o'clock. After being visited by officers from the Custom House, Board of Health, Police, &c., I had to go on shore to the Master of Police, which all passengers have to do. Afterwards we were visited by two officers from the British man-of-war enquiring if we had a mail or could spare them any newspapers. Thermometer at noon 84° in the shade; at eight o'clock in the evening 74°.

Wednesday, September 29th.—Went on shore in the evening with the captain and returned two hours after. Called on John Dickson, Esq., a friend of Mr. Sabine's, and a correspondent of the Society, who received me with great kindness. In the most handsome manner he invited me to stay at his house during my visit and made every preparation to make me comfortable. I showed him my instructions and informed him of the object of my voyage, &c. I learned he had been the host of the late Mr. Forbes two years since. The affectionate manner in which he spoke of him, of his disposition, of the amiable way he conducted himself during his residence, reflects much honour on his memory. Mr. Dickson gave me much facility as to my pursuit by sending his servants and introducing me to the knowledge of his friends.

My movements were greatly frustrated by rains, and as the stay of the vessel was uncertain I could not with propriety make long journeys.

The approach to Rio is particularly grand. The entrance to the bay or harbour is about half a mile broad, at the mouth of which are four or five small islands all covered with wood—on one is a telegraph station. The ground is mountainous, but not rugged, and covered with wood to the summit; and what appears singular, the palm grows more luxuriantly in such a place than in lower situations. On the left is a conspicuous conical hill, known by the name of Sugar-loaf Mountain, of primitive rock, not unlike the Aberdeen granite with which the London streets are paved. A small fort is built at its base, on which are a few guns; on the opposite side stands one of larger dimensions. In general the houses are regularly built, but of coarse workmanship, of freestone; the rooms are lofty, with large doors and windows. Many of the windows are not glazed but have a sort of shutter of lattice-work, with hinges at the top. The only buildings worthy of notice are the churches, among which is the Emperor's private chapel and one adjoining it in the Palace Square, both of Gothic, neat, and reflecting great credit on the architect. I heard service in the latter at midnight; the gorgeous tapestry hung round the saints, the brilliancy of the lamps and candles, with the general neatness of the edifice, impress on the mind of a stranger a pleasing sensation. The palace forms three sides of a square, is of plain rubble-work, and would only do for a potentate of America.

Among those I became acquainted with was William Harrison, Esq., of Liverpool (a brother of Arnold Harrison, Esq., of Aighburgh, who is a Fellow of the Society), who is fond of plants and birds and has introduced many interesting plants to the Botanic Garden of Liverpool; he has a fine garden five miles from town and a collection of African, European and indigenous plants. On an old wall were about seventy species of *Epidendrum* and *Orchideae* in general on a southern aspect, only the branch or stump on which the plant originally grew was nailed on the wall without any earth, many of them were thriving luxuriantly. The number of his live birds in cages amounts to seventy, mostly Brazilian, many of them very beautiful. This gentleman showed me many civilities, and he informed me that most of his relations were Fellows of the Society. On quitting town he gave me a letter of introduction to his friends, Messrs. McCulloch, &c., of St Barbara, New Albion, lest we should put in there or visit it at any future period—tokens of his friendship which at all times I shall think on with pleasure. To have friends in such a remote spot of the globe is of great consequence. Mr. D. made me acquainted also with Mrs. Maria Graham, who writes travels in Chile and Brazil. She is a lady of much information, of very amiable manners, and tolerably conversant in botany—of which she is fond—and some other branches of natural history. Her verbal description of the plants around Valparaiso I heard with delight, and those on Juan Fernandez with equal gratification. I sincerely hope to visit either. Mrs. G. is tutoress to the young princess. The fish, butcher and vegetable markets lie on the east side of the town on the edge of the bay. The varieties of vegetables in the market were few in number and of bad quality—*Convolvulus Batatas*,[1] Yams (*Dioscorea*), both substitutes for potatos, with two peppers. The only cabbage which I saw was purchased by the captain of the Hudson Bay ship, for which he paid half a dollar. Oranges, lemons, and cocoanuts in great abundance were the only fruits. The former finer than any I ever saw before. One hundred oranges or the same number of lemons can be bought for a dollar. Cocoanuts were also cheap, three halfpence each. The culinary vegetables were the dearest, dearer than the same quantity of potatos in England. Beef and pork are the only sorts of meat in the market, both of inferior quality. Poultry is plentiful and much cheaper in proportion than the others. In the fish market there was a very plentiful supply and a very great variety of shell-fish, many of them singular and beautiful. All sorts are had at a very moderate price. In my walks around the city I was much delighted to see many of the plants cultivated in England and

[1] *Ipomoea Batatas*, C. B. Clarke, in Hook. Fl. Brit. Ind. iv. p. 202.

which have but a puny appearance in their exiled state, but here in luxuriant condition. The *Scitamineae* and *Orchideae* are almost endless. I observed *Maranta zebrina*,[1] *Maranta* sp. small, lately figured at Bayswater—several species unknown, *Gloxinia speciosa* [2] and one new species sent in the box; *Passiflora racemosa, P. microcarpa*, two species of *Rhexia*, very fine. In Mr. Harrison's garden is a sort of fruit between a lemon and orange. The form and colour is exactly that of a lemon. It has not the acidity of the former or saccharine of the latter; partakes of both and most assuredly a much pleasánter fruit than either. I am not aware of such being cultivated in England; it might be an acquisition to have it. Young plants of it would have been sent if they had been able to stand the voyage. He informs me that the seeds do not vegetate. I never saw a place that was more inviting; and never laboured under greater disadvantages: during my stay on shore of twelve days only six of them were fair; many of the specimens collected were useless, and more to my disadvantage having to dry them at sea, hampered up in such a small vessel. Collected two boxes of plants, which were to leave for England in a few days, for the Society. This afterwards I thought as useless work, for they would arrive in England in the winter. On the other hand, I regretted to allow any opportunity to pass without endeavouring to fulfil the objects of the Society. I thought it better to pack them in close boxes as there would be no room for them on deck; however, I hope they will reach London in such a state as will at the least compensate for the expense of collecting. Mr. Harrison kindly undertook to see they would be sent with one of the vessels employed by himself. My collection of dry plants, amounting to nearly two hundred species, I have to take with me, not being perfectly dry, and am prevented from ascertaining them for want of a book of general reference.

October 8th.—Wrote to Joseph Sabine, Esq., to Mr. Munro, to Mr. Atkinson, and to my brother, and made preparation for sailing.

Saturday, 9th.—Mr. Dickson gave me a £10 bill on the Society to purchase several articles for the voyage and the country I was to visit. All these things were done for me by his people with his usual politeness. Being the expected day before sailing he invited some of his friends in town, Mr. Louden, the Admiral's Secretary, and Dr. Scott, his physician, to dinner to meet me. The good feeling and harmony that were shown by every guest at table among themselves and good wishes towards my welfare at parting was, I must confess, gratifying to me. I left their agreeable society at eight in the evening. Just as I stepped in the boat it began to rain heavily, with thunder and lightning. I had to take off my coat and vest to keep my specimens dry. I had among the numerous vessels ying in harbour some difficulty in finding the right one, she having hauled out to a more commodious place for sailing since I was last on board.

Sunday, October 10th.—Cloudy, with light rains; at midday the sun broke through. But little wind, and that contrary for sailing. At five o'clock went on shore. Mr. D. was from home; I spent two hours with

---

[1] *Calathea zebrina*, Ind. Kew. fasc. iii. p. 166.
[2] *Sinningia speciosa*, Ind. Kew. fasc. ii. p. 1036.

his assistant, Mr. Gogerty, who was the intimate friend of Mr. Forbes and Mr. Graham. Mr. Gogerty afterward accompanied me to the ship.

October 11th, 12th, 13th.—Being in expectation of having favourable weather for departing I could not leave the vessel. This I regretted exceedingly, the weather being dry and finer than any during my stay.

Thursday, October 14th.—At seven o'clock this morning the anchor was weighed, and with a light air of wind from the north-west I left that interesting country, certainly with regret, but left it with the hope of being enabled to make a longer stay at a future period.

October 14th to 22nd.—The wind for a few days after leaving land continued variable and generally accompanied with rains in the evening. Max. heat 79° to 82° in the shade ; min. 66° to 68°. At four o'clock on Tuesday morning a fine breeze sprang up and we pursued our voyage along the Brazilian coast with pleasure. Day after day passes away almost imperceptibly ; at breakfast inquiries are made how the wind has been during the night, and the like questions. At home among friends this would look ridiculous, but here they appear of great moment. Calculations are made, should the wind be so-and-so we shall be at such-and-such a place at such a time. Without any exaggeration such things are requisite for such places and circumstances. Off the Plate, in Lat. 37° S., Long. 37° W., immense shoals of seaweed passed the ship, only of one species, some of which measured 60 feet in length. (No sounding could be found.) Stem round, three inches in diameter at the thickest part ; leaves alternate, lanceolate, partly serrate and crisped, the young ones all united at the tops of the stem ; vesicles oval, very large ; on the roots were some starfish and bivalve shells ; the earthy matter adhering to the roots was small shells, fine white sand, and lime of a recent formation ; having no fresh water to immerse them in, previous to laying them in paper, I put up in a large jar a portion of the *Fucus* in spirits, which will convey a good idea of its magnitude. When in this latitude, the weather was much like that usually experienced in the Gulf of Florida.

Here the species of petrel which before was scarce was abundant, and could be taken with a hook and line baited with fat of pork. I caught three of the mottled one which may prove to be *P*[*rocellaria*] *capensis*, and preserved them. Two other species were also plentiful ; both are shy. I am not fortunate enough to take any of them ; one large, nearly double the size of the mottled one, of a glossy jet black, another smaller, but larger than *P. capensis*, of a dusky-brown colour. Also two species of Albatross, one large, white, brownish-black on the upper side of the wings ; this may be *Diomedea exulans*, and a black one,[1] smaller, which I take to be *D. fuliginosa* ; the latter I caught off the Falkland Islands in abundance, in the same manner as the petrels, the line of course being stronger and the hook larger. I preserved two of this which have since spoiled with me.

Saturday, November 5th.—Off the Falkland Islands in Lat. 54½° S. We now began to feel the chilliness of Cape Horn and experience the bad

---

[1] In another MS. :—' Of a dark muddy colour.'

weather of its forbidding, dreary climate. It is only when the wind blows furiously and the ocean is covered with foam like a washing-tub that I could take the Albatross. Diametrically opposite to every account I have read of them, they all say calm. Their voice is like the bleating of goats ; on being taken they emit from the mouth an oily matter of different colours, arising no doubt from the great variety of *Physalae, Beroë,* and other zoophytes on which they live.

In all, of the brown ones I caught forty-nine, two of which I preserved ; both males ; no females came under my notice. Off the Cape a third species made its appearance, white on the belly and under the wings, back greyish, blackish-brown on the upper side of the wings ; neck, light azure colour; beak, black upper part and point yellow ; legs and feet black. Two of this I caught, but only one could be preserved (a male) ; it is a much larger and stronger bird than the other : when he attempts to take the bait, or even to light near it, they all, seemingly with fear, leave it to him. He was very ferocious and would bite at sticks held out to him ; one of the sailors in assisting me to lay hold of him was bitten in the thigh through the trousers—the piece was taken out as if cut with a knife. Their flesh is fishy and rancid. On the same day I caught two petrels of a bluish-white colour, beak and legs partly red ; this species on the water is very graceful and by no means very plentiful, their voice is like the chuckling of young ducks ; the two now sent home are males.

During the time (ten days) of rounding the Cape the weather was stormy with generally a fine clear sky. The motion of the vessel was great, the waves frequently breaking over it, and no sleep until completely worn out with fatigue. When the wind blows from the south or south-west the cold is insupportable, and yet the thermometer never was lower than 39°, 45° the greatest ; there is a piercing rawness in the atmosphere (laying aside being so lately in the tropics and of course more susceptible of cold) quite unknown in the northern hemisphere in similar latitudes. Daylight sixteen hours, sky generally clear azure and beautifully tinted in the evenings just as the sun leaves the horizon.

November 16*th.*—We were considered round, and gladly we bade adieu to such inhospitable regions. The weather moderated gradually and we soon found ourselves navigating more pacific water.

November 17*th.*—Caught two of *Diomedea exulans* ; the largest weighed 18 lb. and measured 12 feet 4 inches from tip to tip ; 4 feet from the point of the beak to the tail. Both were moulting and not worth preserving. All the species when sitting on the water raise the wings like the swan— when eating particularly so—but do not shake them like the hawk tribe. When rising from the water they partly run, partly fly, tipping the water with the point of the wings and feet for several hundred yards before they are clear of it. They cannot rise from the deck of the vessel. Their flight is quick but steady ; when fishing they soar with wings in a curved direction.

In the latitude of 54° South 77° West Longitude a curious species of porpoise was seen in abundance with a pure white stripe on each side from the snout to the tail; a much smaller animal than the common one and its motion equally quick; none could be caught.

Until we passed the Straits of Magellan the weather continued variable, the wind was boisterous with a rough sea, rain, and thick fogs.

From this time to the 14th of December nothing worthy to be noticed occurred. Towards noon the Island of Mas-a-Fuera was seen, distant about seven leagues, and appears like a conical black rock. As we drew near the shore it became more like an island. At four in the afternoon of the same day passed within two miles of it ; the surge on the beach prevented the commander from landing. On the whole its appearance is barren, although in the valleys there is herbage and some trees on the hills ; goats were seen in abundance. Our course was then directed towards the island of Juan Fernandez, about eighty miles distant to the north-east.

It afforded me much gratification to see Juan Fernandez on the morning of the second day. At twelve o'clock a boat was sent in search of fresh water, in which I was permitted to embark ; being unsuccessful in some measure, our stay was short and we returned to the ship in a few hours. The following day we went round to the north side to Cumberland Bay, so named by Anson in 1741. The whole island is very mountainous, volcanic, and beautifully covered with wood to the summit of the hills, the tops of which are rarely seen, being enveloped in the clouds. On Friday and Saturday I went on shore and was much gratified with my visit. As we approached the shore we were surprised to see a small vessel at anchor, and on the beach a hut with smoke rising from it. As we were about to step out of the boat a man sprang out of the thicket to our astonishment and directed us into a sheltered creek. He gave me the following account of his adventures. His name, William Clark ; a sailor ; native of Whitechapel, London ; came to the coast of Chile five years ago in a Liverpool ship called *Lolland,* and was there discharged. He is now in the employment of the Spaniards, who visit the island for the purpose of killing seals and wild bullocks, which are both numerous. Five of his companions were on the opposite side, in their pursuit, and came to see him once a week ; he was left to take care of the little bark and other property. When he saw the boat first he abandoned his hut and fled to the wood, thinking us to be pirates. On hearing us speak English he sprang from his place of retreat, and no language can convey the pleasure he seemed to feel. He had been there five weeks and intended to stay five more ; he came from Coquimbo, in Chile. His clothing was one pair of blue woollen trousers, a flannel and a cotton shirt, and a hat, but he chose to go bareheaded ; he had no coat. The surgeon and I gave him as much as could be spared from our small stock, for which he expressed many thanks. His little hut was made of turf and stones thatched with the straw of a wild oat. In one corner lay a bunch of straw and his blanket ; a log of wood to sit on was all the furniture ; the only cooking utensil was a common cast-iron pot with a *wooden bottom,* which he had sunk a few inches in the floor—and placed the fire round the sides ! He longed to taste roast beef (having had none for seven years) and one day tried to indulge with a little *baked,* as he termed it ; but in the baking the bottom gave way, as might reasonably be expected; so poor Clark could not effect the new mode of cooking. I told him under his circumstances roasting beef was an easier task than boiling. He is a man of some information; his library amounted to seventeen volumes —Bible and Book of Common Prayer, which he had to keep in a secret place when his Spanish friends were there; an odd volume of 'Tales of My Landlord' and 'Old Mortality,' some of voyages, Cowper's Poems. He had the one by heart addressed to Alexander Selkirk; but what is still more worthy to be noticed, a fine bound copy of Crusoe's adventures, who himself was the latest and most complete edition. Like all other English sailors he had no aversion to rum; I gave him a single dram, which, not being accustomed to before for a great length of time, made him forget his exile. He was like the heroes of Troy: 'fought his battle over again and slew the slain three times.'

Here a few years ago the Spaniards formed a colony; but it is now abandoned, all the houses are destroyed, and the fort, on which were some very large guns. Twenty-six cannons lay on the shore just below. The vestiges of a church are to be seen; on the lintel of the door the following inscription, 'La casa de Dios puerta del cielo y so colocoesta a 24 de Septiembre, 1811'[1]—'The house of God consecrated 24 September, 1811.' Near this a circular oven of brick, seven feet within, marked on it 1741; probably built by Anson during his residence, it is now occupied by a small species of blue pigeon as their cote; in it I found some eggs, but no young ones. This I told Clark he should use. In the old gardens were abundance of three or four different peaches in a half ripe state, very luxuriant; one apple, a quince, and two pears; a quantity of the last three we took for puddings. Abundance of figs in vigorous state of bearing, and vines, one which thrives luxuriantly; it is just in blossom. The only ripe fruit was a sort of strawberry with large fruit of a pale whitish-red, not unpleasant; leaves, stem, and calyx very downy; dried a paper of seeds of this species lest it may prove indigenous to the island or the coast of Chile. The only culinary vegetable was radish, which grows to a large size. I sowed a small portion of vine, pear, and some other fruit seeds which I had of Mr. Atkinson and some culinary vegetables, and gave some to Clark to sow on various parts of the island. Saturday afternoon was set apart for fishing; a sort of rock codfish and a smaller fish unknown to me were caught in abundance, both good eating, and after such a length of time on salted food were considered a luxury.

Dec. 18*th*. Thermometer on the top of the hill on right of Cumberland Bay, at 2 P.M., 70° Fahr. in the shade; in the valley at 4 P.M., 74°; cloudy and calm, light rain in the evening.

On our quitting the shore Clark presented us with a fine female goat, but not one of Robinson Crusoe's, for it was young. We left him standing on a large stone on the shore on the evening of Saturday, intending to visit him again in the morning. Scarcely had we reached the ship when a strong south-easterly wind set in, which

[1] We cannot vouch for the correctness of the Spanish, but it is given exactly as written in the MS.—ED.

obliged me reluctantly to leave such an interesting speck of the globe and my new acquaintance Clark. The weather continued unfavourable for making the land again ; for three days we were so much driven by its violence that the captain considered it a sacrifice to return. Our course was then directed towards the Islands of Galapagos under the Equator in Long. 80° W. On the morning of Thursday the wind became moderate and we got the south-east trade wind, which we were fortunate enough to carry with us within 1½ degrees of the Equator. Christmas was observed in Lat. 27° S., Long. 84° W. We dined on the goat given to us by Clark ; were comfortable and happy; in the evening we drank the health of our friends in England.

Collected during my visit to Juan Fernandez the following plants :—

(1) Fern, a fine strong plant, plentiful in moist places.

(2) *Aspidium* (?), an elegant strong plant ; abundant in moist places; gives great annoyance in passing through it ; 4 to 6 feet high.

(3) *Asplenium* (?) sp.; small; moist rocks near the sea; abundant on the rocks of a natural arch in Cruz Bay, said by Clark to have been the residence of the hero Crusoe; this I have no doubt will prove *A. marinum*.

(4) *Pteris*, a strong-growing species, frequenting springs and moist ground; under-side of the frond white, upper bright green.

(5) Fern tree, 6 to 10 feet high, branching, pinnated on long footstalks ; a very splendid plant ; abundant in the ravines.

(6) *Aspidium* (?), may prove the same as No. 2 ; the present in a more open, airy situation, and in a young state.

(7) *Adiantum* sp. ; in dry open situation, rare. In dry shady places in rich vegetable soil, looks like *A. pedatum* of N. America.

(8) *Polypodium* sp. ; frond pinnate, dentate, strong nerved, on decayed trees in thick shady woods ; root bright green, covered with thick brown chaffy scales.

(9) *Polypodium*, a very fine plant, leaves root at the point ; on rocks near the summit of the hills ; abundant.

(10) *Asplenium* sp. ; plentiful in moist places near springs.

(11) Fern Flowering, abundant in thick woods.

(12) *Aspidium* sp.; fronds doubly pinnate; footstalks smooth and black; in low shady places among bushes.

(13) *Pteris* sp. ; doubly pinnate ; a fine species, open dry places on the summit of the hills and on rocks.

(14) *Aspidium*, closely allied to (13), but may prove to be a variety of (2). In the same situation.

(15) ? ? Tree Fern, without exception the finest of the kind that came under my notice on the whole island ; stem strong, thick, and rough ; footstalks long and black ; 12 to 15 feet high ; abundant in groups on hilly places ; appearing at a distance like young pines.

(16) Frond large, broad, and entire ; leaves of the spike opposite at the base, alternate at the top, inserted on the upper side of the spike ; very distinct from 1 or 11.

(17) *Vaccinium* sp.; leaves alternate, ovate, minutely serrate, entire at the base; petiole short, smooth, glossy-green on the upper side; leaves on the young shoots obovate, flowers fastigiate, campanulate, revolute, solitary, dingy-white and very fragrant; stamens seven to ten; berry large, globular, compressed at the top; black, tinged with purple; five to seven celled, generally seven. Seeds numerous, small, and vegetate before falling from the bush; peduncle hairy; bractea or bud scales, small, numerous; taste agreeable, like *V. Myrtillus*, only sweeter; young shoots red where exposed to the sun; a most beautiful evergreen species and very variable plant and would form a valuable addition to the numerous group in England; on the low grounds it does not exceed 10 inches or 2 feet high; on the highest peaks of the mountains, in rocky places, 6 to 12 feet; on the high grounds, plentiful in flower and fruit; a paper of this delightful plant is now sent home with specimens of its fruit. Comes near *V. ovatum* of N. America. This curious plant seems to differ a little from *Vaccinium*; the calyx becomes part of the berry, therefore it may come between it and *Gaultheria*; this is indeed the largest shrub on the summit of the mountains.

(18) *Lobelia* sp.; perennial; leaves amplexicaul, ovate, acute at the point, woolly; spike strong; flowers large, bright scarlet (equal to, if not finer than, *L. fulgens*); peduncle long; bractea linear; whole plant covered with a dense wool; a most splendid plant, 4 to 6 feet high, in low damp places; I regret that no perfect seeds of this fine species could be found, being a little too early in the season.

(19) *Berberis* sp.; leaves orbicular; stem smooth; only one small plant, without flowers or fruit, came under my notice on the hills among rocks south of Cumberland Bay.

(20) *Carduus*, annual; in the abandoned fields, probably introduced.

(21) Imperfect, a large tree 40 to 60 feet high with a corresponding thickness; wood hard and apparently durable; may belong to *Eugenia*; this is the principal tree composing the forest.

(22) *Silene*, annual; leaves lanceolate; bractea linear; stem hispid; not in flower; a low species, on the gravelly beach.

(23) *Rumex* sp.; perennial; leaves lanceolate, smooth, undulate; flowers verticillate; valves strongly veined with a strong ciliation; a curious species, found on the mountains.

(24) *Syngenesia* sp. annual; 18 inches to 2 feet high, spreading; leaves imbricate, lanceolate; flowers terminal, yellow; calyx squamous, woolly; stem and leaves pubescent; on dry grounds, abundant; perhaps *Donia*.

(25) *Fragaria* sp. or var.; leaves small, round, pubescent; petiole long; flowers small, white; whole plant very pubescent. This plant is evidently introduced, whether by Lord Anson or the Spaniards from the coast of Chile is uncertain; fruit large, three-quarters of an ounce, whitish-yellow, dry, but has an agreeable acid taste; differs materially from any in England; great numbers of a mealy bug on the plants; in the abandoned gardens at Cumberland Bay, abundant.

(26) *Melilotus* (?) sp., annual; although found on the hills I have no

doubt it has been introduced; however, I did not observe it in the valleys that had been cultivated.

(27) —— (?) calyx (?) corolla none; stamens four; anthers yellow; stigma one; capsule four-celled, four-seeded; seeds small; leaves ovate, lanceolate, denticulate, nearly sessile; a low shrub, on dry elevated situations; plentiful.

(28) *Plantago* sp., perennial; leaves radical, ovate, entire, smooth, seven-ribbed; petiole double the length of the leaf; scape elongated; bractea small; on the high ground; abundant. Can this be *P. major*?

(29) *Umbelliferae*, annual, *Daucus* (?); on the hills, sparingly, 18 inches to 2 feet high.

(30) *Dioecia* (?) female; stem branching in a loose panicle; leaves peltate, 5–6-lobed, six-ribbed (6 feet in diameter); petiole long, scabrous; stem long, 4 inches in diameter, strong, rough, sending out roots 2–3 feet from the ground, like the genus *Arum* and *Ficus*; when broken yields a copious limpid acrid juice; the whole plant has a disagreeable smell; a magnificent plant 10 feet high, covering a space nearly double that in diameter, frequenting the margins of rills; in solitary plants; abundant.

(31) *Cistrum* (?); flowers yellow; a low shrub, in the abandoned fields; plentiful.

(32) *Aralia* (?), calyx in five segments; corolla campanulate; stamens five, inserted on the receptacle; style 1, stigma globular; leaves alternate, ovate, slightly serrate; a beautiful low reclining shrub; flowers bright rose colour, in a large but loose panicle; frequenting dry elevated situations and on rocks.

(33) A large tree, on the high grounds.

(34) *Syngenesia*, perennial; yields a copious clear viscid juice; in the crevices of rocks, in solitary places; rare; 3 feet high.

(35) May prove to be *Spartium*; in dry situations.

(36) *Aster* (?), flowers capitate, white; [leaves] linear-lanceolate, dentate; small plant, in rocky places; perennial.

(37) *Cardamine*, annual; leaves ovate; stem creeping; a small plant, near rills, on moist ground.

(38) *Arenaria* sp., perennial; small, on rocks of the seashore at Cruz Bay.

(39) *Euphorbia* sp. annual; near the deserted houses; introduced.

(40) *Verbena*, male (?) perennial (?); stem square and spreading; leaves alternate, ovate, lanceolate; flowers verticillate, purple; a fine plant, 2½ to 3 feet high in open places.

(41) *Umbelliferae*, biennial; fruit aculeate, oblong, partly solid; a very curious plant; on the hills.[1]

(42) *Malvaceae*, stem reclining; leaves digitate, alternate; flowers solitary; peduncle long; in the valleys,[2] plentiful.

(43) *Melissa* (?), abundant on the old walls of the fort at Cumberland Bay; perhaps introduced.

---

[1] In another MS. :—'abundant in moist places.'
[2] In another MS. :—'on the summit of the hills.'

(44) *Syngenesia, Lactuca* (?) perennial; a curious plant, 2 to 3 feet high, on moist rocks, rare.

(45) *Umbelliferae*, allied to *Daucus*; biennial; fruit oblong, ciliated on the angles, solid; abundant in the valleys.[1]

(46) *Ranunculus* sp.; perennial; stem creeping; leaves round, partly divided; petiole long; corolla inferior, seeds yellow (six to ten), angles beset with bristles; a curious plant, on low moist ground.[2]

(47) Imperfect, plentiful on the old walls.

(48) *Campanula* (?); leaves alternate, sessile, lanceolate, serrate, smooth above, slightly pubescent underneath; flowers capitate, numerous, white; bractea linear; suffruticose; an exceedingly ornamental plant; a low plant, 2 feet high, in tufts on the highest peaks of the mountains.[3]

(49) *Campanula* (var. of 48), corolla striated with blue, like most of this extensive and ornamental genus; has its varieties; this is a smaller plant but by far excels it in beauty, in the same place with the former.

(50) Imperfect, one of the largest trees on the island, 40 to 70 feet high, $3\frac{1}{2}$ to 4 diameter; wood white, hard, and takes a fine polish.

(51) *Gnaphalium*, perennial; leaves amplexicaul, lanceolate, woolly; flowers capitate, yellow; stem glutinous; abundant on dry elevated spots.

(52) *Gnaphalium* sp.; leaves oblong, woolly under; stem slightly pubescent; in the same place with the former.

(53) *Rumex* sp.; perennial; perhaps introduced; plentiful in the abandoned fields.

(54) *Gnaphalium*, probably will agree with 52; in the same situation.

(55) *Ruta* sp.; perennial; flowers yellow; most likely introduced; near the abandoned fields.[4]

(56) *Labiatae*; calyx persistent; segments four[5]; corolla long under-lip, obtuse, bright blue; stamens four, two abortive; fruit small, pulpy, globular, covered with a thin skin, violet-purple colour; two-seeded, seeds angular; leaves opposite, ovate at the base, point acute, smooth, nearly sessile; stem spiny; spines straight; a tree 40 to 50 feet high, very ornamental from the multiplicity of its azure drooping blossoms; the profusion of its fruit of the same colour, and flowering the greater part of the season to all appearance, makes it the most graceful tree in the valleys[6] where it abounds.

(57) *Lythrum* sp.; on the seashore, abundantly.

(58) *Mentha* perennial; introduced, in the abandoned fields.

(59) *Gramineae*, in the mountain valleys, abundant.

(60) *Avena* sp. (?); annual, 6 to 8 feet high, one of the most abundant grasses in the valleys on the island; seeds long, covered with silky hairs; I had some difficulty in procuring the few seeds, being the food of a species of small blue pigeon that devours it before it is ripe.

---

[1] In another MS. :—'abundant on the hills.'
[2] In another MS. :—'on the summits of the hills.'
[3] In another MS. :—' 10 to 18 inches high, in dry light soils on the hills; plentiful; a small portion of seed of this fine plant was had.'
[4] In another MS. :—'abundant on the hills and valleys.'
[5] In another MS. :—'divided into five equal segments.'
[6] In another MS.:—' on the high hills, where it is the only tree of great size.'

This cannot have been cultivated. Clark thatched his hut with the straw.

(61) *Bromus*, perennial, abundant on the hills near the summit.

(62) *Briza*, annual; on the high grounds in dry places; in a bad state.

(63) *Cyperus*, perennial; a fine strong species; abundant on moist ground near rills.

(64) *Gramineae*, perennial; abundant everywhere; a strong grass.

(65) *Gramineae*, perennial; in damp soils[1]; rare.

(66) *Hordeum*, annual; a delightful small species, not more than 8 inches or a foot high; abundant on the highest peaks of the mountains.

(67) *Gramineae*, perennial; a plant 4 to 6 feet high, found everywhere; varying in size, according to situation; seeds long, sharp at the base, with silky hairs; a troublesome plant; the seeds accumulate in masses on the stockings, producing a disagreeable itching sensation. I was under the necessity of stopping every ten minutes to take off the masses of seeds. It gives great pain.

(68) *Gramineae*, perennial; may agree with the preceding No. 67, found in the same places.

(69) *Gramineae*, perennial; in moist, low situations; abundant.

(70) *Phleum*, annual, abundant on the hills.

(71) May perhaps agree with the preceding. In like places.

(72) *Gramineae*, may prove the same as 68; perennial; found on the hills.

(73) *Gramineae*, annual (?); on dry elevated spots; rare.

(74) *Carex*, perennial; in damp places on the margin of rivulets plentiful.

(75) *Carex* sp., perennial; along with the preceding; a fine plant.

(76) *Juncus* sp., perennial; on moist ground in low valleys.

(77) *Medicago* sp., annual; on low grounds, in open situations.

(78) *Umbelliferae*, annual (?), on the hills; tastes like cloves.

Hypnum, three species, on decayed wood in shady places, near water.

Lichen, four, on dry rocks near the summit of the hills.

Fucus, one species at Cruz Bay; *Fucus*, one species, with species of *Confervae*, at Cumberland Bay.

Hymenophyllum, in the deep shady ravines of the mountains; a beautiful species. When I found this I laid myself down on a carpet of it close by a crystal rill descending through the rugged but beautiful hills. Although a plant of humble growth, its delicately veined and crisped foliage contrasts beautifully with the more princely of the tribe by which it is surrounded.

1825.

Saturday, January 1st.—In Lat. 19° S. Weather continued good and nothing occurred deserving to be mentioned. As we approached the Line the heat sensibly increased, but by no means so oppressive as in the Atlantic. Although the difference of the mercury is trifling, there is always a cooling atmosphere which renders it more supportable and

[1] In another MS. :—' in the middle of a small swamp at Cumberland Bay.'

agreeable. At noon on Sunday, 9th, Chatham Island was seen ; we passed
along the east side at 4 P.M. of the same day, fifteen miles from the shore.
It is not mountainous and apparently but little herbage on it. On the
morning of 10th (Monday) I went on shore on James Island, thirty-seven
miles to the west of Chatham Island. It is volcanic, mountainous, and
very rugged, with some fine vestiges of volcano craters and vitrified lava ;
the hills are not high, the highest being about 2000 feet above the level
of the sea. The verdure is scanty in comparison with most tropical
climates, arising, no doubt, from the scarcity of fresh water, although at
the same time some of the trees in the valleys are large, but very little
variety; few of them were known to me. My stay was three days, two
hours on shore each time. Few of its plants were known to me. The
birds are very numerous, and some of them pretty, so little acquainted
with man's devices that they were readily killed with a stick ; a gun
was not necessary except to bring them from the rocks or from the tops
of the trees. Many of the smaller ones perched on my hat, and when I
carried my gun on my shoulder would sit on the muzzle. During my stay
I killed forty-five, of nineteen genera, all of which I skinned carefully,
and had the mortification to lose them all except one species of *Sula* ;
by the almost constant rain of twelve days after leaving the island I could
not expose them on deck and no room for them below. Among them
were two species of pelican, four of *Sula*, four of hawk (one particularly
fine, nearly orange colour), one very curious small pigeon. I was nearly
as unfortunate with plants, my collection amounting to 175 specimens,
many of them, no doubt, interesting. I was able only to save forty.
Never in my life was I so mortified, touching at a place where everything,
indeed the most trifling particle, becomes of interest in England, and to
have such a miserable collection to show I have been there.

With no small labour I dried the few now sent home ; what they may
be I cannot say. The weather during the remaining part of my voyage
was such as did not admit of looking to plants ; since my arrival my time
has been otherwise engaged. In the valleys a very singular large species
of cactus, 20 to 50 feet high, with a trunk 2 to 3 feet in diameter ; comes in
the section of *Opuntia* ; flowers large, bright yellow and proliferous. Seeds
of this are sent home. Also a fine species of *Gossypium* with large yellow
flowers and yellow cotton ; a shrub 4 to 10 feet high ; seeds of this are also
sent, with seeds of a plant which may be found to belong to *Coniferae*. On
the island were a species of tortoise, some of them very large, one weighing
400 lb. ; a lizard, 3 feet long, of a bright orange-yellow ; both good eating.
A fine skin of the lizard I lost, and regret it exceedingly, being not described.
On the shore are abundance of turtle of good quality, probably the green
turtle of the West Indies. No fresh water was found except a small
spring flowing from the crevices of one of the craters. The last day on
shore it ceased to rain for about an hour ; the sun broke through and
raised a steam from the ground almost suffocating. My thermometer
stood at 96°, not a breath of wind.

Left on Wednesday at dusk ; passed along the east side of Albemarle
Island at a short distance. People were on it ; after dark lights were seen.

Some blue lights were let off, to which they answered. The first twelve days, without exception, it was almost constantly raining; light airs of wind but generally calm. On Sunday, 16th, we had a tremendous thunderstorm, vivid lightning lasting from 4 A.M. to 2 P.M.; I never witnessed anything equal to it. Five tons of water was had from the sails and deck; this was a great relief, for our allowance was more and we had our clothes washed. Here we were carried nearly 10° to the westward by the current, a fortunate circumstance, being on our course 45 to 60 miles a day. The remainder of the time within the tropics the wind continued very variable; occasional showers, cloudy sky, with lightning and thunder in the evenings. In the lat. of 34° N. caught two albatross of a blackish colour with a little white about the eyes and beak, resembling D[iomedea] fuliginosa caught off Cape Horn. I was able to preserve only one. The beak of the one at Cape Horn of a lighter colour, feet and legs white. This I have no doubt will prove a very distinct variety, if not a species. Both of them I preserved carefully and are now sent home. Also one agreeing with D[iomedea] exulans, but much smaller than is found in the Southern Hemisphere, 7 feet from tip to tip, colour exactly the same. I was prevented from skinning this one by the violent storm (for as I have mentioned before, I never could take them but when the water was in the most agitated state), during which the second mate fell on the deck and fractured his right thigh. The excruciating pain which this poor man suffered until the termination of our voyage can hardly be expressed. On the 12th of February we were in sight of a river in Long. 134° W., but the weather was so boisterous and frightful that it forbade everything like approaching the coast as useless. We were tossed and driven about in this condition for six weeks, winds prevailing from the south-west. Here we experienced the furious hurricanes of North-West America in the fullest extent a thousand times worse than Cape Horn. In this latitude there is an abundance of a small species of Physalae of an azure transparent colour, which were frequently washed on the mainyard by the spray breaking over the vessel. Prevailing winds from the south-west and north-west. Many efforts were made during this time to reach our destined port in the short intervals of favourable weather. On Saturday, April 2, Cape Disappointment was seen at noon, distant thirty miles. Sail was shortened to wait a new day for entering.

Sunday, 3rd.—Calm in the morning and cold; a keen easterly breeze carried us within four miles of the River, when another violent storm from the west obliged us again to put to sea.

April 5th.—We bore in again for the land, being 170 miles at sea, with weather more inviting.

April 7th.—At daylight on Thursday our course was again directed to the coast, being only 40 miles distant, every person breathing a wish we might be more fortunate than on Sunday. The weather seemingly more steady with a keen north-east wind, such an opportunity was not lost, all sail was set, joy and expectation was on every countenance, all glad to make themselves useful. The Doctor and I kept the soundings. At one o'clock noon we entered the river and passed the sand bank in safety (which is considered dangerous and on which, I learn, many vessels have been injured and some wrecked). At four we came to anchor in Baker's Bay, on the north side of the river.

Several shots of the cannon were immediately fired to announce our arrival to the establishment 7 miles up the river, but were not answered. Thus my long and tedious voyage of 8 months 14 days from England terminated. The joy of viewing land, the hope of in a few days ranging through the long wished-for spot and the pleasure of again resuming my wonted employment may be readily calculated. We spent the evening with great mirth and at an early hour went to sleep, to sleep without noise and motion, the disagreeable attendants of a sea voyage. With truth I may count this one of the happy moments of my life. As might naturally be supposed to enjoy the sight of land, free from the excessive motion and noise of the ship—from all deprived nearly nine months—was to me truly a luxury. The ground on the south side of the river is low, covered thickly with wood, chiefly *Pinus canadensis*,[1] *P. balsamea*,[2] and a species which may prove to be *P. taxifolia*.[3] The north (Cape Disappointment) is a remarkable promontory, elevation about 700 feet above the sea, covered with wood of the same kinds as on the other side.

April 8th.—Constant heavy rain, cold, thermometer 47°. Saturday the 9th in company with Mr. Scouler I went on shore on Cape Disappointment as the ship could not proceed up the river in consequence of heavy rains and thick fogs. On stepping on the shore *Gaultheria Shallon* was the first plant I took in my hands. So pleased was I that I could scarcely see anything but it. Mr. Menzies correctly observes that it grows under thick pine-forests in great luxuriance and would make a valuable addition to our gardens. It grows most luxuriantly on the margins of woods, particularly near the ocean. Pursh's figure of it is very correct. *Rubus spectabilis* was also abundant; both these delightful plants in blossom. In the woods were several species of *Vaccinium*, but not yet in blossom, a species of *Tiarella* and *Heuchera* in flower. In a few hours we returned to the ship amply gratified. On the morning of Sunday the 10th went again on shore and made a short stay; saw nothing different from that seen yesterday, except some *Gramineae* and *Musci*. On our return to the ship we found a canoe with one Canadian and several Indians with intelligence from the establishment who brought some potatos, milk, and fresh butter. The potatos were so much relished that we had some in the evening for tea. The natives viewed us with curiosity and put to us many questions. Some of them have a few words of English and by the assistance of signing make themselves very well understood. The practice of compressing the forehead, of perforating the septum of the nose and ears with shells, bits of copper, beads, or in fact any hardware, gives a stranger a curious idea of their singular habits. They brought dry salmon, fresh sturgeon, game, and some prepared roots with dry

---

[1] *Tsuga canadensis*, Veitch, Man. Conif., ed. 2, p. 463.
[2] *Abies balsamea*, Veitch, loc. cit., p. 492.
[3] *Pseudotsuga Douglasii*, Mast. in Journ. R. Hort. Soc. xiv. p. 245.

berries for sale and soon showed themselves to be a dexterous people at bargaining. On Monday, the 11th, we went up the river to the Company's establishment, distant from the entrance about seven or eight miles. We learned they had nearly abandoned their fort there and had made one seventy miles up the river on the opposite side, to which all persons in their employ were to repair in a few days. I went on shore on Tuesday (12th March,[1] 1825) and was very civilly received by a Mr. McKenzie, the other person in authority; he informed me they were about to abandon the present place for a more commodious situation 90 miles up the river on the north side, also that the chief factor, John McLoughlin, Esq., was up the river at the new establishment, but would be down as soon as he received the news of the ship's arrival. I did not leave the ship until Saturday, but was daily on shore. With respect to the appearance of the country and its fertility my expectations were fully realised. It is very varied, diversified by hills and extensive plains, generally good soil. The greater part of the whole country as far as the eye can reach is closely covered with pine of several species. In forest trees there is no variety or comparison to the Atlantic side, no *Fagus*, *Gleditschia*, *Magnolia*, *Juglans*, one *Quercus*, one *Fraxinus*. The country to the northward near the ocean is hilly, Point Round or Point Adams of Lewis and Clarke on the south side of the river is low and many places swampy. For the distance of 40 miles as far as Cape Lookout there is a ridge of hills that run in a south-west direction and is so named by Vancouver. The breadth of the river at its mouth is about 5 miles not including Baker's Bay which has a deep bend. The current is very rapid and when the wind blows from the west produces a great agitation. The water on the sandbar breaks from one side to the other so that no channel can be perceived; when in such a state no vessel can attempt to go out or come in. Mr. McKenzie made me as comfortable as his circumstances would admit, until he could see the chief factor. My paper being all in the hold, except a very small quantity, and the ship not yet taking out the cargo, I could do but little in the way of collecting. The following is what came under my notice:

Of those marked with an S. seeds are now sent home and a small portion of each kept where they could be divided, which was done; except in a very few instances, those kept are, as I am instructed, to be either taken or sent home across the continent. A dry specimen of each is also kept for reference to the collecting of seeds, and will be sent or taken home by sea, being too bulky to cross the continent, though a few of the most interesting may.

(1) *Cardamine* sp., annual, male; in moist places and margins of creeks.

(2) *Ribes* sp.; leaves large, five-lobed, rough, serrate; flower rising to a very long spike fully 6 to 8 inches long, purplish-yellow, thirty to forty in number; wood white; 6 to 8 feet high, very luxuriant in moist situations in rich vegetable soil; berry about the size of a common currant, round, of a bluish-black and slightly hairy, with a bitter astringent

---

[1] This must be a slip of the pen: April is intended.—ED.

taste; flowers April, fruit in July; habitat banks of the Columbia near the ocean. S.

(3) *Gaultheria Shallon*; called by the natives 'Salal,' not 'Shallon' as stated by Pursh, figure and description good; abundant (as is very correctly observed by Mr. Menzies) in all the pine forests, more luxuriant where partially shaded, particularly so near the ocean. I have since seen it as far as 40 miles above the Grand Rapids of the Columbia, but as it leaves the coast it becomes less vigorous; was in flower when I arrived and continued so till August and in fruit. Bears abundantly, fruit good, indeed by far the best in the country; should the seeds now sent home rise, as I hope they may, I have little doubt but it will ere long find a place in the fruit garden as well as in the ornamental. In my walks I have frequently seen the young plants on the stumps of trees 4 to 10 feet from the ground and on dead wood growing luxuriantly. It might be worth mentioning to Mr. Munro to try it in rich decomposed vegetable soil, being its natural way of propagating. I am sorry that I have it not in my power to send specimens of the fruit in spirits with the ship; that put up when on my last journey to the ocean was by some evil disposed person stolen for the sake of the spirits they were in; I have every reason to think it was some of the Eroque Indians belonging to the establishment. It flowers the whole summer through and the fruit is ripe in July and continues bearing until checked by the frost; thick woods and banks of rivers. S.

(4) *Vaccinium* sp.; flowers small, green tinged with faint yellow; deciduous; 4 to 6 feet high; in shady places; leaves ovate; fruit globular, about the size of a common pea, scarlet, transparent, very juicy with agreeable acid, ripe in June and July. S.

(5) *Vaccinium* sp.; deciduous; corolla faint purple at the base, mouth green; leaves obovate; fruit globular, blackish-purple colour, a little longer than the former, agreeable acid; a shrub 10 to 16 feet high, in the same situation as the former species. This is not so prolific; fruit ripe in June and July. S.

(6) Allied to *Xylosteum* [1]; flowers in pairs, fragrant, golden-yellow; leaves opposite, ovate, acute; berry in pairs surrounded by the calyx; flowers in April, fruit ripe in July; 14 to 18 feet high; in rocky situations and gravelly places, shores of Columbia, near the ocean; a stately and beautiful shrub. S.

(7) *Ribes sanguineum* (?); flowers pink or rose colour, inside of the petals white, anthers white in long racemes; a most beautiful shrub; in open, dry places; 7 to 10 feet high. This exceedingly handsome plant is abundant on the rocky shores of the Columbia and its branches, and in such places produces a great profusion of flowers but little fruit. In the shady woods the flowers are less numerous and beautiful but produce more fruit. I am happy to send a good portion of its seed; flowers in April; fruit ripe in August. S.

(8) *Cardamine* sp., annual; in wet places and creeks of the Columbia, abundant.

---

[1] *Lonicera*, Benth. and Hook. f. Gen. Pl. ii. p. 5.

(9) *Uvularia* (?), perennial ; flowers before expansion green, white, three to five in a cluster ; fruit scarlet ; in rich, shady woods near the ocean. S.

(10) —(?) ; a shrub 6 to 10 feet high ; shores of the Columbia, near the ocean, abundant.

(11) *Alnus* sp. ; a tree 50 to 70 feet high ; may prove *A. glutinosa* ; its size occasioned by the richness of the soil and finer climate ; moist places on the Columbia ; April.

(12) *Ribes* sp. ; allied to *Ribes Grossularia* ; corolla purple ; stamens double its length ; anthers yellow ; flowers in April ; fruit small, black, globular, hairy, and very pleasant ; abundant in rocky places on the shores of the Columbia. S.

(13) *Rubus spectabilis*, Pursh, figure and description good. The same observation is applicable to this delightful plant, relatively to its growth, as to *Gaultheria Shallon*. They are found together in the woods. Bears fruit abundantly, ripe in June, flowers April–May. I was not at the sea during the season of this fruit. That farther up the river I found very pleasant but am informed it is inferior to that on the coast ; fruit oblong, yellow ; abundant on the outskirts of the woods ; the young shoots are stripped of their bark and eaten in a raw state by the natives ; a beautiful plant. S.

(14) *Ribes* sp. ; flowers small, in a raceme, brown and yellow ; appears not to produce fruit ; shores of the Columbia ; abundant ; April. I have been unable to find its fruit, August.

(15) *Tiarella* sp. ; leaves cordate, lobed, viviparous ; calyx five-cleft, petals five, unguiculate, fringed, inserted between the teeth of the calyx, white, after a few days' expansion rose colour, fragrant ; whole plant hispid ; 6 inches to 2 feet high ; perennial ; abundant in shady woods in rich soil, near springs and rivulets ; a fine plant.

(16) *Sambucus*, a shrub or small tree, forms a large part of the under-wood in the forest ; flowers large, white ; abundant near the ocean.

(17) *Equisetum* sp., perennial ; the male stems are eaten by the natives in a raw state and sometimes boiled ; abundant in moist places.

(18) *Tussilago*, in the same place ; this may prove *T. palmata*.[1]

(19) *Juncus* sp., perennial ; in partially shady woods, Columbia river ; a small but fine plant, abundant.

(20) *Phalangium Quamash* [2] (Pursh) ; its roots form a great part of the natives' food ; they are prepared as follows : a hole is scraped in the ground, in which are placed a number of flat stones on which the fire is placed and kept burning until sufficiently warm, when it is taken away. The cakes, which are formed by cutting or bruising the roots and then compressing into small bricks, are placed on the stones and covered with leaves, moss, or dry grass, with a layer of earth on the outside, and left until baked or roasted, which takes generally a night. They are moist when newly taken off the stones, and are hung up to dry. Then they are placed on shelves or boxes for winter use. When warm they taste much like a baked pear. It is not improbable that a very

---

[1] *Petasites palmata*, A. Gray, Syn. Fl. N. Am. i. II. p. 376.
[2] *Camassia esculenta*, Baker, in Journ. Linn. Soc. xiii. p. 257.

palatable beverage might be made from them. Lewis observes that when eaten in a large quantity they occasion bowel complaints. This I am not aware of, but assuredly they produce flatulence : when in the Indian hut I was almost blown out by strength of wind. Flowers large, blue ; abundant in all low alluvial plains on the margins of woods and banks of river. S.

On Saturday, 16th April, the chief factor, John McLoughlin, Esq., came down the river from the new establishment, who received me with much kindness. I showed him my instructions and informed him verbally the object of my voyage, and talked over my pursuit. In the most frank and handsome manner he assured me that everything in his power would be done to promote the views of the Society. Since I have all along experienced every attention in his power, horses, canoes and people when they could be spared to accompany on my journeys. Also in every instance had much assistance from those in authority under him, with all the comfort the country affords—circumstances which I am confident that it will give Mr. Sabine much pleasure in communicating to the committee of the Hudson Bay Company. The same day I had all my articles sent on shore from the ship. Mr. McLoughlin advised me to visit the new establishment as they were shortly to abandon this one on the coast.

Tuesday, 19*th*.—In company with him I left the mouth of the river ; at 8 o'clock morning in a small boat with one Canadian and six Indians ; we made only forty miles, having no wind and a very strong current against us. We slept in the canoe, which we pulled up on the beach. Our supper was a piece of good sturgeon, a basin of tea, and a slice of bread. We had six Indians for paddling the canoe ; they sat round the fire the whole night roasting sturgeon, which they do by splitting a branch and placing the meat in it, twisting a bit of rush at the top to prevent it falling out. They ate a fish weighing about 26 or 28 lb. from ten o'clock at night till daylight the following morning. They had paddled forty miles without any sort of food except the young shoots of *Rubus specta-bilis*, and water. We started at three o'clock the following morning and reached our destination at ten on Wednesday night. The scenery in many parts is exceedingly grand ; twenty-seven miles from the ocean the country is undulating, the most part covered with wood, chiefly pine. On both sides of the river are extensive plains of deep rich alluvial soil, with a thick herbage of herbaceous plants. Here the country becomes mountainous, and on the banks of the river the rocks rise perpendicularly to the height of several hundred feet in some parts, over which are some fine waterfalls. The rocks are chiefly secondary, sandstone and limestone bedded on blue granite. The country continues mountainous as far as the lower branch of the Multnomah river, the Belle Vue Point of Vancouver, about seventy miles from the ocean, when it again becomes low on the banks and rises gradually on the back ground. On the south, towards the head waters of the Multnomah, which are supposed to be in a ridge of snowy mountains which run in a south-west direction from the Columbia, the view is fine. A very conspicuous conical mountain is seen in the distance far exceeding the others in height; this I have no doubt is Mount Jefferson of Lewis and Clarke; two others equally conspicuous are observed, one due east and one to the north, the former Mount Hood, the latter Mount St. Helens of Vancouver. Their height must be very great (at least 10,000 to 12,000 ft.), two-thirds are I am informed continually enwrapped in perpetual snow. I have scarcely perceived any difference in the diminishing of the snow (now August). I was in June within a few miles of Mount Hood. Its appearance presented barriers that could not be surmounted by any person to reach its summit. My residence is on the north bank of the river twelve miles below Point Vancouver (90 from the ocean), the spot where the officer of his squadron discontinued their survey of the river. The place is called Fort Vancouver. In the river opposite my hut lies Menzies Island, so named by Mr. Broughton in honour of Archibald Menzies, Esq., then his companion on the famous expedition. On my arrival a tent was kindly offered, having no houses yet built, which I occupied for some weeks; a lodge of deerskin was then made for me which soon became too small by the augmenting of my collection and being ill adapted for drying my plants and seeds. I am now (August 16) in a hut made of bark of *Thuya occidentalis* which most likely will be my winter lodging. 1 have been only three nights in a house since my arrival, the three first on shore. On my journeys I have a tent where it can be carried, which rarely can be done; sometimes I sleep in one, sometimes under a canoe turned upside down, but most commonly under the shade of a pine tree without anything. In England people shudder at the idea of sleeping with a window open; here, each individual takes his blanket and with all the complacency of mind that can be imagined throws himself on the sand or under a bush just as if he was going to bed. I confess, at first, although I always stood it well and never felt any bad effects from it, it was looked on by me with a sort of dread. Now I am well accustomed to it, so much so that comfort seems superfluity.

(21) *Iris* sp., perennial; flowers blue; a small plant, 6 inches to a foot high; in fertile plains, near the margin of rivulets; abundant. S.

(22) Allied to *Lithospermum*; annual; flowers rose coloured; dry gravelly soil; plentiful.

(23) —— (?); suffruticose; abundant in dry places.

(23 [*bis*]) Allied to *Lithospermum*; perennial; flowers dingy-white; a foot to 18 inches high; plains in rich soils; abundant; a fine plant. S.

(24) *Ornithogalum* (?); flowers yellow; bulbs used by the natives as an emetic; perennial; near Point Vancouver, in open gravelly soils, plentiful in rich plains; a fine plant. S.

(25) —— (?), what Pursh has given as *Lilium pudicum* [1]; stigma three-cleft, which removes it from that genus; perennial. Bulbs of this plant are eaten in a boiled state by the natives. Abundant on the plains and near the outskirts of woods. Bulbs of this are sent home in a jar among dry sand. S.

[1] *Fritillaria pudica*, Baker, in Journ. Linn. Soc. xiv. p. 267.

(26) *Umbelliferae*, biennial; whole plant downy, leaves tripartite; abundant in dry open places with the following. S.

(27) *Umbelliferae*; biennial; leaves pinnate; abundant in dry places. April. S.

(28) *Ranunculus* sp., annual; flowers small, yellow; same place.

(29) —— (?) perennial; sandy soils; abundant.

(30) *Trientalis americana*, in shady pine-woods among moss; abundant.

(31) *Smilacina racemosa* (?), perennial; shady woods, in rich vegetable soil; flowers white; plentiful.

(32) *Smilacina* sp., perennial; a small plant 6 inches to a foot high in the same situations as the former. S.

(33) *Cynoglossum* sp., perennial; flowers fine blue; a strong plant 2 to 3 feet high; in thick shady woods. S.

(34) —— (?); shady woods among moss; flowers white; 4 to 6 inches high; annual. Can this belong to *Trientalis*; April.

(35) *Syngenesia* —— (?) *acaulis*, perennial; 6 inches to a foot high; flowers yellow; April; open places in dry soil.

(36) *Syngenesia* (?); different species from the former but found together; flowers also yellow; from the roots of both, when cut, a thick gum exudes having a smell like turpentine; two fine plants; the seeds of both are pounded and made into a sort of bread. They may prove to be *Helianthus*. Seeds of both are sent.

(37) *Berberis nervosa*, Pursh; figure and description good. Plentiful in all mountainous situations among rocks and woods; 2 to 3 feet high in low, moist woods in rich vegetable soil where it flowers beautifully but rarely produces fruit in such places; in more open and elevated situations it bears fruit in abundance. Berry large, round, 3 to 7 seeded, purple-black; it is not found on the coast; at the mouth of the Columbia on the mountains 50 miles from it and is seen as far as the Great Falls. Lewis and Clarke say it is in the valleys of the Rocky Mountains. Seeds of this are sent.

(38) *Acer macrophyllum*, of Pursh; one of the largest and most beautiful trees on the Columbia River. Its large foliage and elegant racemes of yellow fragrant flowers contrast delightfully with the dark feathery branches of the lordly pine; 6 to 16 feet in circumference; 60 to 90 feet high. Banks of the Columbia as far as a few miles above the Grand Rapids to the ocean, also in several of its branches. A jar of seed is put up with a large paper. Correctly noticed by Pursh to have the largest foliage of any.

(39) *Acer* sp.; flowers red, different from the former; will prove *Acer circinnatum* of Pursh; leaves smooth on both sides, ciliate. This Acer forms part of the underwood in the pine forests; 20 to 40 feet high; seldom has a trunk of great size, 6 to 10 stems rising together which are twisted and crooked in all directions forming growing arches. Is called by the voyageurs *Bois de diable* from the obstruction it gives them in passing through the woods. The wood is very tough and is used by the natives for making hooks with which they take the salmon. Wood white; bark white, smooth, green on the young shoots. April. S.

(40) *Calypso borealis*, plentiful in thick shady pine-woods, among moss ; April.

(41) *Fragaria* sp. ; flowers white, small, resembles *F. sterilis* [1] of Britain ; in shady woods ; Point Vancouver ; abundant ; April ?

(42) *Mespilus* sp. ; 10 to 15 feet high ; flowers white and fragrant ; forms part of the underwood ; abundant on the banks of the Columbia ; April. S.

(43) *Cornus* sp. ; a tree 30 to 40 feet high, with a smooth bark ; the wood is hard and very tough, and much used by the Canadian voyageurs for masts and spars for their canoes ; April ; very abundant in the pine-forests ; its great profusion of large white flowers makes it one of the most ornamental trees of the forest ; fruit red.

(44) *Acer* sp. ; a tree 40 to 50 feet high, but never attains a thickness in proportion to its height ; the bark on the stems is white, on the young shoots green ; used for canoe masts or spars ; flowers red ; leaves glabrous ; a fine tree. May agree with 39.

(45) *Umbelliferae*, perennial ; in dry sandy soils, on the plains of the Columbia near Point Vancouver ; April ; plentiful.

(46) *Arabis* sp., annual ; in moist places on the banks of rivulets and springs ; Columbia ; April.

(47) —— (?) ; shady pine-woods, in dry soil ; April ; near Point Vancouver.

(48) *Cymbidium* sp. ; in shady woods, among moss ; a very beautiful plant ; plentiful.

(49) *Lupinus* sp., perennial ; flowers small, numerous, mostly red, with a little mixed purple ; abundant in sandy soils on the plains ; a fine species ; Columbia ; April. S.

(50) *Syngenesia*, may prove a species of *Senecio* ; perennial ; 2 to 3 feet high ; flowers yellow ; in dry, open places.

(51) *Aquilegia*, probably *canadensis* ; abundant in partially shaded woods in dry soil ; April. S.

(52) *Claytonia* sp., annual ; flowers white and pink ; plentiful on elevated situations in sandy soil. [2]

(53) *Vicia* sp., perennial ; abundant on the margins of rivulets and in the woods ; a fine plant.

(54) *Aronia* [3] sp., of Nuttall (?) ; forms part of the underwood in the pine-forest, but more principally the outskirts ; 6 to 16 feet high ; a slender shrub ; flowers white ; abundant at Point Vancouver ; April.

(55) *Taxus* sp. ; a tall tree, 20 to 60 feet high ; the natives on the Columbia prefer this wood to any other for making their bows ; branches pendulous ; a handsome tree ; plentiful in dark low valleys.

(56) *Aspidium* sp. ; the natives eat the roots boiled ; probably *marginale* [4] ; abundant everywhere in all low moist woods.

(56*) *Vaccinium ovatum*, evergreen, flowers white, tinged with pink

---

[1] *Potentilla Fragariastrum*, Hook. f. Stud. Fl. ed. 3, p. 126.
[2] In another MS. :— 'roots of dead trees where they have been burned.'
[3] *Amelanchier*, Benth. and Hook. f. Gen. Pl. i. p. 628.
[4] *Dryopteris marginalis*, Christensen, Ind. Fil. p. 276.

or rose; berries small, black, ripe in August; a very handsome shrub 4 to 10 feet high; on rocky places near the ocean; evergreen; April; Cape Disappointment; abundant. S.

(57) *Celtis* sp., a small tree, flowers of a rusty colour; berry ripe in August, 3-seeded, seeds flat; abundant on the banks of Columbia, near the ocean. S.

(58) *Fragaria* sp., perennial; flowers white, large, spreading; underside of the leaves very pubescent; flowers in April; fruit ripe in June, scarlet, large and very fine; this may be found to be worth cultivating for its fruit as I am informed this species produces food; get specimens of fruit and seed; frequents sandy shores of Columbia from the ocean as far as the Falls; plentiful.

(59) *Salix* sp., male and female; a tree 20 to 40 feet high; sides of rivers and damp woods; plentiful; April.

(60) *Carex* sp., perennial; plentiful on the alluvial plains of the Columbia; April.

(61) *Gramineae*, perennial; in the same situation; also found on the summit of the high hills in moist places.

(62) *Juncus* sp.; in partially shady, low woods.

(63) *Poa annua*, abundant everywhere in moist plains.

(64) *Juncus* sp., perennial; same places.

(65) *Gramineae*, sp., perennial; in dry elevated situations, in rich soil; April; mouth of Columbia river.

(66) *Cardamine*, margins of creeks and low woods; abundant; annual.

(67) *Oxalis* sp., perennial; flowers white,[1] with two red striations in each petal; a small plant; abundant in open woods in rich earth.

(68) *Populus* sp.; a very large tree, 18 to 24 feet in circumference, 70 to 100 high; usually found in the low grounds on the Columbia River as the chief trees on its numerous islands.

Chrysosplenium sp.; in marshy places near the mouth of the Columbia·

(69) *Claytonia* sp.; may prove *sibirica*; annual; plentiful in open woods, growing very luxuriantly in decayed vegetable soil; flowers white; April.

(70) *Sinapis* (?), annual, plentiful on the sandy shores of the river.

(71) *Cardamine*, sp., annual; plentiful in low grounds.

(72) *Lycopodium* sp.; perennial, on hills and dry ground[2]; on trees it grows very long and has a fine effect; generally found on *Alnus*; get it in fruit.

(73) *Trillium grandiflorum*; flowers turn pale red when old; April; close to shady pine-woods.

(74) *Caryophylleae*; perennial; root granulous; radicle leaves twenty-three, palmate cauline; alternate; two to three flowers, pink, fringed, fragrant; 6 inches to 1 foot high; dry grassy plains[3]; a fine plant; rare.

(75) *Bartsia* sp.; perennial; 8 to 14 inches high, flowers deep red; one of the most ornamental plants on the plains; April; there is a yellow variety of it; in dry gravelly soils; abundant. S.

---

[1] In another MS. :—' yellow.'
[2] In another MS. :—' on rocks and dead wood in moist places.'
[3] In another MS. :—' on dry hilly soils.'

(76) *Viola* sp.; perennial; flowers blue; sandy soils near rivulets, on elevated grounds; April; comes near *V. canina*.

(77) *Antirrhinum* (?) sp.; annual; flowers blue, numerous; a beautiful plant; get seeds of it; on the plains and hill situations; April; may prove a second species to Nuttall's *Collinsia*.

(78) *Alnus* sp.; male and female; may prove *A. glutinosa*; the trees in size far exceed the accounts given of them on the Atlantic side, arising no doubt from the richness of the soil and superior climate; may be the same as 11.

(79) *Valeriana* (?) sp.; annual; in wet meadows; a small plant; Point Vancouver. S.

(80) *Valeriana* or *Fedia* sp.; annual; in dry light hilly soils; abundant at the same place, April 20th. S.

(81) *Caryophylleae*, annual; flowers white; dry sandy places.

(82) *Syngenesia*, annual; flowers yellow; same place.

(82 [*bis*]) *Pinus* sp.; exceeds all trees in magnitude; I measured one lying on the shore of the river 39 feet in circumference and 159 feet long; the top was wanting, but at the extreme length 2½ in diameter, so I judge that it would be in all about 190 feet high if not more, girth 48 feet; they grow very straight; the wood is softer than most of the *Pinus* except *P. canadensis*,[1] and easily split. This species, although I have not yet seen the cones, I take to be *P. taxifolia*.[2] The most common tree in the forest.

(83) *Dodecatheon* sp.; 6 inches to a foot high; in the plains [3]; found a beautiful white variety of it; April.

(84) *Viola* sp.; perennial; leaves radical, lanceolate, acute, pubescent; flowers numerous, yellow; abundant on the plains in dry soils [4]; a fine plant 6 to 8 inches high. S.

(85) —— (?) in shady woods.

(86) *Rubus* sp.; sides of rivulets, woods, and in rich soil [5]; branches slender; flowers white; calyx hispid; fruit oblong, large, black and well tasted.

(87) *Veronica* sp.; annual; a plant 6 to 8 inches high; flowers solitary, white; near springs and rivulets on the alluvial plains of Columbia River, plentiful.

(88) *Ranunculus* sp.; plains near the river, and on moist ground; abundant; flowers small, yellow.

(89) *Sisyrinchium* sp.; perennial; flowers blue; in fertile meadows; abundant; April.

(90) *Erigeron* sp.; flowers white; in the same place.

(91) —— (?); annual; flowers white, solitary, marked with minute black dots in the inside; calyx five-partite, acute, ciliated, with a small bractea inserted between the teeth; corolla five-lobed, obtuse; shady

---

[1] *Tsuga canadensis*, Mast. in Journ. R. Hort. Soc. xiv. p. 255.
[2] *Pseudotsuga Douglasii*, Mast. *loc. cit.*, p. 245.
[3] In another MS. :—' in hilly situations.'
[4] In another MS. :—' on dry rich soils near woods.'
[5] In another MS. :—' banks of rivers in sandy soils.'

woods, April; stem and leaves hispid; capsule two to four seeded; seeds angular.

(92) May prove the same as the preceding; inside of the corolla marked with minute black dots: on dry elevated places; annual.

(93) *Ribes* sp.; flowers brownish-green[1]; leaves overrun with a species of *Uredo*. I have been unable to procure fruit of this species; side of the Columbia, Point Vancouver.

(94) *Geum*, perennial; in low moist plains with the three following plants; scarce.

(95) *Rumex* sp., perennial; 2 to 4 feet high; plains; plentiful.

(96) *Carex* sp., perennial; alluvial soil, Columbia.

(97) *Juncus* sp., perennial; same situations.

(98) *Quercus* sp.; may prove *Q. agrifolia*; all the trees which have yet come under my observation are generally low and scrubby; seldom is seen to exceed 40 to 60 feet in height, but thicker in proportion, sometimes 10 to 15 diameter; in many places, particularly in the upland soils, it dwindles to a mere scrubby bush; some in fertile spots attain the height of 70 feet; the wood is white when dry, hard, close-grained, and takes a good polish; it is used for building part of the Fur Company's apartments; the Canadian voyageurs informed me that it is only found at the Grand Rapids, thirty miles downwards; abundant on all elevated grounds to the Rapids of the Columbia, and on a bend of the river 45 miles from the ocean called by Vancouver 'Oak Point.' Seeds not yet ripe.

(99) *Berberis Aquifolium*; this fine plant and *B. nervosa* are sometimes seen growing together; this however is a rarer species; it grows much stronger than the latter. It is very seldom seen in the low woods, but on the banks of rivers; it is correctly stated by Pursh to be found on the falls and rapids of the Columbia, but is there of humble growth, scarcely more than a foot or 18 inches high; on the banks of the Multnomah river I have seen it 4 to 10 feet high, with stems 5 to 8 inches round, growing in light brown, sandy loam, with a livid green luxuriant foliage. When there I could not preserve specimens of its fruit in spirits, but as I am shortly to make a journey towards the head waters of that river it may be in my power to accomplish it. Except when in blossom it can hardly be called a handsome plant; it sends out but few branches and these soon become destitute of leaves, leaving only a few at the extremities; different from *B. nervosa*, which covers the whole ground in the upland woods just like the common heath in the forests of England. A large portion of seeds of both species are sent; seeds of this smaller than the other, 2 to 5 seeded; flowers golden-yellow with faint green, red before expansion; stamens irritable. Nuttall's division into *Mahonia* is trifling. The other species (37) thrives best in close pine-woods, in rich soil; seldom seen more than 2½ to 3 feet high; leaves long; flowers yellow, tinged with green, rises to a spike; this is by far the more beautiful plant; recollect you must get seeds of them both.

(100) *Orobanche* sp.; roots of trees in shady woods.

---

[1] In another MS. :—'brownish-yellow.'

(101) *Siliculosa*, may prove *Thlaspi*; annual; 6 inches; margins of Columbia River; abundant.

(102) *Sisymbrium* (?), annual; same place; 8 to 14 inches high.[1]

(103) *Cardamine* (?), annual; flowers small, white; same situations.

(104) *Sisymbrium* (?) sp.; annual; 18 inches to 2 feet high[2]; flowers yellow; a fine plant, abundant in moist places.

(105) *Cerastium* sp., annual; plains in dry sandy soils; plentiful.

(106) *Claytonia* sp., annual; flowers white; probably not different from *C. sibirica*; very abundant in damp shady places, in rich vegetable soil near woods and Indian villages; very luxuriant.

(107) *Geum* sp., annual; flowers small, yellow; 6 inches to a foot high; in dry light soil near rivulets, &c., at Point Vancouver; plentiful. S.

(108) Does this differ from the other species of *Sisymbrium*? May prove the same as 104.

(109) *Rubus* sp.; stem smooth, erect, white; petiole long; two linear bracteæ at the foot of each; leaves simple, three to five lobed, pubescent on both sides; fruit scarlet, flat like *R. odoratus*; flowers large, white, fragrant; petals round or nearly so; plentiful on the banks of Columbia river, near the ocean. The whole plant differs but little from *R. odoratus* except flowers white and fruit red. This, I have no doubt, will prove a new species; fruit rarely to be had, but of a very fine flavour. S.

(110) *Rosa* sp.; stem slender, twiggy, brown, thickly set with slender straight prickles; leaves nine to eleven, ovate, serrate, smooth; flowers small, faint pink, very fragrant; a shrub 2 to 5 feet high; frequent in dry elevated situations. May be *R. blanda*, a species plentiful on all elevated gravelly and rocky soils; fruit not yet ripe.

(111) ——(?), allied to *Phlox*; annual; small, low plant; flowers pink and handsome; on dry sandy soil. S.

(112) *Carex* sp.; alluvial plains,[3] one of the most conspicuous objects; perennial.

(113) *Hordeum*, banks of river, in light soil.

(114) *Gramineae*, annual; same place.

(115) *Actaea* sp., shady close woods in dry rich soil.

Fraxinus, sp.; scarce, only two trees near the river at Fort Vancouver.

(116) *Lupinus* sp., perennial; stem hirsute; leaves digitate, leaflets fourteen to seventeen, lanceolate; flowers partly verticillate, purple and faint red; appears to differ materially from *nootkatensis*, to which it comes nearer than any other species; one of the most magnificent herbaceous plants which have yet come under my notice; 2 to 4 feet high, frequently having a spike of 18 inches. In rich alluvial plains, abundant. S.

(117) *Lupinus* sp., perennial; leaves digitate, leaflets seven to nine, lanceolate, tomentose; this comes near *Lupinus sericeus* of Pursh; both species are found together in plains and banks of rivers; on the outskirts of woods, where partially shaded, in light rich soil, they thrive in the

---

[1] In another MS. :—' 6 to 8 inches high.'
[2] In another MS. :—'18 inches to a foot high.'
[3] In another MS. :—'a strong plant in swamps.'

I

highest perfection and make two of the most beautiful plants I ever beheld; April 29th. S.

(118) *Saxifraga* sp., perennial; dry rocky situations; abundant.

(119) *Myosotis* sp., annual; flowers white; in dry situations; plentiful.

(120) *Phalangium Quamash*[1]; flowers white; sometimes also a pure white tinged with blue; plains.

(121) *Vicia* sp., perennial; small, flowers purple and red; stem slender; in dry shady woods, in open places; abundant.

(122) *Astragalus* sp., perennial; tall, 2 to 4 feet high; flowers dirty yellow colour; abundant in rich soil, on the side of woods and rivers, or shady damp situations; April 30th.

(123) *Asarum* sp., perennial; calyx invariably three-cleft, bright brown; probably *A. canadense*; close shady pine woods, in vegetable soil; April.

(124) *Viburnum* sp.; a fine strong shrub 14 to 20 feet high; flowers white; forms part of the underwood in most of the forests; very abundant in the islands and on the banks of the Columbia, from the ocean to the Grand Rapids.

(125) *Urtica* sp., perennial; tall, strong-growing species 6 to 10 feet high; very striking; very plentiful in all situations.

(126) *Galium* sp., perennial; long, creeping; edges of woods; plentiful among bushes and dead wood.

(127) *Cruciferae*, annual; stem erect, smooth; leaves alternate, lanceolate, amplexicaul, glabrous; flowers small, faint yellow; a singular plant; alluvial plains[2]; April. S.

Rosa sp.; agrees with the one found a few days since at Point Vancouver; April 30th.

(128) *Pyrus* sp.; a tree 10 to 35 feet high, sometimes 18 inches in circumference; few spines on it; flowers white, very fragrant; from the wood of this tree the natives make their wedges for splitting the pine; fruit small, not yet ripe; plentiful on the south side of the Columbia, near the Multnomah river.

(129) *Crataegus* sp.; a very strong shrub, sometimes attains to the size of a considerable tree; on the plains, where no other tree comes in contact with it, its branches extend over a large space, it grows to the height of 30 to 40 feet with a proportionable thickness; like most of the species it is very fragrant, particularly in moist weather.

Mespilus sp.; does this agree with that collected a week ago?

(130) *Carex* sp., perennial; one of the most luxuriant of the genus, 4 to 6 feet high; in low marshy places; the natives use it for making their couches (sometimes mats are made of it), thatching their huts, &c.

Sunday, May 1st.—Early in the morning left the fort for the purpose of visiting an extensive plain seven miles below on the same side of the river. Passed several Indian steaming huts or vapour baths; a small hole is dug about 1 foot deep, in which hot stones are placed and water thrown on them so as to produce steam; the bather then goes in naked

---

[1] *Camassia esculenta*, Baker, in Journ. Linn. Soc. xiii. p. 257.
[2] In another MS. :—'in all dry soils.'

and remains until well steamed; he immediately plunges into some pool or river, which is chosen so as not to be far distant. They are formed of sticks, mud, and turfs, with a small hole for means of entering. They are most frequently used when the natives come from their hunting parties, after the fatigues of war, and also before they go on any expedition which requires bodily exertion. My curiosity was not so strong as to regale myself with a bath. Saw plenty of *Fraxinus*, which a few days since I said was scarce. Also very fine tree of *Quercus*. It seldom grows to a great height, not more than 60 or 70 feet at most. Some of them measure 6 feet diameter 6 feet from the ground. The bark is very singularly crossed and rugged, and at a distance resembles the ash; could get only male flowers. Collected the following:

(131) *Dodecatheon* sp., var. *alba*; this fine variety, among the beautiful blossoms of the species, makes a fine contrast.

(132) *Ranunculus* sp., perennial; small still pools; plentiful.

(133) *Corallorrhiza* or *Cymbidium*, perennial; lip crenulate, obtuse; white with pink spots; flowers erect; differs from the one (48) gathered some days since; the whole plant is smaller, lip the same colour as the petiole, with three longitudinal streaks, the middle one of a deeper hue than the marginal ones; two altogether distinct species, both found in shady close pine-woods in rich soil; plentiful.

Phalangium Quamash [1]; flowers white; sometimes varying from blue to purplish-blue, pink and white.

Rosa sp.; open places, near the sides of rivulets; abundant; flowers blush colour, faint odour.

Gaultheria Shallon. I find it thrives more luxuriantly near the ocean than elsewhere; the further removed from it, it gradually becomes less vigorous.

Gnaphalium sp., perennial; abundant on dry sterile plains.

(134) *Pentandria, Monogynia*; flowers blue, calyx five-partite; corolla monopetalous, five-lobed; stigma globose; a single specimen of this plant I found near the ocean on the south side of the river in an unexpanded state; the tree 17 to 20 feet high; since found abundantly on dry soils. *Scabiosa* (?) S.

(135) *Vicia* sp.; tall luxuriant plant, found near woods in rich soil; flowers blue and purple; perennial; probably different from any gathered before.

(136) *Populus*, female, same as before.

(137) *Cerastium* sp., annual; flowers white; may be the same as collected before.

(138) *Claytonia perfoliata* (?); near Indian houses; plentiful.

(139) and (140) omitted.

May 2nd.—Made a visit to Menzies Island, in the Columbia river, opposite the Hudson Bay Company's establishment at Point Vancouver, seventy-five miles from Cape Disappointment. The island is low, sandy shores, rich vegetable soil in the middle, frequently inundated when the river is much swollen. Collected the following plants:

[1] *Camassia esculenta*, Baker, in Journ. Linn. Soc. xiii. p. 257.

(141) *Leontodon* sp., perennial; on the sandy shores of river.

(142) *Achillea* sp., perennial; flowers white; abundant, probably *A. Millefolium*.

(143) *Artemisia* sp., perennial; on shores of river.

(144) *Erigeron*; perennial; the same as collected before, but in a more luxuriant state.

(145) *Syngenesia*; annual; in open dry places and on the shores of river; flowers yellow; a very beautiful low plant, 1 foot to 18 inches high.

(146) *Stellaria* sp., perennial; flowers white; in partially shaded places on the edges of woods; not uncommon.

(147) *Trifolium* (?); (this is by no means *Trifolium*; May 5); a very singular small plant; seed vessels small, short, on very short stalks, which with the seeds are buried in the sand; on Menzies Island; obtain seeds of this curiously interesting plant. S.

(147 [*bis*]) *Claytonia* sp., may prove *Claytonia perfoliata*; in all rich open spots, particularly where the elk or other animals scratch the ground so as to destroy the stronger herbage, this with the other species are sure to take possession of the soil.

Myosotis sp., more perfect specimens than those I had before; agrees in every other respect with them.

Cerastium sp.; agrees with the one collected some days ago.

(148) *Valeriana* sp., annual; a small plant, 6 to 10 inches high; in open dry soils on the plains; different from the other; flowers small, faint pink.

(149) *Populus*, female; a very large tall tree, 60 to 100 feet high, to 12 diameter; all the low banks of the river are covered with it.

(150) *Scrophularia* sp., perennial; 4 to 6 feet high; margins of rivers, pools, and all moist situations, plentiful.

(151) *Myosotis* sp., annual; hirsute, branching; leaves long, entire; linear-lanceolate; flowers bright yellow; tube long; mouth of the corolla spreading, with a dark spot opposite the teeth; seeds not yet known; this very interesting species was found on Menzies Island in company with Mr. Scouler, who agreed with me to call it *Myosotis Hookeri* after Dr. Hooker of Glasgow; scarce, only three specimens of it were found, two of which are in my possession. I have since found it in abundance near all the Indian lodges above the Rapids of the Columbia. S.

(152) *Phlox* sp., annual; cauline leaves lanceolate, smooth; floral stem minutely pubescent; floral leaves lanceolate, somewhat pubescent and ciliated; flowers small, beautiful pink, in large clusters at the top; corolla-tube long; whole plant branching, 6 to 16 inches high; seeds mucilaginous; very abundant on Menzies Island, in open places and in sandy light soils on the banks of rivers; prefers rich moist places; will prove *Collomia* of Nuttall, which is also a trifling generic distinction. S.

(153) *Veronica* sp., perennial; moist places; abundant; may prove *V. Beccabunga*.

(154) *Salix* sp., male; a large tree; banks of rivers.

(155) *Salix* sp., female; both species found on the sides of rills; plentiful.

Mimulus luteus, on the edge of pools and rills; very abundant. S.

(156) *Rubus* sp.; petals long, white; leaves ternate, glabrous, serrate; stems covered with short hooked prickles, red where exposed to the sun; fruit longish, black and not unpleasant; a strong rambling shrub, abundant on dry sandy places near the river. S.

May 5th.—Obtained the following *Gramineae*, &c.

(157) *Bromus* sp., annual (?); in sandy soil, banks of rivers.

(158) —— (?), a fine grass, whole plant glaucous; 18 inches to 2 feet high; perennial; plentiful in the same places.

(159) *Gramineae*; perennial; same situation.

(160) *Scirpus* sp., perennial; in and on the margins of still ponds.

(161) *Melica* (?) sp.; same place.

(162) *Scirpus* sp., perennial; moist plains.

(163) *Carex* sp., perennial; abundant on alluvial plains.

(164) *Pentandria, Monogynia*; stigma trifid; a very handsome shrub, 6 to 12 feet high, abundant in all sizes on the edges of woods and rivers. May prove a *Rhamnus*, not unlikely *R. alnifolia*.

(165) *Potentilla* sp., perennial; flowers yellow; a fine plant; 12 to 18 inches high; plentiful on dry gravelly soils. S.

Geum sp., perennial; probably the same as before.

(166) *Chrysanthemum* (?) sp., perennial; flowers yellow; dry, light, or gravelly soils; plentiful.

(167) *Juncus* sp., perennial; wet natural meadows; plentiful.

(168) What I formerly suspected, before expansion, to be a *Spiraea*, now *Pentandria, Monogynia*; 6 inches to a foot high; plentiful in dry gravelly soils.

Friday, May 6th.—Rain during the night and early in the morning. Employed drying paper, turning specimens, and fixing my new tent; in the evening laid in the following plants, collected two miles east on the north side of the river:

Cynoglossum sp., perennial; same as before.

(169) *Gramineae*, annual; abundant in open places, in rich soil at edges of woods, &c.

(170) *Rhus* sp., a small upright shrub 2 to 4 feet high in low woods and plains, in dry gravelly soil; flowers faint yellow.

(171) *Diadelphia*, annual (?); dry places, under the shade of *Pteris aquilina* [1]; abundant.

(172) *Geranium* sp., perennial; flowers purple; dry meadows; plentiful.

(173) *Umbelliferae*, perennial; partially shaded woods in dry soil.

(173 [*bis*]) *Pentandria, Digynia*; leaves radical, orbicular; petioles and nerves hirsute; calyx five-cleft, green at the teeth; petals small, white, revolute; filaments long; perennial; Menzies Island; plentiful.

(174) *Geum* sp., annual; flowers small, yellow; 6 to 10 inches high; banks of river, in sandy soil; abundant.

(175) *Rumex* sp., perennial; a strong plant, plentiful in all moist grounds in open places; a very strong species.

[1] *Pteridium aquilinum*, Christensen, Ind. Fil. p. 591.

May 7th.—(176) *Caprifolium* [1] sp. ; corolla golden-yellow before expansion, after, tinged with a bright red ; style hirsute ; tube hirsute half-way up inside ; leaves sub-amplexicaul; floral leaves connate ; this agrees perfectly with *C. ciliosum* [2] of Pursh. Nuttall supposes this to be a mere variety of *C. parviflorum*,[3] which is incorrect, more in them [*sic*] than his generic description of *Mahonia* from *Berberis* ; abundant on the rapids of the Columbia. Seeds of this fine plant are sent.

Geum marked 174 and 107 are the same.

Lupinus sp., 49 ; var. flore albo ; this I doubt not will prove *L. villosus* of Pursh ; only two plants of this beautiful variety.

(177) *Galium* sp., *boreale* (?), on dry gravel and rocky places near rivers ; abundant.

(178) *Caulophyllum* (?) *thalictroides* (?) ; shady dry woods ; very plentiful.

(179) *Arenaria* (?), annual ; dry natural meadows ; plentiful.

(180) *Oxalis* sp., perennial ; a small species, scarcely 4 or 5 inches high ; flowers yellow ; on the edges of woods, in light soil.

(181) *Pyrola aphylla* ; in thick shady dry woods among moss, in soil composed of the leaves of *Juniperus* and *Taxus* ; frequents the same situations as *Monotropia.* It was with much pleasure I found this curious little plant, which will I think on careful examination be found to differ from the description, it has a profusion of bracteate leaves, is parasitic on decayed wood like *Monotropa* or *Orobanche.* Style declining ; stamens bent upwards, about 70 specimens of this singular plant I collected which will I am confident find admirers ; I am very glad to be enabled to confirm this plant, as it is still by some persons said not to exist ; I have little doubt but it agrees with that I saw in the possession of Mr. Menzies ; his specimen, however, was imperfect.

May 10th.—Made a journey down the river as far as the ocean for the purpose of transporting the remainder of my property ; returned on Monday 16th. Collected the following plants on my journey :

(182) *Cheiranthus* sp. (?), biennial or perennial ; flowers bright yellow ; on the banks of the river ; only one plant.

(183) *Sedum* sp., perennial ; flowers yellow ; rocks ; very plentiful.

(184) *Arbutus* sp. ; leaves ovate, oblong, entire, glabrous on the upper side, glaucous on the under ; raceme axillary ; peduncle pubescent ; an obtuse bractea at the foot of each ; calyx minute, five-cleft, white ; corolla nearly entire, partially revolute, white and very fragrant ; pistil double the length of the corolla, filaments at the base, very pubescent ; wood hard and brittle, of a white colour ; bark smooth, yellow, not unlike the second bark of *Betula.* This very ornamental tree, which attains the height of 60 feet and sometimes 2 in diameter, differs in some measure from the description of *A. laurifolia* and does not agree with *A. Menziesii.* The leaves on the young shoots where there are no flowers answer the description of *A. laurifolia.* I have little doubt it is that noble species ; the fruit is not

---

[1] *Lonicera*, Benth. and Hook. f. Gen. Pl. ii. p. 5.
[2] *Lonicera ciliosa*, A. Gray, Syn. Fl. N. Am. i. II. p. 16.
[3] *Lonicera glauca*, A. Gray, *loc. cit.*, p. 18.

yet ripe. In the collection are abundance of specimens in flower and fruit; on the rocky shores of the Columbia and its branches; plentiful. Do not fail to put up a treble supply of its seeds; being evergreen it is the more desirable.

May 13*th*.—(185) *Juncus*, is it the same as before, in seed ?

(186) *Cheilanthes* sp.; very abundant on rocks on the south bank of Columbia river, forty miles from the ocean; I think will prove *C. dealbata*.[1]

(187) *Polypodium*; probably *P. vulgare*; on rocks and trees.

(188) *Mimulus* sp., annual or perennial ? stem creeping, jointed; leaves subrotundate, partially dentate, three-nerved, glabrous; flower leaves obovate-lanceolate; flowers small, very fine bright yellow, in the lower lip two streaks of a purplish-crimson colour with two minute dots of the same in the upper; a very beautiful little plant. This interesting species I call *M. Scouleri*,[2] after John Scouler, who has been the agreeable companion of my long voyage from England and walks on the solitary Columbia, who first noticed it when in company with me on one of our walks on the Columbia. Plentiful on a conspicuous neck of land called ' Tongue Point,' on moist rocks; I have since seen it on the rocks at the Falls with *M. luteus* and another species. I was rather late in making a second visit to it to procure abundance of seed, a small portion is now sent in the earth it grew in; a fine plant for cultivation.

(189) *Smilacina*, probably *S. bifolia*,[3] perennial; plentiful in open woods, among rich soil.[4]

(190) *Galium* sp., perennial; a small inconspicuous plant, plentiful in soils among rocks and stones.

(191) *Umbelliferae* (?); perennial; flower purple; abundant, in shady places, at Tongue Point.

(192) *Viola* sp., perennial; flowers small, yellow; in open ground in light rich soils; plentiful.

May 14th.—(193) *Valeriana* sp., perennial; flowers white; banks of the river, on moist rocks and near springs.

(194) *Ranunculus* sp., perennial; probably *R. sceleratus*; in marshy ground or near rivulets.

(195) *Geum* sp., perennial; on sandy banks of the river, sixty miles from the ocean.

(196) *Saxifraga* sp. (?); perennial; dry shady rocks, very plentiful.

(197) *Rosa*; leaves ovate, obtuse, serrate, under side rough, upper smooth; flowers beautiful pink, after expanded a few days a faint colour; fragrant; a strong climbing plant, 6 to 10 feet high; near the junction of the Multnomah river with the Columbia. I have since seen this beautiful plant in abundance on the Multnomah in great perfection. S.

(198) *Anemone* sp. (?), perennial; flowers white; a very fine small plant; abundant on the north.[5] I have in vain sought for it in seed.

[1] *Pellaea dealbata*, Christensen, Ind. Fil. p. 479.
[2] Hook. Fl. Bor. Am. ii. p. 100.
[3] *Maianthemum bifolium*, Baker, in Journ. Linn. Soc. xiv. p. 563.
[4] In another MS. :—' among stones.'
[5] In another MS. :—' in shady woods, rare.'

The *Rubus* with large white flowers has notched petals in shady places.

(199) *Mespilus* sp.; a large tree; flowers white; on the banks of river.

(200) *Salix* sp., male and female; sides of lakes and rivers; abundant; a handsome species.

(201) *Gramineae* (?), perennial; sides of woods; plentiful.

(202) *Scirpus* sp., perennial; margins of pools, shores of rivers and marshes.

(203) —— (?); same situations as the preceding.

(204) *Claytonia* sp., annual; a beautiful small slender creeping plant; frequents moist rocks; plentiful.

(205) *Heuchera* sp., may prove *H. caulescens* [1]; a splendid plant; plentiful on elevated dry soils.[2] S.

(206) *Vaccinium* sp., *ovatum* (?); fine evergreen shrub; flowers delicate white; where exposed to the sun, faint red; in open, dry, hilly woods, in conjunction with *Gaultheria*.

(207) May prove a species of *Nyssa* in fruit; a small tree; on edges of woods and rivers; abundant.

(208) *Cucurbitaceae*; perennial; flowers white, no male ones expanded; leaves somewhat three-lobed, rough; fruit round, yellow, hairy, three to eight seeded; seeds orbicular, flat, large; a strong free-growing plant, with a large tap root 2 to 6 feet long in the ground; has a bitter pungent taste; plentiful in sandy soils. S.

(209) *Astragalus* sp., perennial; edges of woods and rivers; may be the same as collected already.

(210) *Pulmonaria* sp. (?); perennial; flowers blue; in shady moist woods.

On my way down the river passed some trees of *Acer macrophyllum* of immense size, 60 to 90 feet high, 3 to 8 inches in diameter; the large deeply lobed leaves, with the fragrant raceme of flowers, make a fine contrast with the pines by which it is surrounded. Measured some trees of *Pinus taxifolia* [3] the roots of which were laid bare by the water, blown down by the wind; 100 to 140 feet, at the top 18 inches in diameter; many of them exceed 15 and some measure 20 and 25 feet in circumference; the only two species except *P. taxifolia* [3] are *P. balsamea* [4]; and *canadensis* [5]; both far exceed those on the Atlantic side in size.

(211) *Caltha palustris*; rare; in marshes and sides of rivers.

(212) *Fumaria* sp., perennial; flowers red and purple; in very shady woods near springs; two feet high.

(213) *Aspidium* sp., perennial; abundant in all low moist woods; *marginale* ? [6]

(214) *Carex* sp., perennial; a very strong species; plentiful in marshy ground near the ocean.

[1] *Heuchera villosa*, S. Wats. Bibl. Ind. N. Am. Bot. p. 326.
[2] In another MS.:—' plentiful on moist rocks where partially shaded.'
[3] *Pseudotsuga Douglasii*, Mast. in Journ. R. Hort. Soc. xiv. p. 245.
[4] *Abies balsamea*, Veitch, Man. Conif. ed. 2, p. 492.
[5] *Tsuga canadensis*, Veitch, *loc. cit.*, p. 464.
[6] *Dryopteris marginalis*, Christensen, Ind. Fil. p. 276.

(215) *Rumex Acetosella*; common in dry pasture; does this agree with the European?

(216) —— (?) *Triandria* Order? *Digynia* (?); flowers yellow, calyx brown; common near springs and rivulets in shady woods. S.

(217) *Asperifoliae* (?); perennial; flowers dark purple; stamens exserted; a low plant 4 to 10 inches high; common on the outskirts of woods.

(218) *Menziesia ferruginea*; this handsome shrub is very abundant in rocky and all elevated grounds near the ocean and a few miles up the banks of the river, but not more than ten miles from the ocean; the further from the sea the less vigorous it becomes; thrives best in a light brown earth. Seeds not yet ripe.

(219) *Potentilla* sp., perennial; abundant on the shores of the river and all dry sandy soils.

(220) *Veronica*, perennial, may prove *V. officinalis*; common in spring [1]; flowers blue.

(221) *Pyrus*; probably different from that found higher up the river; flowers larger; leaves more acute; a large tree, 20 to 40 feet high.

(222) —— (?), common in shady woods; a pretty vivid green foliage.

(223) *Acer macrophyllum*, in a further advanced state than that collected before further up the river.

(224) *Rubus spectabilis*, with fruit in a half-ripe state; this very beautiful and abundant-bearing species would above all others be a valuable addition to the garden; frequents the edges of woods; thrives best in partially shaded situations, in light rich loam; do not fail to procure a very large supply of seeds; devise means to send fruit home.

The above are what were collected on my way up and down the river, from Point Vancouver to the ocean. Returned on Wednesday evening, having experienced rather a fatiguing journey; had only one dry day.

May 17th, 18*th*.—Employed drying paper, arranging what plants I had brought with me on the present occasion and changing the paper of the last laid in before leaving; frequent showers of rain for the last ten days, which greatly interrupted my excursions.

May 20th.—Collected as follows near Fort Vancouver, making an excursion of ten to thirty miles:

(225) *Delphinium virescens* [2] of Nuttall; a very fine early flowering plant; frequents all dry light soils near the outskirts of woods and under the shade of trees in the plains; abundant. S.

(226) May probably turn out *Trifolium*; a very fine plant in the like places as the preceding plant, 10 to 20 inches high; flowers pink colour; annual; plentiful. S.

(227) *Syngenesia*, perennial; 1 to 2 feet high; flowers yellow; on dry gravelly plains; abundant.

(228) *Linnaea borealis*; this beautiful fragrant plant is very plentiful in most upland or hilly forests, in great profusion in the close pine or

---

[1] In another MS. :—' near springs rare.'
[2] *Delphinium azureum*, S. Wats. Bibl. Ind. N. Am. Bot. p. 12.

cedar woods among moss; I have not yet seen it nearer the ocean than forty miles. I am unable to find its fruit.

May 22nd.—(229) *Brodiaea grandiflora* [1]; Pursh is correct as to it being hexandrous; anthers united in pairs; one of the finest and earliest flowering plants; will the bulbs keep to England? (try). This splendid bulbous plant is plentiful in all dry elevated light soils on the banks of rivers, rarely to be found in seed; 18 inches to 2 feet high. Seeds are sent with a quantity of bulbs in a jar among dry sand. I hope they may reach England in a good state; May.

(230) *Gramineae*, perennial; plentiful on the dry plains.

(231) Probably *Lychnis*; annual; calyx five-toothed; three-celled; seeds flat; on the plains. S.

(232) *Eriogonum* sp.; suffruticose; leaves partly cordate, long, ovate; petioles very downy, particularly the under side; stem smooth, green, red where exposed to the sun, hollow; calyx none; corolla five petals obtuse; stamens eight to nine, mostly nine, nearly as long as the corolla, green colour; flowers white, stamens and ovary pubescent; this curious plant I cannot refer to its proper species; whether included in the American flora, I am unable to say; a splendid plant with large flowers, 2 to 3 feet high; plentiful on the southern declivities of the hills, in dry soil.

(233) *Viburnum* sp., may turn out *V. pubescens*, a strong shrub which generally inhabits margins of rivers, marshes, moist woods, &c.; abundant.

May 23rd.—(234) *Lupinus*, perennial; var. *alba* of what I suppose *L. nootkatensis*.

(235) *Lupinus* sp.; perennial; different from 49; flowers very bright light blue and white, the other purple-rose; lip of the present more acute; leaves 5 to 7 digitate, lanceolate, silky underneath; a small beautiful species which I cannot refer to any yet described; spike 8 inches to 14 high, not more than one cauline leaf on a stem; in light gravelly soils, on elevated ground, near the Grand Rapids. S.

(236) *Symphoria* sp. [2] (?); flowers pink; berry white, in clusters; plentiful on the high banks of rivers, plains, and edges of woods; May. S.

(237) *Dracocephalum* sp. (?), perennial; flowers blue, pubescent within; a small plant, plentiful in the plains, in dry open soils.

(238) *Erigeron* sp. (?), perennial; flowers large, light blue and purple; in the plains, very abundant.

(239) *Mimulus* sp., perennial; roots creeping; stem spreading; leaves opposite, nearly sessile, ovate-lanceolate, somewhat dentate, five-nerved; flowers large, in pairs at the foot of each leaf; peduncle double the length of calyx and corolla; flowers bright yellow, brown streaks in the inside; a very beautiful plant; whole plant very woolly, from which I call it *M. lanatus*. In moist rocks on the Multnomah and the Columbia at the falls of both, with *M. luteus* and *M. Scouleri*; will be a great addition to the garden; May. S.

(240) *Circaea* sp.? perennial; in dark, shady woods, plentiful.

(241) *Cyperus* sp., perennial; a strong species, in marshes.

---

[1] *Brodiaea lactea*, Ind. Kew. fasc. i. p. 340.
[2] *Symphoricarpos*, Benth. and Hook. f. Gen. Pl. ii. p. 4.

(242) *Carex* sp., perennial ; 18 inches high ; a very singularly handsome species, plentiful in moist meadows. S.

(243) *Hordeum* sp., perennial ; plentiful in dry alluvial plains[1] ; 2 to 4 feet high.

(244) *Tephrosia* sp., perennial ; flowers bright yellow ; plentiful in dry gravelly meadows.

(245) *Allium* sp., perennial ; 6 to 10 inches high ; flowers purple ; bulb mild ; plentiful in open woods. S.

May 24th.—(246) *Epilobium* sp., annual, biennial (?) ; a small plant with minute white flowers ; on dry sandy soil near the banks of the Columbia ; rare, only one plant of it.

(247) *Pentstemon* sp. ; flowers large, pink, tinged with rose ; a curious plant ; *grandiflorus* (Nuttall), perennial ; rare on Menzies Island ; I found it at the Rapids in great abundance, on moist rocks and gravelly soils. S.

(248) *Oenothera albicaulis* (Nuttall) ; flowers white and rose colour ; abundant on the sandy shores of the Columbia, seven miles from the ocean. S.

(249) *Senecio* sp. ; perennial ; 18 inches high ; flowers yellow ; rare, near rivers.

(250) *Polygonum* sp. ; annual ; flowers white ; on the sandy shores of the Columbia, plentiful.

(251) *Pentandria, Digynia* ; annual ; calyx and corolla none ; stamens five, short ; anthers globose, yellow ; stem villous, partly succulent, branching ; leaves opposite, sessile or partly amplexicaule, linear ; flowers situate between the leaf and stem ; capsule one-celled, one-seeded ; seeds flat, orbicular ; on the sands of the Columbia within the tide mark ; a small plant ; allied to *Salsola.*

(252) *Campanula perfoliata*[2] (Pursh), plentiful in moist soils near springs.

(253) *Epilobium* sp. ; annual ; flowers small, white ; a small plant, abundant in all moist ground.

(254) *Pentandria, Monogynia* ; calyx five-leaved, acute ; corolla five-petalled ; linear, obtuse ; filaments longer than the corolla ; style the same length ; flowers blue, capitate ; leaves sessile, alternate, pinnate ; annual ; of this very beautiful little plant I am for the present unable to procure more than one specimen ; on the banks of the Columbia, near the Grand Rapids. (Found later very abundantly in dry, sandy soils, near rivers.) *Scabiosa.* S.

May 25th.—(255) *Verbena*, perennial ; leaves opposite, sessile ; flowers small, blue[3] ; on Menzies Island and at the falls of the Columbia. S.

(255 *) *Equisetum hyemale* ; in marshes and sides of rivers, plentiful.

(256) *Erigeron* sp., annual ; flowers pale rose colour ; rare ; on Menzies Island, in sandy soil.

(257) *Gramineae*, perennial ; tall, 3 to 4 feet high ; in light sandy soil ; abundant.

---

[1] In another MS. :— in all alluvial soils near woods.'
[2] *Specularia perfoliata*, DC. Prod. vii. p. 491.   [3] In another MS. :—' pink.'

(258) *Gramineae*, perennial ; belongs to the same genus as the former plant ; found together in the same situation.

(259) *Gramineae, Arundo* (?), perennial ; the strongest grass which has yet come under my notice ; 6 to 10 feet ; in all alluvial grounds ; very luxuriant near springs and on the shores of rivers.

(260) *Gramineae*, perennial ; a handsome, slender grass, with fine leaves ; plentiful near rivers in sand.

(261) *Sisymbrium* (?) annual ; pod short, curved ; in sands plentiful ; probably collected before.

(262) *Lupinus* sp.[1] ; annual ; leaves small, hairy, digitate, leaflets five to seven, villous on both sides, oblong ; calyx half the length of the corolla, with small linear bractea ; upper lip bidentate ; flowers alternate, lower lip lanceolate, blue, upper obtuse, white, with two rows of black dots in the centre ; bright purple after a few days' expansion ; legume very villous, six-seeded ; seeds small, grey colour, with a longitudinal black streak on each side ; stem branching, with dense long hairs ; 6 inches to a foot high ; on gravelly, dry grounds ; rather rare ; this fine little plant has some resemblance to *L. pusillus* of Nuttall, but that is only two-seeded ; corolla bright blue ; both sides pubescent alike ; this, glabrous on the upper side ; near the Grand Rapids, Columbia River. S.

(263) *Lupinus* sp.[2] ; perennial ; leaves digitate, leaflets seven to nine, lanceolate, obtuse ; flowers faint blue and purple, striated, with blue of a deep hue ; whole plant glabrous, branching ; the flowers are rarely to be seen in a perfect state, being attacked by insects which devour them in the bud[3] ; a strong plant ; in the same place a variety (*alba*, 264) frequently to be found equal in size, and agreeing in every other respect with this specimen, except its white flowers ; both abundant on elevated grounds in dry soil, banks of rivers, in sand, Grand Rapids, Columbia River. S.

Tuesday, May 31st.—Made a journey of three days on the north banks of the river towards the Rapids, accompanied by one Indian belonging to a tribe called Kyuse ; his name was 'Yes,' I mean his Indian name ; he had no good qualification except being a good huntsman. Collected the following plants, a few miles below the Grand Rapids :—

(264) *Antirrhinum* sp., annual or biennial ; leaves glabrous, alternate, linear, obtuse ; flowers bluish-purple, flower-stems erect, 18 inches high, others creeping ; abundant on dry ground in light soil. S.

(265) *Pentstemon* sp., perennial ; radical leaves glabrous ovate-lanceolate, cauline partly amplexicaule, all entire ; flowers numerous, verticillate ; purplish-blue ; sterile filament bearded ; a handsome tall plant, 2½ to 3 feet high ; on rising bluffs, in marshy grounds ; plentiful (different from 247). S.

[1] In the MS. at the side of 262 *L. bicolor* is written. This is *L. micranthus* var. *bicolor* according to S. Wats. Bibl. Ind. N. Am. Bot. p. 238.
[2] In the MS. at the top of 263 *L. albicaulis* is written.
[3] In another MS. :—'being destroyed by a species of aphis which confines itself to this plant.'

(266) *Diadelphia, Decandria*, perennial; calyx two-lipped, upper three-cleft or toothed, spreading, lower two-cleft and only half as deeply; corolla yellow, wings white; whole plant smooth; leaves pinnate; roots creeping, covered with a white, soft, spongy substance; abundant in marshy ground. S.

(267) *Ranunculus* sp., perennial; flowers yellow; in marshy grounds; plentiful.

(268) *Orchis* sp.; tall, beautiful plant; flowers white, very fragrant; root palmate; plentiful in wet meadows; this species does not seem to have been noticed by Pursh.

(269) *Trifolium* sp., annual; leaves nearly orbicular, minutely ciliated and pubescent; flowers small; tube of the corolla red; wings and orifice white; this may prove *T. microcephalum* of Pursh; abounds in moist soil where wood has been burned; indeed in every place where stronger herbage has been checked it readily makes its appearance; wild animals, most likely elk, seem to be fond of it, it being only under brushwood where they cannot get to it that a perfect specimen can be had. S.

(270) *Myosotis* sp., annual; plentiful in all wet ground and natural ditches; *M. palustris*?

(271) *Philadelphus* sp. (?); a shrub 4 to 12 feet high; on the banks of rivers and marshy ground; very plentiful on most of the branches of the Columbia from the Rapids to the ocean.

(272) *Arbutus Uva-ursi* [1]; abundant on all rocky dry situations; the leaves, when dried over a fire, are smoked by the natives.

(273) *Umbelliferae*, perennial or biennial; leaves large, lobed; flowers white; plentiful in marshes and moist soils with the following. S.

(274) *Umbelliferae*, perennial; leaves smooth, pinnate; flowers dark purple; same places. S.

(275) *Umbelliferae*, perennial; leaves glaucous; seed-vessel two-seeded; seeds flat, oval; same situations.

(276) *Umbelliferae*, perennial; leaves not unlike 274; flowers white; in dry sandy soils; plentiful.

(277) *Lupinus* sp., perennial; leaves digitate, leaflets four to seven, pubescent; stems and footstalks purple; flowers purple-rose, branching stems black; a handsome plant; 1 to 2 feet high, very plentiful on open grounds in rocky and gravelly soil near the Grand Rapids. S.

(278) *Ornithogalum* sp.; flowers umbellate, white with a green stripe in the middle of each petal; plentiful in alluvial, rich plains among strong grass; may prove *O. umbellatum*. S.

(279) *Carduus* sp., biennial (?); flowers yellow; a very tall strong-growing plant; abundant in all meadows on the Columbia in all dry soils.

(280) *Carduus* or *Cnicus* sp., biennial; 4 to 10 feet high; flowers red; in dry meadows; abundant.

(281) *Geranium* sp., annual; a small plant, not more than 6 inches to a foot high; frequents edges of woods and mountain springs; flowers

---

[1] *Arctostaphylos Uva-ursi*, A. Gray, Syn. Fl. N. Am. ii. i. p. 27.

faint pink, nearly white; places that have been burned are its favourite situation. S.

(282) *Umbelliferae* (?), probably *Chaerophyllum*; when young, the stems and leaves are eaten by the tribes on the Columbia; abundant in all marshy grounds; perennial; flowers white.

(283) *Prunella* sp., perennial; flowers purple; a low plant abundant in all meadows.

(284) *Lilium* sp., perennial; flowers orange, with dark brown spots in the centre, generally only one flower, when in rich soil two, but never more; sweet-scented, like *Daphne odorata*[1]; plentiful in open woods and on the banks of rivers in sandy soil.

(285) *Hypericum* sp. (*Trigynae, Herbaceae*, Pursh); flowers yellow; 18 inches to 2½ feet high; perennial; plentiful in rocky and all dry soils.

(285 [*bis*]) *Gramineae*, perennial; plentiful in moist ground, Point Vancouver.

(286) *Oenothera* sp., annual; stem erect, rarely branching except when in very rich soil, slender; leaves sessile, alternate, linear, smooth, entire; flowers very large, rose colour; petals obtuse, round, nearly entire, with a beautiful dark purple spot in the centre of each; anthers white, stigma yellow; capsule sessile, long; this exceedingly beautiful species I call *O. Lindleyana*[2] after Mr. John Lindley the secretary; abundant on elevated gravelly plains and rising grounds. S.

(287) *Oenothera* sp., annual; stem slender, erect, slightly pubescent, rarely branching; leaves alternate, entire, linear, sessile, pubescent on both sides; flowers small, faint rose colour, fainter than the preceding; petals round, entire, a pink spot near the tip of each petal; capsule short, very pubescent, sessile; distinct from the former capsule, not only shorter but eight-grooved, while that is only four, and then but slightly; this fine although by no means equal to the former species will also form a valuable addition to this handsome tribe of plants; found together in hilly and rising grounds, abundant. (Both species are now in flower among some turnips sown by me, are very strong, 2 feet high and *very branching* from cultivation.) S.

(288) *Euphorbia* sp., annual; in dry gravelly meadows; common.

(289) *Syngenesia*, annual; flowers yellow; dry soils, on elevated plains, near the edges of woods; abundant round individual trees in the plains. S.

(290) *Lysimachia* sp., perennial; flowers small, yellow; plentiful in all alluvial meadows.

(291) May prove an annual species of *Silene*; in dry soils. S.

(292) *Spiraea tomentosa* (?); in all low grounds that are inundated, sides of rivers, and mountain rills; common.

(293) *Pteris crispa*[3]; amongst rocks at the Grand Rapids, plentiful.

(294) *Saxifraga* sp., perennial; flowers white; on the dry rocks between the Rapids and the Falls, rare.

(295) *Alnus*; a large shrub, plentiful on the gravelly hills, twenty miles

---

[1] *Daphne Cneorum*, Ind. Kew. fasc. i. p. 717.
[2] *O. amoena*, Ind. Kew. fasc. iii. p. 334.
[3] *Cryptogramma crispa*, Christensen, Ind. Fil. p. 187.

above the Rapids; the only tree which survives the intense heat and drought on the high grounds!

June 20th.—Towards midday left my residence for a journey up the river in company with the canoes going to the different posts in the interior, a few miles above the Great Falls, about two hundred miles from the ocean. I was at a loss to decide whether my time would be better employed there or between here and the ocean. In the latter, from what I have already seen, I should reap a rich harvest, and leave it probably for a less fertile one; although, on the other hand, I might obtain some interesting objects peculiar to the plains and mountains of the interior, as John McLoughlin, Esq. (the chief factor), from whom I have experienced every attention and assistance as to the furthering of my pursuit and comfort which he has in his power to show, assures me there will be no obstacle to my crossing the continent, and that he will use every means to make my journey beneficial to the Society and agreeable to myself. Before the vessel left the river for Nootka and thereabouts, I had some thought of going there. But as he informed me, that my opportunities of collecting, arising from the turbulent disposition of the natives, would be so limited—persons being under the necessity of meeting them armed and in a large party—in unison with his opinion I thought my time would be devoted to the best advantage by remaining on the Columbia, and to make journeys in various directions as opportunities would occur. My ascent was slow, the current at this season being exceedingly powerful, so that I had many excursions on the banks and adjoining hills. The water ran with such rapidity that when the wind blows from a contrary direction it produces a swell like an inland sea; frequently we had to take shelter in the creeks, and although our canoes were considered good, yet we could not see each other except at a short distance, so great was the swell. The Grand Rapids, as they are termed by Lewis and Clarke, are formed by the river passing through a narrow channel about 270 yards broad in a south-west direction, very rocky, the fall of water about 147 feet above which stand three small islands; one of them is the burial-place of the natives who inhabit the southern banks of the river. The extreme length of the Rapids may be about two miles, but for only a short space (about 600 yards where the river makes a turn S.W.) the water passes with great agitation. At this season they are seen to a disadvantage, the river being 9 feet higher water than in May (from May 24 to July 16 the river rose 12 feet 8 inches); I am informed it is lower this season than generally. The banks are high, steep, and in many places rugged; limestone, sandstone, on blue and grey granite. Many large trees in a petrified state are to be seen lying in a horizontal position between the layers of rock, the ends touching the water in many places. There seem to be two kinds, a soft wood and a hard; one I take to be *Pinus balsamea*,[1] the other a species of *Acer*, which must be *A. macrophyllum*, being the only hard wood of large dimensions on the place; some of both measure 5 feet in diameter. This being the fishing season, the natives are numerous on the banks of the river; they come several hundred miles to their favourite fishing grounds. At the

---

[1] *Abies balsamea*, Veitch, Man. Conif. ed. 2, p. 492.

Rapids an almost incredible number of salmon are caught. They are taken in the following manner : before the water rises on the approach of summer, small channels are made among the stones and rocks, 2 feet broad and running out into various branches, over which is placed a platform for the person to stand. Several channels are made, some higher, some lower, so as to suit the water as it falls or rises. A scoop net or net fastened round a hoop at the end of a long pole, 12 to 15 feet, is all that is used ; the person stands on the extremity of the stage or platform and places his net at the top of the channel, which is always made to fit it exactly, and it is carried down with the current. The poor salmon, coming up his smooth and agreeable road as he conceives it to be, thrusts himself in the net and is immediately thrown on the stage ; the handle or pole of the net is tied to the platform by a rope lest the pressure of water or strength of the fish should snatch it out of the hands of the fisher. The hoop is made of *Acer circinnatum* of Pursh, which is very tough and not unlike *A. rubrum*. The pole is balsam pine, which after drying is light. The net is made from the bark of a species of *Apocynum*, which is very durable. The fish are of good quality, much about the same size as those caught in the rivers of Europe, 15 to 25 lb. generally, some more. I measured two, the one 3 feet 5 inches from the snout to the extremity of the tail, 10 inches broad at the thickest part, weighing about 35 lb. ; another 3 feet, and 9 inches broad, a little lighter. Both were purchased for 2 inches of tobacco (½ oz.) value twopence, or one penny each. How little the value from that in England, where the same quantity would cost £3 or £4, and not *crisped* [1] *salmon* as it is termed by those acquainted with refinement of dishes, as I have it, cooked under the shade of a lordly pine or rocky dell far removed from the abodes of civilised life. It is very wonderful the comfort, at least the pleasant idea of being comfortable in such a place surrounded by multitudes of individuals who, perhaps, had never seen a white person before, and were we to judge by their appearance are very hostile, viewing us narrowly with surprise. The luxury of a night's sleep on a bed of pine branches can only be appreciated by those who have experienced a route over a barren plain, scorched by the sun, or fatigued by groping their way through a thick forest, crossing gullies, dead wood, lakes, stones, &c. Indeed so much worn out was I three times by fatigue and hunger that twice I crawled, for I could hardly walk, to a small abandoned hut. I had in my knapsack one biscuit ; the third and last time I was not so bad with hunger, but very weak. I killed two partridges an hour before I camped, which I placed in my little kettle to boil for supper. The Canadian and the two Indians had eaten their dry salmon and were asleep. Before my birds were cooked Morpheus seized me also ; I awoke at daybreak and beheld my supper burned to ashes and three holes in the bottom of my kettle. Before leaving my resting place I had to make a little tea, which is the monarch of all food after fatiguing journeys. This I did by scouring out the lid of my tinder-box and boiling the water in it ! I have oftentimes heard that ' Necessity has many inventions,' which I now know and partly

---

[1] *Sic* MS., but query ' crimped.'—ED.

believe. The natives are inquisitive in the extreme, treacherous, and will pillage or murder when they can do it with impunity. Most of the tribes on the coast (the *Chenooks*, *Cladsaps*, *Clikitats*, and *Killimucks*) from the association they have had with Europeans are anxious to imitate them and are on the whole not unfriendly. Some of them are by no means deficient of ability. Some will converse in English tolerably well, make articles after the European models, &c. They are much prejudiced in favour of their own way of living, although at the same time will not fail to eat a most inordinate quantity if offered to them. My canoe-men and guides were much surprised to see me make an effervescent draught and drink it boiling, as they thought it. They think there are good and bad spirits, and that I belong to the latter class, in consequence of drinking *boiling* water, lighting my tobacco pipe with my lens and the sun, and they call me *Olla-piska*, which in the Chenook tongue signifies *fire*. But above all, to place a pair of spectacles on the nose is beyond all their comprehension : they immediately place the hand tight on the mouth, a gesture of dread or astonishment.

Salmon are also caught on sandy shores, where free from large stones, with a draught net in the same manner as the salmon fishing in Britain. The net is made of *Apocynum* bark, floated by pieces of wood in lieu of cork. This mode is only practised where there are no rapids or projecting rocks, or places to make channels for the scoop net. From the Rapids to the Great Falls, distant about fifty-eight or sixty miles, the banks are steep and in many places rugged. Some of the hills are very high but all destitute of trees or large shrubs. The wood becomes smaller the further the river is ascended. *Acer* is not found above the Rapids ; *Thuya*, *Pinus balsamea*,[1] and one species of *Populus* on the edges of creeks, all of which gradually diminish into low scrub-wood. Sixteen miles below the Falls we are no longer fanned by the huge pine stretching its branches in graceful attitude over a mountain rivulet or deep cavern, or regaled by the quivering of the aspen in the breeze. Nothing but extensive plains and barren hills, with the greater part of the herbage scorched and dead by the intense heat. I had to cross a plain nineteen miles without a drop of water, of pure white sand, thermometer in the shade 97°. I suffered much from the heat and reflection of the sun's rays ; and scarcely can I tell the state of my feet in the evening from the heat in the dry sand ; all the upper part of them were in one blister. Six miles below the Falls the water rushes through several narrow channels, formed by high, barren, and extremely rugged rocks about two miles long. It is called by the voyageurs *The Dalles*. On both sides of the river very singular rocks of a great height are to be seen, having all the appearance of being water-worn ; not unlikely they have been the boundaries of the river at some former period. The present bed of the river is more than 6000 feet lower. The Falls stretch across the whole breadth of the river in an oblique direction, which may be about 400 yards, about 10 or 12 feet of a perpendicular pitch. At present its effect is somewhat hid, the water being high, but I am told it is fine when the river is low. The ground on both sides is

---

[1] *Abies balsamea*, Veitch, Man. Conif. ed. 2, p. 492.

K

high, destitute of all sorts of wood or shrub except *Berberis nervosa* and *B. Aquifolium, Tigarea tridentata*,[1] and one species of *Ribes* with small red smooth berries.

During my journey I collected the following plants, some very interesting and will, I am sure, amuse the lovers of plants at home :—

(296) *Lupinus* sp., perennial; a tall strong plant, all parts alike hairy; flowers faint rose colour, with a tint of yellow, very beautiful; on the plains near the Falls of the Columbia. S.

(297) *Lupinus* sp., perennial; a small plant, seldom more than 10 inches or a foot high; flowers bright purple; in the same situations as the preceding. S.

(298) *Coreopsis tinctoria* (?) Nuttall; annual; within the sides of the Columbia and its branches that are inundated; plentiful. S.

(299) *Myosotis* sp., perennial; flowers white, with a yellow orifice; seed-vessels very hirsute; rocky situation, near mountain rivulets. S.

(300) *Syngenesia*; perennial; flowers blue; on dry sandy soil, near the Falls; rare.

(301) *Malva* (?) sp., perennial; flowers purplish or pink; a strong plant 3 to 4 feet high; margin of springs and streams.

(302) *Helianthus* sp., annual; 4 to 8 inches high; flowers yellow; sandy plains; rare.

(303) *Diadelphia*, perennial (?); small and creeping; flowers purple; on the sandy banks of rivers.

(304) *Spiraea* sp.; plentiful on the Rapids; grows very luxuriant in low damp shady woods.

(305) May prove *Trifolium*; annual; flowers white; plentiful in all alluvial soils.

(306) —— (?), imperfect; on all sandy soils.

(307) Small annual plant, on the shores of the river.

(308) *Syngenesia*, perennial; flowers yellow; on the plains; rare.

(309) *Polygonum* sp., annual (?), perennial; on the banks of rivers.

(310) *Polygonum* sp., annual; small, 8 to 14 inches high; flowers small, white; same situations as the former.

(311) May prove a *Psoralea*; sandy places; stems creeping; rare.

(312) *Onosmodium* of Michaux, perennial; dry soils and among rocks; plentiful. S.

(313) *Achillea* sp., perennial; among rocks, on the banks of all the rivers.

(314) *Asclepias* sp., perennial; flowers brown, fragrant; rare; in dry sandy soil.

(315) *Santolina* (?); perennial (?), biennial; summit of barren dry hills in sandy soil; rare.

(316) *Syngenesia*, (?) biennial; fine plant; leaves spinous; whole plant white and woolly; flowers yellow; plentiful; same situations as the former. S.

(317) *Ribes* sp.; leaves round, partially three-lobed, serrate; fruit globular, small, smooth, scarlet, transparent; seldom more than one or

[1] *Purshia tridentata*, S. Wats. Bibl. Ind. N. Am. Bot. p. 309.

two at most perfect seeds in a berry ; taste rather insipid ; very luxuriant among stones and in fissures of rocks where there is scarcely a particle of earth. S.

(318) *Hypericum* sp., perennial ; probably an alpine variety of that found on the plains near the ocean ; on the hills.

(319) *Syngenesia* ; flowers yellow, glutinous, and scaly ; at the Falls ; rare.

(320) *Diadelphia*, perennial ; seed-vessel one-seeded ; among rocks ; only found one plant.

(321) *Stellaria* sp., (?) perennial ; on dry barren rocks ; plentiful.

(322) —— (?) in seed ; perennial ; 12 to 18 inches high ; abundant in sandy dry hills. S.

(323) *Syngenesia*, perennial ; flowers yellow ; a low handsome plant ; abundant in rocky places.

(324) *Plantago* sp., annual ; leaves linear, pubescent, shorter than the scape ; a fine species ; abundant on the gravelly hills, near the rivers.

(325) *Clematis* sp., perennial ; flowers white ; edges of rivers and woods.

(326) *Epilobium* sp., annual (?), small; on moist rocks and mountain springs ; abundant at the Rapids. S.

(327) *Cheiranthus* sp., annual or biennial ; flowers orange-yellow, leaves and stem rough ; siliqua long.

(328) *Cheiranthus*, annual ; flowers dark yellow ; may not be different from the former ; has also long pods.

(329) *Clarkia pulchella* (Pursh), annual ; description and figure very good ; flowers rose colour ; abundant on the dry sandy plains near the Great Falls ; on the banks of two rivers twenty miles above the Rapids ; an exceedingly beautiful plant. I hope it may grow in England. S.

(330) *Gaura* (?) sp., annual, perennial (?) ; in the same situations.

(331) *Erigeron* sp., perennial; flowers white ; abundant in all low grounds.

(332) *Phlox* sp. new ; annual ; root fibrous ; stem round, somewhat pubescent, rarely branching ; leaves alternate, sessile, linear, lanceolate, upper side minutely pubescent, under scabrous ; calyx cyathiform, small ; corolla large, funnel long, spreading at the mouth ; lobes ovate, obtuse, yellow-cream colour, surrounded by bractea broader than the leaves but shorter ; seeds mucilaginous, like Nuttall's *Collomia*, which I already have ; 6 to 18 inches high [1]; the flowers of this one, five times the size ; it will be a great addition ; this exceedingly beautiful species I name *P. Sabinii*,[2] in honour of Jos. Sabine, Esq., for the zeal he has taken in illustrating this beautiful genus of plants, when described or figured I beg it will be adopted ; this interesting species I found on the subalpine ground between the Rapids and Falls of the Columbia, July 6th ; I regret much that only a very few seeds of it could be found sufficiently ripe on the Columbia and its branches. When I found this fine plant in June, I regretted that only about a dozen perfect seeds could be found. Since I have seen

[1] In another MS. :—' 1 to 2½ feet high.'
[2] *Phlox speciosa* var. *Sabini*, A. Gray, Syn. Fl. N. Am. ii. 1. p. 134.

it in abundance and now send about a pound; I hope it will ere long decorate the garden at Chiswick. S.

(333) *Hedysarum* sp., (?) perennial; suffruticose, decumbent; flowers white; on the sandy shore of rivers. S.

(334) *Brodiaea* sp.; flowers blue; this is very different from and has larger flowers than what I took to be *B. grandiflora*; this I have no doubt will be a second species to the genus; 8 inches to a foot high; like the other anthers united, white, hexandrous; roots of this fine plant are sent in a jar among dry sand with bulbs of the other; on all hilly sandy soils among stones with the following plant; this is a very fine plant. S.

(335) *Hexandria, Monogynia*; calyx three-leaved, lanceolate; corolla three petals, roundish, same length as the calyx; anthers as long as the filaments; style shorter than the stamens; capsule three-sided, three-celled; seeds angular; stem three or four jointed; leaves sheathing, gramineous; flowers violet colour seldom more than two on a stem; nectary surrounded by yellow hairs half-way up the petals, seldom more than a brown streak up the middle of them; inside of the calyx same colour as the flower; plant 1 to 2 feet high; this splendid plant, which equals *Ferraria Pavonia*,[1] I cannot refer to any in the American flora; I have no doubt it is not yet included; seeds were not ripe when there, and since I have not been able to make a second journey; roots are sent in a jar of dry sand, I hope they will keep and vegetate. For the present I call this *Munroa speciosa*, after Mr. Munro. Found on elevated grounds; rare.

(336) —— (?) a beautiful branching plant in thick clumps, with white flower and very fragrant; abundant on elevated sandy dry ground, forming elegant round groups, at a distance appearing like Laurustinus; I had not time at the moment to examine it, and cannot say to what it belongs; perennial; near the Falls. S.

(337) *Tigarea tridentata*[2] (Pursh); I regret very much to see this very interesting plant in an imperfect state, being just out of blossom; plentiful on the plains near the Falls and rocks below them; it thrives luxuriantly in sandy soil, and so dry and loose that a stick may be pushed down 2 or 3 feet. I hope to have seeds of it in the Fall; the only shrub to be seen on the barren plains.

(338) *Pentstemon* sp., perennial; radical leaves opposite, lanceolate, deeply dentate, on short petioles, cauline sessile, somewhat cordate, dentate, smooth; flowers large, fine purple, inside of the margin blue; upper lip bearded; this new species I call *P. Richardsonii*,[3] after that distinguished traveller Dr. John Richardson, now on his second hazardous journey to the Polar Sea; plentiful in rocky soils on the Columbia and its branches. S.

(339) *Polyandria, Di-Pentagynia*, annual; calyx five-parted, segments equal, lanceolate; petals five, superior; stamens numerous, partially united, 3, 4, 5 filaments broader than the rest; capsule glutinous, longish,

---

[1] *Tigridia Pavonia*, Baker, Handb. Irid. p. 67.
[2] *Purshia tridentata*, S. Wats. Bibl. Ind. N. Am. Bot. p. 309.
[3] Dougl. in Lindl. Bot. Reg. t. 1121.

imperfect ; leaves alternate, lanceolate, dentate, sessile, scabrous ; stem branching, round, smooth, white ; flowers bright yellow ; 2 to 3 feet high ; this very handsome plant I found on the dry banks of the Columbia, a few miles below the Falls ; it is by no means plentiful ; what it belongs to I cannot say ; I regret that seeds of it cannot be procured for the present.

(340) *Syngenesia*, annual ; flowers yellow ; in all dry grounds ; abundant, particularly near the edges of woods and clumps of trees.

(341) *Campanula* sp. ; perennial ; flowers blue ; rocky situations, near the Rapids.

(341*) *Arabis* sp. ; annual ; leaves oblong-ovate, hispid ; moist rocks ; rare.

(342) *Umbelliferae*, annual ; on dry rocky places, near the Rapids ; abundant. It is found also in low meadows, where it grows much larger than on the rocks at the Rapids. S.

(343) *Syngenesia*, perennial ; flowers yellow ; root large and bitter ; on dry sandy grounds, banks of rivulets.

(344) *Polygonum* (?) sp., perennial ; leaves radical, lanceolate, on long petioles, pubescent ; flowers small, white ; a strong branching plant, 2 to 4 feet high ; in subalpine ground between the Rapids and Falls ; plentiful.

(345) *Syngenesia*, perennial ; flowers yellow ; in dry soil at the Falls. S.

(346) —— (?) is this allied to *Sanguisorba* ; annual ; on dry soil, among bushes and woods, where they are partially shaded ; plentiful. S.

(347) *Echium* sp., annual ; leaves alternate, upper ones hastate, sessile ; segments of the calyx linear, ciliated ; flowers blue, tinged with rose colour ; a fine small plant ; in all dry open plains, in light soil. S.

(348) *Lobelia* (?) sp., annual ; flowers blue and white ; a small plant, grows to a foot high ; in edges of pools and mountain rills. S.

(349) *Polygonum* (?) annual ; a small plant, 2 to 6 inches high ; on moist ground.

(350) —— (?) *Pentandria, Monogynia* ; allied to *Phlox* ; flowers white ; a curious little plant, growing in great abundance with the former ; I had not time to examine it when gathered. S.

(351) *Polypodium fragile*[1] ; moist rocks.

(352) *Pteris crispa*[2] ; same situations.

(353) *Asplenium melanocaulon*[3] ; same situations.

(354) *Maclura* (?) in an imperfect state, on the dry plains between the Rapids ; a small tree.

(355) *Nyssa* (?) in fruit ; in the like places ; a shrub 6 to 10 feet high.

(356) *Bartsia* sp.[4] ; annual ; dry open subalpine grounds.

(357) *Chara* sp. ; in small still pools.

(358) *Ranunculus* sp., annual ; very small, in the like places.

---

[1] *Cystopteris fragilis*, Christensen, Ind. Fil. pp. 203, 528.
[2] *Cryptogramma crispa*, Christensen, *loc. cit.*, pp. 187, 595.
[3] *A. trichomanes* var. Christensen, *loc. cit.*, pp. 121, 136.
[4] In another MS. he has *Pedicularis*.

(359) —— (?); in the same situation, very minute.

(360) —— Three small plants in a bad state, found on dry rocky places; shrubs.

(361) *Hordeum* sp., annual; in dry plains; common.

(362) *Triticum* sp., perennial; a strong plant, 6 to 10 feet high, in the plains and edges of rivers.

(363) *Juncus* sp., perennial; in moist soils, mountain springs; abundant.

(364) *Juncus* sp., perennial; this I think will agree with one collected some time since in the low grounds; this can only be an alpine.

(365) *Alopecurus* sp., perennial; same place.

(366) —— (?) *Gramineae*, perennial; a very strong grass, in all alluvial soils; 4 to 8 feet high.

(367) *Gramineae*, perennial; in moist grounds.

(368) *Gramineae*, perennial; on the dry sandy plains, near the Great Falls.

(369) *Gramineae*, perennial; in the same place.

The above seventy-three species are what I collected during my journey of ten days up the river; my time I consider well spent. Returned on the 6th of July, and from this time to the 19th employed putting into order what were collected, writing, and making journeys in the neighbourhood of my residence.

July 6th–19th, collected the following :—

(370) *Stachys* sp., perennial; flowers purple; edges of woods, rivers, pools, and moist soil, in shady places; plentiful.

(371) *Lithospermum* sp., annual; on the sandy banks of rivers, among bushes; 2 to 4 feet high; flowers white.

(372) *Apocynum* sp., perennial; flowers white; stems red; leaves opposite, sessile; the bark of this plant affords the flax from which the natives make their nets, &c.; probably this will be *A. cannabinum*; the flax is collected from the withered stems in autumn; 2 to 5 feet high; abundant in rich soil; grows luxuriantly in a low moist place.

(373) *Sambucus* sp.; flowers yellowish-white; a large shrub; fruit not yet ripe; plentiful in all low woods, near the outskirts and sides of pools.

(374) *Mentha* sp., perennial; in all low grounds; flowers white, tinged with red.

(375) *Orchis* or *Satyrium* sp. *viride*[1] (?); flowers greenish-white; in dry subalpine plains, under the shade of individual trees; has a fine effect.

(376) *Gramineae*, perennial; on all moist low fields and woods; most animals are very fond of this grass.

(377) *Campanula* sp., perennial; flowers white; teeth of the corolla reflexed after a few days' expansion; a small curious species; in thick shady woods, among moss; plentiful; plant reclining. S.

(378) *Didynamia, Gymnospermia*, perennial; calyx five-cleft, campanulate; corolla bilabiate, upper lip two segments, lower three; flowers

---

[1] *Habenaria viridis*, Ind. Kew. fasc. iv. p. 808.

white; leaves nearly round, partly entire, opposite, on short petioles; stem creeping; in shady woods.

(379) *Philadelphus* sp.; flowers white, fragrant; I had not time to examine it; a fine small shrub, plentiful on hilly banks of the Columbia above the Rapids, which may prove *P. Lewisii* of Pursh; plentiful in all outskirts of woods; forms part of the underwood in the pine forest.

(380) *Sanicula* (?); perennial; in dry gravelly woods, in partially shady situations; flowers white; peduncle and calyx beset with brown glands. S.

(381) *Neottia pubescens* [1]; in all shady pinewoods, in dry decomposed vegetable soil; abundant.

(382) *Sonchus* sp., annual; flowers bright blue; plentiful in moist meadows.

(383) *Gramineae*, perennial; in the edges of pools, moist meadows, rivers, &c.; a strong grass.

(384) *Gramineae*, perennial; same situations; a strong coarse grass; no animal will eat of this species.

(385) *Gramineae*, perennial; in the like places as the preceding; no animal will eat of it.

(386) *Orobanche ludoviciana* [2] (Nuttall); on the alluvial plains of the Columbia; *parasitic* on the roots of various grasses which have been burned by the natives in the autumn for the purpose of affording a tender herbage in spring for their horses; flowers white, mixed with rose colour, sometimes sulphur-yellow where very much shaded; in very shady moist places; abundant.

(387) *Spiraea* sp. *capitata* [3] (?) Pursh; a handsome shrub, 5 to 10 feet high; flowers white, pendulous, very large; thrives luxuriantly in all low places, moist woods, low valleys, near rivulets, in shady woods; this is one of the most magnificent plants of the wood; seeds not yet ripe; get a large package of seeds.

(388) *Prunella* sp. var. *alba*; only one plant in low marshy grounds.

(389) *Bartsia* (?) sp.,[4] annual; flowers rose colour; abundant in low marshy ground.

(390) *Verbena* sp., perennial; stem erect, square, covered with rigid hairs; leaves opposite, ovate, acute, serrate, strongly nerved; flowers small, purple.

(391) *Lythrum* sp., annual; flowers rose colour; margins of pools and rivers.

(392) *Campanula* sp., perennial; in all dry meadows and rocks, on the banks of rivers. *C. rotundifolia*? abundant.

(393) *Scutellaria* sp., annual; flowers blue; in low, swampy grounds; abundant.

(394) *Veronica* sp.; *scutellata*? annual; in the like places; also abundant; flowers blue.

---

[1] *Goodyera pubescens*, Ind. Kew. fasc. iii. p. 304.
[2] *Aphyllon ludovicianum*, A. Gray, Syn. Fl. N. Am. ii. I. p. 313.
[3] *Neillia opulifolia* var. *mollis*, S. Wats. Bibl. Ind. N. Am. Bot. p. 290.
[4] In another MS. he has *Pedicularis sp.*

(395) *Solidago* sp., perennial; flowers yellow; in all elevated grounds; plentiful.

(396) *Silene* (?), perennial; flowers small, pale purple; stem and calyx glutinous; abundant on little rising grounds in swamps. S.

(397) *Gramineae*, perennial; in all dry places over plains or gravelly or light soils.

(398) *Syngenesia*, perennial; flowers yellow; probably *Helianthus*, will form a third species with 36 and 37; plentiful in low, wet grounds. I am not aware if this species is used in making bread like the others. S.

(399) *Diadelphia*, perennial; in seed; in dry bluffs in the middle of swamps among trees. It occurs to me that I gathered this plant in May in the same place in flower near the Rapids (No. 10); near the Rapids it is very plentiful and grows to 35 to 40 feet high. S.

(400) *Rumex*, perennial; a strong plant; plentiful in moist soils.

(401) *Scirpus*, perennial; 4 to 10 feet; in marshy ground; the natives make mats of it by weaving them together with tissue from the roots of the *Cyperus*, or with twisted strings made from the leaves of *Typha angustifolia*; the tender part of the stem next the root, which is white, is eaten by them and considered a luxury.

(402) *Oenothera* sp.; biennial; *biennis* (?); sandy banks of rivers and edges of woods; plentiful.

(403) *Aster* sp.; perennial; flowers blue; in all alluvial soils.

(404) *Gnaphalium* sp.; annual; flowers yellow; on sandy and all light soils.

Collected the following plants on Menzies Island, on sandy banks of the Columbia river:

(405) *Artemisia* sp. *longifolia* (?), Nuttall.

(406) *Artemisia* sp.; stems straight, partly suffruticose; leaves entire, sessile, linear-lanceolate; smells strong like Tansy.

(407) *Artemisia* sp.; 1 to 3 feet high; has no smell; all these three species have yellow flowers; the two former from 4 to 6 feet high.

(408) *Syngenesia*, annual; flowers small, white; 1 to 3 feet high; shores of the river; abundant.

(409) *Gnaphalium* sp.; perennial; flowers yellow; whole plant tomentose, with a sweet but overpowering scent; plentiful.

(410) *Chenopodium* sp.; annual; plentiful in all rich soils; fertile banks of rivers; grows very strong and abundant around Indian villages and camps; among the numerous vegetables used by them, it is rather singular they should omit one which is almost universally used in every country; even the tender shoots of several species of *Rubus*, and sprouts of a species of *Equisetum* and *Scirpus* are greedily sought after, used and considered good, while a wholesome plant is left untouched.

(411) *Syngenesia*, annual; may prove a second species of No. 145, collected in May on Menzies Island; very plentiful on all dry, open, elevated ground; a beautiful plant.

(412) *Hieracium* sp.; perennial; abundant in the like situations as the above; flowers yellow.

(413) *Epilobium angustifolium*; edges of woods and rivers among decayed vegetable and sandy soils.

(414) —— (?) small plant, resembles *Polygonum*; annual or perennial; sandy shores of rivers; abundant.

(415) *Gramineae*, perennial; a very fine silky grass; abundant in all meadows.

(416) *Gramineae*, perennial; a strong beautiful grass; in similar places as the former, banks of rivers.

(417) *Gramineae*, perennial; in the like places; a fine grass; also plentiful.

(418) *Syngenesia*, perennial; in all dry barren plains.

July 19th.—Early in the morning I left my residence in a small canoe, with one Canadian and two Indians, for a journey to the shores of the ocean, principally for the purpose of searching for and inquiring after the tuberous-rooted *Cyperus* mentioned by Pursh in his preface, the root of which is said to afford the natives food something like potatos when boiled. After a laborious route of twelve days along the shore north of Cape Disappointment, I was obliged reluctantly to return without being fortunate enough to meet with it. I observed several dead roots, washed on the shore by the surge and agreeing exactly with the description given by Lewis and Clarke, which I conjecture to be it. My guide, who is tolerably conversant with many of the tongues spoken by the inhabitants of the coast, learned that it is very abundant along the shore from Point Adams, the southern entrance of the river, at no great distance. I am for the present prevented from prosecuting my journey in that direction, several of the tribes being at war with each other. I laboured under very great disadvantage by the almost continual rain; many of my specimens I lost, and although I had several oilcloths, I was unable to keep my plants and blanket dry or to preserve a single bird; saw many pelicans of one species, but could not obtain any, (I believe it to be the same as one I killed in the Galapagos,) one albatross, some petrels which did not come under my eye during my voyage out, one large brown gull, and a smaller white with bluish wing on the upper side. This one I have since seen on the sandbanks of the river as far up as the Great Rapids. Now I have a little idea of travelling without the luxuries of life. Only two nights were dry during my stay on the shore; before I could lie down to sleep my blanket drying generally occupied an hour. In the creeks I caught plenty of a small trout and young salmon. With a basin of tea, a small piece of biscuit, and now and then a duck, I managed to live very well. On my return I visited Cockqua, the principal chief of the Chenooks and Chochalii[1] tribes, who is exceedingly fond of all the chiefs that come from King George—words which they learn from Broughton, of Vancouver expedition, and other commanders of English ships. His

---

[1] In another MS. :—'Chits.'

acquaintance I previously had. He imitates all European manners;
immediately after saluting me with ' clachouie,' their word for ' friend,'
or ' How are you ? ' and a shake of his hand, water was brought immediately for me to wash, and a fire kindled. He then carried me to one
of his large canoes, in which lay a sturgeon 10 feet long, 3 at the thickest
part in circumference, weighing probably from 400 to 500 lb., to choose
what part should be cooked for me. I gave him the preference as to
knowledge about the savoury mouthfuls, which he took as a great com-
pliment. In justice to my Indian friend, I cannot but say he afforded
me the most comfortable meal I had had for a considerable time before,
from the spine and head of the fish. A tent was left here, which could
not be carried further, in which I slept. He was at war with the Cladsap
tribe, inhabitants of the opposite banks of the river, and that night expected an attack which was not made. He pressed me hard to sleep
in his lodge lest anything should befall me : this offer I would have most
gladly accepted, but as fear should never be shown I slept in my tent
fifty yards from the village. In the evening about 300 men danced the
war dance and sang several death songs. The description would occupy
too much time. In the morning he said I was a great chief, for I was
not afraid of the Cladsaps. One of his men, with not a little self-conse-
quence, showed me his skill with the bow and arrow, and then with the
gun. He passed arrows through a small hoop of grass 6 inches in diameter, thrown in the air a considerable height by another person ; with
his rifle he placed a ball within an inch of the mark at the distance of
110 yards. He said no chief from King George could shoot like him,
neither could they sing the death song nor dance the war dance. Of
shooting on the wing they have no idea. A large species of eagle, *Falco
leucocephala*, was perched on a dead stump close to the village ; I charged
my gun with swan shot, walked up to within forty-five yards of the bird,
threw a stone to raise him, and when flying brought him down. This
had the desired effect : many of them placed their right hands on their
mouths—the token for astonishment or dread. This fellow had still a
little confidence in his abilities and offered me a shot at his hat ; he threw
it up and I carried the whole of the crown away, leaving only the brim.
Great value was then laid on my gun and high offers made. My fame
was sounded through the camp. Cockqua said ' Cladsap cannot shoot
like you.' I find it to be of the utmost value to bring down a bird flying
when going near the lodges, at the same time taking care to make it
appear as a little thing and as if you were not observed. In the lodge
were some baskets, hats made after their own fashion, cups and pouches,
of very fine workmanship ; some of them made with leaves of *Typha
angustifolia* and leaves of *Helonias tenax*[1] ; some with the tissues of *Thuya*
roots, and of the inner bark, and some with a small linear-leaved *Fucus*
with leaves of the stronger *Carices*. I received from him an assemblage
of baskets, cups, &c., and his own hat, with a promise that the maker (a

[1] *Xerophyllum asphodeloides* var., Baker, in Journ. Linn. Soc. xvii. p. 467.

little girl twelve years of age, a relation of his own) would make me some hats like the chief's hats from England. I made a short stay, always collecting what came under my notice; on the 5th August I made some small presents of tobacco, knives, nails, and gun flints, and then left for my residence up the river to Fort Vancouver, which occupied two and a half days. Collected the following plants:

(419) *Santolina* sp.; perennial; a strong plant, 2 to 3 feet high; flowers yellow; on the seashore; abundant.

(420) *Lathyrus* sp.; perennial; flowers large bluish-purple; a splendid strong-growing plant; the roots are large, run deep in the sand, and are eaten by the natives in a raw state; abundant. S.

(421) *Rosa*; a very strong plant, 10 to 14 feet high, with large globular fruit; probably I have it in blossom; abundant on the banks of rivers; very luxuriant in light rich dry soils. S.

(422) *Samolus Valerandi* (?); in shady places near the edges of creeks and mountain springs; plentiful.

(423) *Sagina procumbens* (?); in the same place.

(424) *Prunella* sp.; probably only a variety of what was collected before, only more luxuriant; leaves more acute; in moist ground.

(425) *Rhinanthus Crista-galli*; in sandy, open, dry meadows; abundant.

(426) *Lupinus* sp.; perennial; stem slender, creeping; leaves digitate, leaflets five to seven, linear, silky on both sides; peduncles long; seed-vessel pubescent, eight to twelve-seeded, generally ten seeds, small and mottled; root large, spreading, with small tubercles; roots when large are roasted by the natives and sometimes chewed raw; plentiful on the seashore; *Lupinus tuberosus*. S.

(427) *Trifolium* sp.; perennial; flowers large, light red; plentiful near the sea.

(428) *Lonicera* sp.; shoots small, slender, hirsute; leaves opposite, ovate, obtuse, nearly sessile, pubescent on both sides; this little plant I could not find in flower or fruit; this I regretted the more as I have no doubt of its not being described; abundant on rocky and gravelly soils near the ocean, and on the banks of rivers.

(429) —— (?); *Leguminosae*, in an imperfect state, on the seashore; obtained only one specimen.

(430) *Umbelliferae*, perennial; in marshes near the sea; strong growing. S.

(431) *Umbelliferae*, perennial; flowers white; sides of river and mountain springs; plentiful.

(432) *Carex* sp., perennial; a low-growing plant; dioecious, with very large fruit; on the seashore north of Cape Disappointment; rare.

(433) *Scirpus* sp.; perennial; on the seashore; plentiful.

(434) *Gramineae*, perennial; plentiful.

(435) *Juncus* sp.; perennial; small, 3 to 6 inches high; in seed; a fine delicate plant.

(436) *Juncus* sp.; perennial; also small and resembles the preceding; both grow together; the former rather rare; this is found in abundance.

(437) *Juncus* sp. ; 8 to 18 inches high ; in flower and seed ; a hand some species ; very plentiful in all sandy soil near the sea ; viviparous.

(438) *Triticum* sp. ; very handsome ; abundant on the shore ; I regret that all the seed of this beautiful species was gone.

(439) *Juncus* sp. ; small, 4 to 6 inches high ; in seed ; abundant.

(440) *Gramineae* sp. ; annual (?) or perennial (?) ; a very singular handsome grass, 1 to 2½ feet high ; in a low marsh near the sea ; plentiful. S.

(441) *Gramineae* sp. ; perennial ; strong in the like places as the former one ; also abundant.

(442) *Rumex* sp. ; annual ; small ; on the shores ; plentiful.

(443) *Spergula*, annual ; creeping ; leaves linear, verticillate ; in seed ; on the grounds around the Company's apartments. S.

(444) *Spiraea* sp. ; perennial ; a fine plant, 2 to 4 feet high ; this I expect will prove the same as one I saw in the possession of Mr. Menzies with red flowers ; I saw it only in seed ; plentiful in the moist rocks near the ocean, and at the Great Rapids.

(445) *Pteris aquilina* [1] ; abundant and grows exceedingly strong in rich open woods and meadows ; the roots are dried and eaten by th natives, when boiled or roasted.

(446) *Aspidium* sp. ; strong growing in all low moist woods ; very plentiful.

From the 6th to the 18th of August employed drying what plants had been collected on my journey ; making occasional short routes in the neighbourhood of my residence, collecting seeds of plants already collected. Nothing of interest occurred during the time ; my labours were frequently retarded by rainy weather.

August 19th.—Towards afternoon left in a small canoe with one Canadian and two Indians, in company with a party of men going on a hunting excursion to southward, on a visit to the Multnomah River, one of the southern branches of the Columbia. The distance I was enabled to go was about fifty-six miles. The river is large, nearly as large as the Thames ; thirty-six miles from the Columbia are very fine falls, about 43 feet high, across the whole river, in an oblique direction ; when the river is low they are divided into three principal channels, all of which have a perpendicular pier ; when the water is high it rushes over in an unbroken sheet. This season, in July, which is the time it is at its greatest height, it rose 47 feet. From the Columbia to the Falls there is but little or no current ; gorged back by the waters of that river. The banks are covered with *Pinus taxifolia*,[2] *P. balsamea* [3] *Quercus*, and *Populus*. The soil is by far the richest I have seen. Above the Falls, as far as I went, at many places the current is rapid. I had considerable difficulty in making the portages at the Falls, having to haul the canoe up with ropes ; this laborious under-taking occupied three hours, and one hour on my return. This at one time was looked on as the finest place for hunting west of the Rocky Mountains.

[1] *Pteridium aquilinum*, Christensen, Ind. Fil. pp. 591, 592.
[2] *Pseudotsuga Douglasii*, Mast. in Journ. R. Hort. Soc. xiv. p. 245.
[3] *Abies balsamea*, Mast. *loc. cit.*, p. 189.

The beaver now is scarce; none alive came under my notice. I was much gratified in viewing the deserted lodges and dams of that wise economist. Abundance of a species of deer (which probably may be the one spoken of by Mr. Sabine in his description of the animals observed by Franklin's party) are to be had. During my stay (ten days) seventeen were killed, both males and females. It grieved me exceedingly I was so placed that only a small one could be preserved: a young male, which I killed at 115 yards with ball. The flesh is very fine, of a beautiful fine delicate white. Unfortunately I lost my note of their proportions, colour, &c., but as I am very shortly to make a second visit I shall not fail to preserve a pair; horns 3 to 4, branched, short, about 15 inches long; light brown, white on the belly, young ones white spotted until six months old. Near my tent was a small salt-marsh, to which in the morning they daily resorted. Killed 2 females and 3 males of a fine species of pigeon; feet, legs, and part of the beak yellow, a white ring round the neck. I was only able to skin one, a male, and that is a miserable specimen; female same colour except neck and breast, which is of a darker hue. I was very fortunate having good weather all the time, except one day.

Collected the following plants and obtained seeds of several very important plants already collected:

(447) *Nicotiana pulverulenta* [1] (?) of Pursh, correctly supposed by Nuttall to exist on the Columbia; whether its original habitat is here in the Rocky Mountains, or on the Missouri, I am unable to say, but am inclined to think it must be in the mountains. I am informed by the hunters it is more abundant towards them and particularly so among the Snake Indians, who frequently visit the Indians inhabiting the head-waters of the Missouri by whom it might be carried in both directions. I have seen only one plant before, in the hand of an Indian two months since at the Great Falls of the Columbia, and although I offered him 2 ozs. of manufactured tobacco he would on no consideration part with it. The natives cultivate it here, and although I made diligent search for it, it never came under my notice until now. They do not cultivate it near their camps or lodges, lest it should be taken for use before maturity. An open place in the wood is chosen where there is dead wood, which they burn, and sow the seed in the ashes. Fortunately I met with one of the little plantations and supplied myself with seeds and specimens without delay. On my way home I met the owner, who, seeing it under my arm, appeared to be much displeased; but by presenting him with two finger-lengths of tobacco from Europe his wrath was appeased and we became good friends. He then gave me the above description of cultivating it. He told me that wood ashes made it grow very large. I was much pleased with the idea of using wood ashes. Thus we see that even the savages on the Columbia know the good effects produced on vegetation by the use of carbon. His knowledge of plants and their uses gained him another finger-length. When we smoked we were all in all. S.

---

[1] This must be a slip of Douglas's, as the only specific name in *Nicotiana* for which Pursh is the authority is *quadrivalvis*, Pursh, Fl. Am. Sept. i. p. 141.

(448) *Oxalis* sp. ; perennial; flowers yellow ; abundant in low moist woods.

(449) *Papaveraceae*, perennial ; calyx two-leaved, deciduous ; petals four, ovate or nearly round ; siliqua superior, two-celled ; seeds small, globular, numerous ; flowers yellow ; leaves alternate, pinnate, glaucous, partly succulent ; stems when broken yield a milky juice ; a very hand-some plant, 8 inches to a foot high. I cannot refer it to any plant in the Flora. S.

(450) *Triglochin maritimum* .(?) ; in a small salt-marsh near an en-campment on the Multnomah ; plentiful. One rose in fruit, probably in flower in my collection.

(451) *Eryngium* sp. ; perennial in low plains ; 4 to 12 inches high.

(452) *Lycopus* sp. ; in all wet meadows.

(453) *Euonymus* sp. ; a small tree ; 6 to 15 feet high ; on the banks of rivers; It gave me much pleasure to meet with *Phlox Sabinii*,[1] with abundance of seed ; a very small portion of which I had from the first place I found it on the Columbia ; here it was much stronger, 2 feet to 30 inches high, in light rich brown loam in open places of the wood.

(454) *Eriogonum* sp. ; different from the other species ; in dry sandy soils and on rocks ; rare.

(455) *Malva* sp. ; perennial ; flowers faint rose colour ; a fine plant, 18 to 30 inches high ; on dry plains. *Quercus* sp. ; in fruit, which is very rare to be had ; I have nowhere seen it, except on the tree my specimens are from. S.

(456) *Prunus* sp. ; a small tree ; fruit small, red, and sour.

(457) *Viburnum* sp. ; fruit black ; a strong shrub ; sides of rivulets ; plentiful. *Alnus* ; probably different from any already had. (458) *Didynamia*, perennial ; flowers blue ; stamens and upper lip reflexed ; very curious ; rare. (459) *Hypericum*, annual ; in wet meadows ; plentiful. (460) *Didynamia*, annual ; flowers blue ; on high dry grounds ; plentiful. S. *Alnus* sp. ; probably different from that collected before. *Mentha* sp. ; perennial ; may be different from any already had.

Returned on the 30th of August. From that time till Thursday, September 1st, employed drying, arranging, putting up seeds, and making up my notes. Early on Thursday went on a journey to the Grand Rapids to collect seeds of several plants seen in flower in June and July. Went up in a canoe accompanied by one Canadian and a chief (called Chum-talia) of the tribe inhabiting the north banks of the river at the Rapids. I arrived on the evening of the second day and pitched my tent a short distance from the village. I caused my Canadian to drench the ground well with water to prevent me from being annoyed with fleas, although I was not altogether exempt from them, yet it had a good effect. I found my Indian friend during my stay very attentive and I received no harm or insult. He accompanied me on some of my journeys. (They were only a few years since very hostile. The Company's boats were frequently pillaged by them and some of their people killed.) My visit

[1] *Phlox speciosa* var. *Sabini*, A. Gray, Syn. Fl. N. Am. ii. 1. p. 134.

was the first ever made without a guard. On Saturday morning went on a journey to the summit of the mountain near the Rapids on the north side of the river, with the chief's brother as my guide, leaving the Canadian to take care of the tent and property. This took three days, and was one of the most laborious undertakings I ever experienced, the way was so rough, over dead wood, detached rocks, rivulets, &c., that very little paper could be carried. Indeed I was obliged to leave my blanket (which, on my route is all my bedding) at my first encampment about two-thirds up. My provision was 3 oz. tea, 1 lb. sugar, and four small biscuits. On the summit all the herbage is low shrub but chiefly herb plants. The second day I caught no fish, and at such a great altitude the only birds to be seen were hawks, eagles, vultures, &c. I was fortunate enough to kill one young white-headed eagle, which (then) I found very good eating. I roasted it, having only a small pan for making tea. On the summit of the hill I slept one night. I made a small fire of grass and twigs and dried my clothes which were wet with perspiration and then laid myself down on the grass with my feet to the fire. I found it very cold and had to rise four times and walk to keep myself warm, fortunately it was dry and a keen north wind prevented dew. On Monday evening at dusk I reached my tent at the village much fatigued and weak and found all things going on smoothly. Made a trip to the opposite side two days after, also to the summit of the hills, which I found of easier ascent, the only steep part near the top. My food during my stay was fresh salmon, without salt, pepper, or any other spice, with a very little biscuit and tea, which is a great luxury after a day's march.

Collected the following, which did not come under my observation before, with many of the seeds—the object of the journey:

(461) *Pinus* sp.; in fruit; a large tree on the hills, I have nowhere seen before.

(462) *Pinus* sp.; a tall splendid tree; leaves glaucous. Probably I am mistaken as to *P. taxifolia*,[1] which I supposed to be plentiful on the banks of the river; this I think more likely to be it. The cones being on the top, I was unable to procure any. All the trees were too large to be cut down with my small hatchet, and as to climbing, I have already learned the propriety of leaving no property at the bottom of a tree. I went up one, but the top was too weak to bear me; the height was so great that I could not bring down any cones with buck-shot. Make a point of obtaining it by some means or other.

(463) *Pinus* sp.; different from the former; leaves longer and not so glaucous; branches drooping, while the other has few branches and straight. It is not near so large.

(464) *Pinus* sp.; what I supposed to be *P. taxifolia*[1] on the mountains; it is considerably lower than in the valleys; very abundant, more than any other, the trunk very straight with a very rough rotten bark, and yields a great quantity of gum.

(465) *Helonias tenax*,[2] of Pursh; perennial; on the summit of

---

[1] *Pseudotsuga Douglasii*, Mast. in Journ. R. Hort. Soc. xiv. p. 245.
[2] *Xerophyllum asphodeloides* var., Baker, in Journ. Linn. Soc. xvii. p. 467.

the hills. This very interesting plant gave me at first great pleasure when it struck my eye; in an imperfect state, being neither in flower nor seed. I looked on it as certain to obtain either, but after a search of three days I had the mortification to be content with strong plants of this year's growth, and decayed stalks and capsules of last year's growth. No seeds could be found. Probably only flowers every other year. Is not eaten by any animals; its beautiful green verdure bids defiance to the intense cold of its high habitation. My thermometer stood at 43° in the shade at noon 8th September; at Fort Vancouver on the same day 85°. The natives at the Rapids call it ' Quip Quip.' Pursh is correct as to their making water-tight baskets of its leaves. Last night my Indian friend Cockqua arrived here from his tribe on the coast, and brought me three of the hats made on the English fashion, which I ordered when there in July; the fourth, which will have some initials wrought in it, is not finished, but will be sent by the other ship. I think them a good specimen of the ingenuity of the natives and particularly also being made by the little girl, twelve years old, spoken of when at the village. I paid one blanket (value 7s.) for them, the fourth included. We smoked; I gave him a dram and a few needles, beads, pins, and rings as a present for the little girl. Faithful to his proposition he brought me a large paper of seeds of *Vaccinium ovatum* in a perfect state, which I showed him when there, then in an unripe state. I have circulated notices among my Indian acquaintances to obtain it for me.

(466) *Monoecia*, annual; curious plant; stamens short; anthers superior yellow; near Indian lodges abundant.

(467) *Poa* sp.; annual; small, creeping; on the sandy banks of rivers, plentiful.

(468) *Cyperus* sp.; annual; very small fine plant; this and the former grow together in abundance.

(469) *Euphorbia* or *Polygonum* sp.; annual; small creeper; in the same place as the two preceding; also abundant.

(470) *Diadelphia*, perennial; leaves alternate, lanceolate, compoundly serrate; flower white, tinged at the tips of its wings with rose.

(471) *Arbutus tomentosa*[1]; a shrub 2 to 14 feet high; abundant on the mountains in rocky places, in light soils; fruit perfect, brownish-yellow; this is a plant which, if the seeds rise, will be a valuable addition; it will stand the English climate, being from a high altitude.

(472) *Bidens* (?) sp.; annual; flowers yellow; abundant on the sides of springs and rivulets in the mountains.

(473) *Polypodium* sp.; wet places on the mountains.

(474) *Oxalis* sp.; perennial; leaves large; flowers white; near mountain springs, abundant.

(475) *Monoecia*, (?) *Umbelliferae*; berry, 1, 2, 3, 5-seeded; seeds red, flat, oval; leaves lobed, serrated; stem prickly; bark white; on moist ground, in the shade of pines; 6 to 10 feet high; this I take to be what Dr. Fischer of Petersburg mentioned to Mr. Lindley, who spoke to me

---

[1] *Arctostaphylos tomentosa*, A. Gray, Syn. Fl. N. Am. ii. I. p. 28.

about a *Panax* or *Aralia* in North-West America, which grows 20 feet high; abundant.

(476) *Dianthus* sp.; perennial; in seed, on the hill.

(477) *Chelone* sp.; perennial; leaves opposite, ovate, dentate; petals short, ovate, acute, round, at the base dentate; flowers fine rose; also seeds of this fine plant on the mountains. S.

(478) *Pterospora andromedea* (Nuttall); this curious plant I found two years ago near Albany, State of New York, in the low valleys; this I found in the thick shady woods, among moss, in rich earth, at a very great height; in flower and seed.

(479) *Vaccinium* sp.; deciduous; leaves alternate, ovate, serrate; berries bright brown shining, pulpy and fine taste; very fine flower, large and abundant; young shoots red, old white; 4 to 8 feet high; on the summit of the mountains, plentiful. One of the three shrubs on the top of the hills, *Sorbus*, and a *Monoecious* one which to me is unknown.

(479*) *Monoecia*; leaves lanceolate, glabrous, sulphur-yellow below; in flower and half-ripe globular fruit; a shrub 4 to 10 feet high; on the summit of the hills; abundant.

(480) *Rhododendron* sp.; imperfect; 4 to 6 feet high; on the hills; plentiful under the shade of pines.

(481) *Rubus* sp.; small; probably *pistillatus* [1]; on the tops of the hills; rare.

(482) *Lycopodium denticulatum* [2]; in moist places on the hills; abundant.

(483) *Pyrola secunda*; in woods; plentiful. *Pyrola umbellata* [3] plentiful.

(484) *Pyrola* sp.; small; same places; plentiful.

(485) *Pyrola rotundifolia*; only one species of it could be found.

(486) *Monotropa* sp.; annual; in thick shady woods; only one specimen.

(487) *Pyrola* sp.; small; rare; in the like places.

(488) *Cornus canadensis*; in fruit.

(489) *Syngenesia* perennial; flowers yellow; rare; on the side of the river; only one plant.

(490) *Gramineae*, perennial; on the sands; plentiful.

(491) —— (?) small plant; bifoliate; fruit globular, bright azure colour; in shady upland woods; abundant. S.

(492) *Vaccinium* sp.; imperfect; different from any which I have seen; I regret that no seeds or flowers of it could be found.

(493) *Artemisia* sp.; perennial; on the sandy banks of river, with the following species, all in abundance:

(494) *Artemisia* sp.; small species.

(495) *Artemisia* sp.

---

[1] *Rubus arcticus* var. *grandiflorus*, S. Wats. Bibl. Ind. N. Am. Bot. p. 314.

[2] *Selaginella denticulata*, Baker, Handb. Fern. Allies, p. 37, but this is a native of the Mediterranean region, so Douglas must have made a slip of the pen. Did he intend to write *Lycopodium dendroideum* ?—which is referred to *L. obscurum*, Baker, *loc. cit.*, p. 24.

[3] *Chimaphila umbellata*, A. Gray, Syn. Fl. N. Am. ii. 1. p. 45.

(496) *Artemisia* sp.

(497) *Artemisia* sp. ; perennial ; tomentose underneath.

(498) *Artemisia* sp. ; very fine species. The strongest of any.

(498 [*bis*]) *Gentiana* sp. ; perennial ; leaves opposite, ovate-lanceolate ; a species ; S.

(499) *Sorbus* sp. ; a low shrub ; on the tops of the high hills at the Rapids.

Returned on 13th, and on my arrival I found that Mr. Scouler had taken possession of my house and learned that the vessel had returned from the north and would be despatched for England without delay. My time must now be taken up packing, arranging, and writing for a short time. From that time till October 3rd employed dividing my seeds and specimens and finishing transcribing my Journal. Wrote to-day to Jos. Sabine, Esq., to Dr. Hooker and Mr. Murray of Glasgow, to A. Menzies, Esq., and to my brother. I am to-morrow morning to leave here to see my boxes safely placed in the vessel.—Fort Vancouver, Columbia River, October 3rd, 1825, D. D.

October 4th to 22nd.—In consequence of receiving a wound on my left knee by falling on a rusty nail when employed packing the last of my boxes, I am unfortunately prevented from proceeding with my collection to the ship. In the meantime I wrote a note to Captain Hanwell, requesting he would have the goodness to place them in an airy situation, particularly the chest of seed, and, if possible, above the level of the water. I gave him also a note to Joseph Sabine, Esq. He kindly answered my note immediately on receipt and assured me that as far as was in his power he should feel glad in complying with my request, and that he should make a point of calling on Mr. Sabine on his arrival in England. On the 7th my leg became violently inflamed and a large abscess formed on the knee-joint which did not suppurate until the 16th. It is needless to observe that I was unable to continue my journeys or increase my collection during the time. This very unfortunate circumstance gave me much uneasiness, being my harvest of seed.

October 22nd to November 15th.—Learning the ship had been detained by contrary winds, and finding myself much recovered, I left for Vancouver in a small canoe, with four Indians, for the purpose of visiting my old shipmates on my way to Whitbey Harbour on the Cheecheeler River in latitude 48°, near to which were several plants that had not come under my notice or of which I had only obtained imperfect specimens and a supply of seed, among them *Helonias tenax*,[1] a very desirable plant for cultivation. I camped at the junction of the Multnomah River at sundown, having made only twenty miles, a strong wind setting in from the sea. On Sunday at daylight I embarked, but before leaving my encampment the canoe had to be fresh gummed. I had not proceeded many miles when it struck on the stump of a tree, which split it from one end to the other, and I had to paddle to shore without loss of time, the water rushing in fast. During the time my Indians

---

[1] *Xerophyllum asphodeloides* var., Baker, in Journ. Linn. Soc. xvii. p. 467.

were repairing it I occupied the office of cook. I made myself a small basin of tea and boiled some salmon for them. At ten o'clock I proceeded on my route. At eight the same evening I put ashore at the village of Oak Point to procure some food, where an Indian gave me a letter from Mr. Scouler, the surgeon of the ship, who informed me in his note they would not yet leave for a few days, and as the vessel was seen that same day in the bay I was desirous of writing to Mr. Sabine up to that date. After obtaining a few dried salmon and a wild goose, I went on four miles further down the river, where we took some supper, and continued my journey at ten o'clock, expecting to reach the sea before daylight, being only forty-three miles distant. At four in the morning of Monday a strong westerly breeze set in, which produced a very angry swell on the river and obliged me to cast along the shore. Indeed this was almost necessary under any circumstances, my canoe being so frail. I landed at the mouth of the river at 9 A.M., where I was informed by the Indians the ship had sailed an hour before. I felt no little disappointed, having my letter ready to hand on board. After breakfast my canoe-men lay down to sleep, and I took my gun and knapsack and walked along the bay in quest of some seeds. In the evening I returned to the lodge of Madsue or Thunder, one of the Chenook chiefs, where I found his brother Tha-a-muxi, or the Bear, a chief from Whitbey Harbour. As he was then going to his home he offered to accompany me, to which I agreed. On Tuesday the 25th Com Comly or Madsue ferried us across the Bay. Our canoe being small, and as I found his so much more commodious, I negotiated with him to lend it to me, which he did in Baker's Bay at the entrance of Knight's River. In the evening I gave the two chiefs a dram of well-watered rum, which pernicious liquor they will make any sacrifice to obtain. I found an exception in my guide Tha-a-muxi; he would not taste any. I inquired the reason, when he informed me with much merriment that some years since he got drunk and became very quarrel-some in his village ; so much so that the young men had to bind his hands and feet, which he looked on as a great affront. He has not tasted any since. In lieu of that I found him an expensive companion in the way of smoking : so greedily would he seize the pipe and inhale any particle of smoke in the lungs, that he would regularly five or six times a day fall down in a state of stupefaction. Smoking with them being the test of friendship, it is indispensable. I was of course compelled to join. I found my mode gave him as much sport as his gave me. He observed, " Oh, why do you throw away the food ? Look at me, I take it in my belly." On Wednesday I made a portage of four miles over Cape Disappointment, on the north point of the Columbia, to a small rivulet which falls into the ocean twelve miles to the north. I found it very laborious dragging my canoe through the wood, over rocks, stumps, and gullies. On reaching the bay I proceeded along the coast a few miles ; two hours before dusk a thick fog with a drizzly rain obliged me to encamp for the night under a shelving rock a little above the tide-mark, overshadowed by large pines. In the evening I felt my knee more troublesome and very stiff, arising from the exertion I had to make in transporting the canoe, or probably with the cold and rain.   After a comfortless night's rest I resumed my route at daylight, and as I was disappointed in not procuring salmon at the village I passed yesterday, it being abandoned, I had nothing to eat except a small cake of chocolate about two ounces, so with as much speed as possible I proceeded to Cape Foulweather, which I gained in the evening, forty miles being made that day along the coast.   As I had here a portage of sixteen miles to make, too great an undertaking to be done by so few, I sent two of the Indians with the canoe to the Columbia. As we had not this day had any food, they preferred leaving that same evening in hopes of obtaining some fish ; my guide and the other two remained with me.   They had not been away more than two hours when a most violent hurricane set in from the west, producing an agitation on the shoal water frightful in the extreme.   I was much alarmed for their safety, but learned on my return they happily put into a sheltered creek at the commencement of the storm and remained till it abated.   I was very hungry in the evening, and went out and gathered a few berries of *Arbutus Uva-ursi*,[1] being the only thing which could be found at the place. The wind was so high, with heavy rain, that scarcely any fire could be made.   Long ere day I was ready to leave Cape Foulweather, which name it merits ; being in a very bad state for walking.   All the wild-fowl had fled to the more sheltered parts ; not a bird of any description could be seen.   Being the two days without food, I resolved to endeavour to walk over the portage to the north side of Whitbey Harbour, where I was informed by my guide he expected a fishing party from his village to be.   On my arrival there at six o'clock, being on my legs from four in the morning, I hardly can give an idea of my afflicted state.   The storm continued with equal violence, which prevented the fishing party from leaving their village, which increased my misery.   While my guide and the Indians were collecting fuel, I made a small booth of pine branches, grass, and a few old mats ; my blanket being drenched in wet the preceding day, and no opportunity of drying it, the night raining heavily, I deemed it prudent not to lie down to sleep.   Therefore I spent this night at the fire.   On Saturday I found myself so much broken down and my knee so much worse that I did not stir out for the whole of the day.   A little before dusk the weather moderated, when I crawled out with my gun ; providentially I killed five ducks with one shot, which, as might be expected, were soon cooked ; one of the Indians ate a part raw, the other did not take time to pluck the feathers off but literally burned them to save time.   I was certainly very hungry, but as soon as I saw the birds fall my appetite fled ; it had brought such a change over me that I could hardly persuade myself I had been in want.   I made a basin of tea, on which, with a bit of duck, I made a good supper.   Very little sufficed me.   At midnight my guides arrived ; our fire had attracted their notice, and, as the chief was expected, they had come to wait on him.   I was asleep, and did not know until Tha-a-muxi roused me in the morning to embark.   He would not allow them to wake me or make any noise, having had no sleep last night and very little the two nights before.

[1] *Arctostaphylos Uva-ursi*, A. Gray, Syn. Fl. N. Am. ii. 1. p. 27.

Crossing the Bay I killed two large gulls : one white, bluish on the wings, with black feathers with points ; one of an equal size, of a mottled grey and a species of *Colymbus*. I had no opportunity at the time to preserve them. The Cheecheeler River is a large stream, nearly as large as the Thames, very rapid, with numerous cascades. I reached my guide's village a little before dusk, where I had every kindness and all the hospitality Indian courtesy could suggest, and made a stay of several days at his house, during which I was fortunate enough to procure a little seed of *Helonias*[1] ; being so late in the season, I was unable to procure as much as I should have done if earlier. Abundance of seeds of that splendid *Carex* (432) and *Lupinus* (426) the roots of which are gathered and roasted in the embers and eaten. This is the wild liquorice spoken of by Lewis and Clarke. There is in the root a large quantity of farinaceous substance, and it is a very nutritious wholesome food. I procured several other seeds not in my possession before. The *Lupinus* is called by them *Somŭchtan* ; seed-vessel one-celled, seed angular ; calyx none ; corolla five-petalled, lanceolate ; stamens five to nine ; style three-cleft ; flowers faint white ; leaves alternate, linear, sessile, revolute ; stem suffruticose, covered with chaff scales.

On the 7th November, I proceeded up the river in a canoe with my guide ; made halts at places such as presented anything different from what I had seen before. On the 11th, I reached sixty miles from the ocean, where I found my canoe too large to pass in many places by reason of cascades and shallowness of the water. I abandoned the idea of proceeding further in that direction. I therefore made my guide such presents as were adequate for the service and kindness I had experienced from him. Before leaving me he requested I would shave him, as he had pretensions to civilisation and aped with nicety European manners. I accordingly did so, and invited him to come at the New Year to see me, when I would give him a dram, a smoke, and shave him again. He told me before he left, to let all King George's chiefs know of him, when I spoke to them with paper. This river is a large stream nearly as large as the Thames, very rapid in many parts with cascades. The banks are rocky, steep, and covered with the like woods as are found on the Columbia. At the village where I put up I bargained with an Indian to carry my baggage on his horse to the Cow-a-lidsk River, one of the northern branches of the Columbia. I had some difficulty with this fellow in accomplishing my end ; he was the most mercenary rascal I have seen. I had to give him twenty shots of ammunition, two feet of tobacco, a few flints, and a little vermilion. The following day rained so heavily that I could not proceed. Early on the 13th I set out with my two Indians on foot, the horse carrying my little baggage with the owner. The distance may be about forty miles, and a very bad road owing to the late heavy rains ; much water was in the hollows, and the little creeks and rivulets so much swollen that my clothes were often off three times swimming across some of them. In the afternoon the rain fell in torrents, and as the country was

[1] No doubt refers to *H. tenax* which is *Xerophyllum asphodeloides* var., Baker, in Journ. Linn. Soc. xvii. p. 467.

an entire plain and no commodious place for camping, I was urged to exert myself to endeavour to reach the Cow-a-lidsk, which I accomplished at sundown, being greatly fatigued. My track was along the foot of Mount St. Helens of Vancouver, which lay a little to the north-east. At Schachan-away's or the chief of the Chenook tribe's house I learned he had just returned from a trading visit from other parts and had brought with him a bag of potatos, flour, a little molasses, and rum, of all of which I had a portion and a comfortable night's lodging. A small boat had been lent to him, which I considered fortunate, as it enabled me to proceed without delay.

On the 14th I had breakfast and was on my route before five o'clock in the morning. This is a large river, 150 to 200 yards wide in many parts, very deep and rapid, the current running more than six miles an hour in many parts. At mid-afternoon camped on a small woody island at its mouth, where it joins the Columbia, fifty miles from the ocean. Being high water when I put in, the boat grounded at ebbtide; not having strength enough to slide her along on the sand, I had to wait longer in the morning than I would have otherwise done. At six in the morning of the 15th I proceeded up the Columbia with a freshening breeze of wind, my blanket and cloak serving as sails. I arrived again at Fort Vancouver at half-past eleven at night, being absent twenty-five days, during which I experienced more fatigue and misery, and gleaned less than in any trip I have had in the country.

Collected the following plants:—

(500) *Monoecia*, annual; male flowers small; anther yellow; leaves alternate, ovate, entire, scabrous; capsule two-celled, two-seeded; seeds oblong; this plant I found early in spring, destitute of leaves, with perfect seeds on the withered stalks of last year; plentiful near villages and banks of rivers.

(501) *Cerastium* sp.; annual; stem prostrate; flowers small, white; on the sandy shores of rivers; plentiful.

(502) *Artemisia* sp.; perennial; tall, 3 to 4 feet high, erect and rarely branching; a fine species; banks of the Cow-a-lidsk River; rarer than most other species.

(503) *Oenothera* sp.; biennial (?); leaves lanceolate, sessile, some-what dentate; calyx pubescent; flowers very large, yellow; capsule large, four-grooved; different from (402); on the banks of rivers in sandy soil; rather rare.

(504) *Portulaca* sp.; annual; creeping; on sandy shores of rivers and all low soils.

(505) *Syngenesia* sp.; perennial; flowers yellow; abundant on the margins of lakes and rivers and in all low damp grounds.

(506) *Gramineae*, annual; a curious fine grass in seed; rare; on the banks of rivers.

(507) *Donia* (?) sp.; annual; abundant on all dry elevated gravelly plains; this I think I found a few starved specimens of early in the season at the Falls of the Columbia.

(508) *Mimulus* sp.; leaves opposite, sessile, linear-lanceolate, min-

utely dentate ; peduncle bibracteate ; flowers small, white ; tube yellow ; stem nearly round ; a beautiful, erect-growing plant, 4 to 6 inches high ; on the shores of the Columbia ; rare. A fine species. ? *Mimulus albus.* I have frequently sought in vain to find seeds of this valuable addition to that interesting genus.

(509) —— (?) Stem suffruticose covered with chaffy scales ; leaves alternate, sessile, linear, revolute ; flowers faint white ; calyx none ; corolla of 5 petals, lanceolate ; stamens 5–9 ; stigma 3-cleft ; capsule 1-celled, 1-seeded ; seeds angular ; on the sea sands.

(510) *Pinus* sp. ; a low tree rarely more than 20 to 40 feet high and seldom thicker than 10 or 18 inches in diameter ; on the barren grounds between Cape Foulweather and Whitbey Harbour ; plentiful.

November 16th till December 31st.—The rainy season being set in, with my infirm state, totally banished every thought from my mind of being able to do much more in the way of botany for a season. It is with serious regret that I am compelled to resign my labours, so much sooner than if that accident had not befallen me. At midday on the 18th the express, consisting of two boats and forty men, arrived from Hudson's Bay which they left on the 21st of July. They were observed at the distance of some miles, rapidly descending the stream. In this distant land, where there is only an annual post, they were by every person made welcome guests. I hastened to the landing-place, congratulating myself on the news from England. I learned with much regret there were no letters, parcel, or any article for me. I was given to understand they left Hudson's Bay before the arrival of the ship which left London the May before, so that if Mr. Sabine wrote to me, the letter will remain on the other side of the continent till next November. I was exceedingly disappointed. A Mr. McLeod, the person in charge of the party, told me he met Captain Franklin's party on Cumberland Lake on their way to Bear Lake, their winter residence, early in July ; their stay being only a few minutes, Dr. Richardson did not write to me.

I learned there was a Mr. Drummond attached to them as botanist ; he accompanied Mr. McLeod as far on his route as the foot of the Rocky Mountains, and is to pass the summer in the country towards Peace and Smoky Rivers. This I take to be Drummond of Forfar, from the description given of him. Mr. McLeod, whom I find an agreeable gentlemanly man, and from whom I have had much kindness, informs me that he spent the last five years at Fort Good Hope, on the McKenzie River, and of course possesses more knowledge of that country than any other person ; that (if the natives can be believed, with whom he was well acquainted and perfectly conversant in their language) there is a river nearly equal to the McKenzie to the westward of it, running parallel with it, and falls into the sea near Icy Cape. At the mouth of the said river there is a trading establishment on a woody island, where ships come in the summer. The people have large beards and are very wicked ; they have hanged several of the natives to the rigging and have ever since been in much disrepute. Much stress may with many be laid on this statement with all safety, as he showed me several articles of Russian manufacture, among which were small copper Russian coins, metal combs, &c.

But the most convincing proof that the difficulty of transportation by land or water is trifling, is large four, five, and six gallon malleable-iron pots of very coarse workmanship and very different from anything in the trade of the British Fur Company. He exchanged some of his for theirs. The sea to the west of the McKenzie River is said to be open after July, so that there is little difficulty in going either by water or land to Icy Cape. Mr. M. had the Indians assemble for the purpose of extending their territory in that direction, when he had to leave and proceed to Hudson's Bay. In him there is a great example of perseverance, visiting the Polar Sea, the Atlantic and Pacific Oceans, in the short space of eleven months. In the short spells of fair weather, when able, I crawled out, either with my gun collecting birds or other animals, or picking up Musci or any Cryptogamic plants in the woods. As yet (15th December) there has been scarcely any frost. When dry, weather generally very pleasant during the day; the nights invariably cold and damp. On the 24th December the rain fell in such torrents, without the least intermission, that my little hut of *Thuya* bark, which stood in rather a low situation, was completely inundated; 14 inches of water was in it. As my lodgings were not of the most comfortable sort, Mr. McLoughlin kindly invited me to a part of his house in a half-finished state. Therefore on Christmas Day all my little things were removed to my new dwelling. After the morning service was performed, they took an airing on horseback. I was prevented from joining them in their pleasant excursion by my troublesome knee.

1826.

Sunday, January 1st.—Commencing a year in such a far removed corner of the earth, where I am nearly destitute of civilised society, there is some scope for reflection. In 1824, I was on the Atlantic on my way to England; 1825, between the island of Juan Fernandez and the Galapagos in the Pacific; I am now here, and God only knows where I may be the next. In all probability, if a change does not take place, I will shortly be consigned to the tomb. I can die satisfied with myself. I never have given cause for remonstrance or pain to an individual on earth. I am in my twenty-seventh year.

January 2nd to March 1st.—As my Journal would be of little consequence containing a statement of the weather and so on, I do not transcribe it. The following birds came under my notice during the season : Silver-headed Eagle is abundant all over the country where there are rivers containing fish. They perch on dead trees and stumps over-hanging the water, and are invariably to be found near falls or cascades. It is a very wary bird and difficult to obtain; although powerful, it is overcome by several other species. Its voice is a weak whistle; called by the natives ' chuck, chuck,' which name they give from its own call. They build their nests on large trees on the banks of rivers, and seem to prefer a point, for on every conspicuous eminence or neck of land are nests. I have not seen the egg; has two, three, or four young at a time. They keep the nest on the branches of the trees for several weeks, and seldom leave the place where they were hatched any considerable distance. The colour of the first plumage is a brownish-black. The first spring they assume a mottled-grey, the head and tail of a lighter cast; the second, the head and tail become perfectly white, and the body black. When returning from the Grand Rapids last September, I observed one take a small sturgeon out of the water and come over my head. I lifted my gun and brought him down. The claws were so firmly clenched through the cartilaginous substance of the back, that he did not let go until I introduced a needle in the vertebræ of the neck. The sturgeon measured 15 inches long, weighing about 4 lb. Common Magpie; is a rare bird in the low country. The first I observed was in November. I am informed they are very abundant in the upper country at all seasons, whither they probably migrate in the summer. They appear not to differ specifically from the European species except in size, and the tail feathers of the male a brighter azure-purple. The American variety has the same trait in his character as the European of annoying horses that have any sores about them. I killed a pair, male and female, in January. Wood Partridge; is not a rare bird although they are by no means seen in such numbers as many of the tribes on the other side of the continent. They frequent dry gravelly soils on the outskirts of woods, among hazel bushes and other brushwood; are very shy. The breaking of a small twig is sufficient to raise them, and as they very generally are in the low thicket, it is only by a chance shot on wing they can be secured. I preserved two pair of this fine species, but a villainous rat mutilated one of the males so much that I had to throw it away, and I had no opportunity to replace it, and there is in the collection one male and two females. On the Multnomah there is one of very diminutive size not so large as the English thrush, with a long azure crest; the whole bird is pease-grey except the neck and head which are azure-purple. I have not seen it myself. I have furnished one of the hunters with a small quantity of fine shot to procure it for me. In the upper country there are two or three species of grouse, one, a very large bright grey bird as large as the smaller size of turkeys, it is very plentiful and easily procured; another of the same colour, about the size of the English black-cock, inhabits the same place, and is abundant. In addition to these there is a very beautiful species of pheasant, a little on this side of the Rocky Mountains, about the size of the common hen, of a blackish colour. It cackles exactly like a hen, it was never seen to fly, but runs with great speed. The large grouse I have never seen alive, only tail feathers, and parts of the skin forming war capes in the possession of Indians from the interior. Small Blue Jay; a very distinct bird from C[uculus] cristatus of Wilson. Indeed I do not remember any species that will agree with it in his work. If I recollect rightly, the common blue jay is rather a shy bird, and in the autumn is seen in great flocks, seldom near houses. This one is also very plentiful, but seldom more than thirty or forty together; it is very tame and visits dunghills of Indian villages, the same as the English robin. I preserved three, sex unknown. It is of a darker blue than the other, with a black crest. Large Brown Eagle; is a less plentiful bird than most species of the tribe, and not so shy as many. Is not so ferocious as the Silver-headed, of which he stands in great awe. I was able to procure only one of this species in February; the sex is unknown to me. Appears not to live on fish, as wild-fowl was in the stomach. Small Eagle; this appears to be a rare bird; only one pair have I seen, one of which I killed. It flies with amazing speed and pursues all other species although far inferior in strength and much smaller; the legs and feet of a bright light blue colour. What food it lives on I cannot say, as the stomach was empty. Large Horned Owl; seems not to be very abundant; I have not seen more than twelve or fourteen. One I killed by the light of the moon, after having watched for six successive evenings. It was not the species I was in quest of; I am given to understand there is a species here much larger than the Snowy Owl, of a yellowish-brown, but although I have been constantly in search of him, I am as yet unsuccessful. Two species of Crow, one large and one small; the small one is less abundant and more shy, generally seen on the sides of rivers; both frequent old encampments and live on carrion; one of each is in the collection; killed in February. Of the Hawk tribe I have seen four species; only two males of different species I have been able to kill, which are both preserved. I have seen one nearly a pure white about the size of a Sparrowhawk, a very active bird and continually on the chase after all the other species which all shun its society. I am sorry that this with the other two species I am unable to kill. In Wildfowl there appears to be little difference from those found in most parts of uninhabited America. The common Canadian Wild Goose, the Grey or Calling Goose, and the Small White Goose, are very plentiful in all lakes, low plains, and on the sandbanks of the Columbia. They migrate northward in April and return in October. A pair of each are in the collection, the male of the grey is a fine mottled bird. Of Swans there appear to be three species or varieties: one large, the Common Swan; one small, of the same colour (probably age may account for that); a third, equal in size to the largest, bluish-grey on the back, neck, and head, white on the belly. All three are seen together in flocks frequenting the same spots as the wild geese and migrate at the same time. The third species, of which there is a female in the collection, differs I think specifically from the others and is not so plentiful. In Ducks there are ten or twelve species; I have been able to kill only three. On the Columbia there is a species of Buzzard, the largest of all birds here, the Swan excepted. I killed only one of this very interesting bird, with buckshot, one of which passed through the head, which rendered it unfit for preserving; I regret it exceedingly, for I am confident it is not yet described. I have fired at them with every size of small shots at respectable distances without effect; seldom more than one or two are together. When they find a dead carcase or any putrid animal matter, so gluttonous are they that they will eat until they can hardly walk and have been killed with a stick. They are of the same colour as the common small buzzard found in Canada, one of which was sent home last October. Beak and legs bright yellow. The feathers of the wing are highly prized

by the Canadian voyageurs for making tobacco pipe-stems. I am shortly
to try to take them in a baited steel-trap. I learn from the hunters
that the Calumet Eagle is found two degrees south of the Columbia in
the winter season; two were killed by one of them.

The variety of species of quadrupeds is not I think so great as in many
other parts of America. The Elk (which the hunters say agrees with
the Biche of the other side) is plentiful in all the woody parts of the
country; is particularly abundant near the coast. Two species of Deer,
one called by the hunters *le Chevreuil* or *Jumping Deer*, is found in most
parts of the Columbia; it is of a light grey and white on the belly and
inside of the legs, with a very long tail, a foot to 15 inches; it is very
small, a little longer than the English Hart; the horns are 15 inches long,
much curved inwards, round and small, not exceeding one or two branches;
great numbers are killed on the Multnomah or Willamette River, one of
its southern branches. During my route in that quarter last July, with
a party of hunters, seventeen of this species were killed; only one young
one could be preserved at the time. The other, the Black-tailed Deer,
is not so abundant as the former, grows larger, darker grey on the back,
bluish-grey and yellow on the belly; the ears are remarkably large, resembling the ears of an ass, being much longer and a little broader than
those of the other species, brownish-black on the outside. The other's
ears are in colour the same as its body. The tail of the Black Deer is
shorter, not exceeding 8 inches to a foot. It is a much larger animal
than the *Chevreuil*. Both are found in the upland countries all through
that extensive range of mountains and plains in the Snake and Flathead
Indians' lands. The one sent home last October is a young *Chevreuil*
which I killed in August on the Multnomah where they are found in
abundance. As nothing would be more interesting than some knowledge
of this genus I have instructed several of the hunters in the mode of
preparing the skin and furnished them with a small portion of preserving
powder. I hope to get a pair at least of each. There are two species
of Rabbits, one of Hare, but neither have I seen alive; the Hare is only
found in the interior and is said to be very large. On the Multnomah
there is a most singular species of Fox, smaller than any other, except
the White Fox of the other side. The extreme length 33 to 40 inches.
The hair is remarkably short, very coarse, and, what is singular, brown
at the base, white in the middle, and black at the points, which gives
a light grey; the belly white; sides, and sides of the neck and forehead
of a light brown, ears and nose somewhat black with a grey beard, with
a longitudinal black stroke on the back from the shoulders to the point
of the tail. It differs from most of the genus in its propensity for climbing
trees, which he mounts with as much facility as a squirrel. The first
that came under my notice were two skins forming a robe for an Indian
child, belonging to the Calapooie tribe, inhabitants of the higher reaches
of the Multnomah. In August 1825, I was desirous of purchasing some
for the purpose of showing at the establishment, but too great value
was put on them. On a hunting excursion in February, in company
with Mr. McLeod, we raised a large Lynx, a small bull and terrier dog

immediately seized the lynx by the throat and killed it without much trouble. It was a full-grown female. This skin I preserved, being in a good state. On the banks of the rivers and lakes are some curious species of mice and rats. I have never been able to procure any more of that singular species with pouches, which troubled me so much during the autumn. The Ground Rat, or a species of *Arctomys*, the skin of which the Chenooks and other tribes of Indians near the coast make their robes, I have been unable to procure. They are plentiful in the upper parts of the Cow-a-lidsk River. When there in November my broken-down state prevented me from a day's hunting after them. I hope soon to have specimens of them. On the 20th February, Jean Baptist McKay, one of the hunters, returned to the establishment from his hunting excursion on the Multnomah; he brought me one cone of a species of *Pinus* which I requested of him last August when there. The first thing that gave me any knowledge of it, was the very large seeds and scales of the cones which I saw in an Indian's shot-pouch; after treating him to a smoke, which must be done before any questions are put, I enquired and found it grew a little to the south on the mountains. As McKay was going in that direction I asked him to bring me twelve cones, a few twigs, and a small bag of seeds and some of the gum. He informed me that the seed was all gone before he went in the autumn, and he only brought one cone to show me. The cone measures $16\frac{1}{2}$ inches long, and 10 inches round at the thickest part. The pine is found on the mountains two degrees south of the Columbia in the country occupied by the Umptqua tribe of Indians. He is in a few days to start for the same quarter, and as he has left orders with some of the Indians to collect seed cones and twigs, I am certain of obtaining it. It belongs to Pursh's second section. The trees, 20 to 50 feet in circumference and 170 to 220 feet high, are almost destitute of branches till within a short distance of the top which forms a perfect umbel. The trunk is remarkably straight, the wood is fine and yields a large quantity of resin. Growing trees that have been burned by the natives to save the trouble of collecting other fuel yield a sugar-like substance which they gather and use in seasoning, in the same manner as sugar in civilised society. At the end of summer the seeds are gathered, dried, and pounded and baked into a sort of cake, which is considered a great treat amongst them. As I have offered McKay a reasonable compensation to bring it to me, lest it may be impossible for me to visit that quarter myself, I am pretty certain of gaining more information of this very desirable tree. From the same person I obtained an elk snare—a netted purse, made of a desirable sort of grass, I think a different species of *Helonias*. This being also a plant worthy of inquiry, I am to receive a quantity of seed and grass of the plant; from the little I have seen I have no doubt it will prove *Helonias*. I have furnished him with a few paper bags, a little paper, and some fine shots to procure me some small birds which will be sent to me in the course of the summer. I was much indebted to Mr. McLoughlin for the trouble he took to explain to him what I wanted, and at the same time enjoined him to obtain them for me by some means or other.

From what I have seen in the country, and what I have been enabled to do, there is still much to be done ; after a careful consideration as to the propriety of remaining for a season longer than instructed to do, I have resolved not to leave for another year to come. From what I have seen myself of the upper country towards the head-waters of this river and the boundless track contiguous to the Rocky Mountains, I cannot in justice to the Society's interest do otherwise. However, I am uncertain how far I may be justified in so doing. If the motive which induces me to make this arrangement should not be approved of, I beg it may at least be pardoned. In doing so, two considerations presented themselves : first, as I am incurring very little expense ; second, being laid up an invalid last autumn during my seed harvest, I lost doubtless many interesting things which I would have otherwise had. Lest the former should be made any objection to, most cheerfully will I labour for this year without any remuneration, if I get only wherewith to purchase a little clothing. I could have crossed the continent this season to Montreal, and most gladly would I have availed myself of such an opportunity, but could never for a moment forget myself so far as to pass over unnoticed a country deserving the strictest research. Lest it should be impossible for me to cross in the spring of 1827, I shall without loss of time embrace the first opportunity of reaching London by sea after that period. That, however, I should be sorry to do, as so much time is lost, and as George Simpson, Esq., the Governor of the Western District, will be on the Columbia early in September, most likely I shall not lie under the disagreeable necessity of undertaking such a long voyage. My headquarters will be either at Wallawallah the lowest, Spokane the middle, or Kettle Falls the highest, establishment on the Columbia, and its branches as may appear most interesting ; I shall make such stays at each of the establishments as shall appear necessary, and as the extreme distance does not exceed more than 800 miles, frequent journeys can be made to and from each in the course of the season. I shall probably reach the Rocky Mountains in August. In all probability a vessel will soon arrive in the river, in which it is expected I will return, but as I shall not be on the coast till November, if then, I will pack the whole of my collection up to this time, to be transmitted in her to England ; also send my package of seeds which I intended to carry across myself to Hudson's Bay to the care of J. G. McTavish, and make extracts of my Journal, although at this season it can be of but little interest to the Society.

March 1st to 20th.—During this period I was employed packing the residue of my plants, birds, and other things in two boxes, to be sent to England by the first ship, which is soon expected to arrive in the river. Also making preparations for my journey to the interior. The whole of this time was very rainy—not so much as one dry day—thermometer 40° to 45°, minimum 28° to 34° ; winds westerly. By the kindness of Mr. McLoughlin I was enabled to pack up thirty quires of paper weighing 102 lb., which, with the whole of my other articles, is by far more than I could expect when the difficulty and labour of transportation is taken into consideration.

I packed in a small tin-box 197 papers of seeds, being a portion of all that have been collected, to be sent across the continent to Hudson's Bay. Dry plants are too bulky to be sent over such a tract of country, and indeed in my own opinion they will reach England by sea in a better state than any other mode of conveyance, as all the chests would require to be lined with tin or some such article to preserve them against water, things which cannot be had in this country. On the afternoon of Monday, the 20th, at four o'clock, I left Fort Vancouver in company with John McLeod, Esq., a gentleman going across to Hudson's Bay, and Mr. Francis Ermatinger, for the interior, with two boats and fourteen men. The day was very rainy, and we camped on a low piece of ground among poplars and willows, on the north side of the river, a few miles from the establishment, at dusk. The following morning at daylight we proceeded up the river. As there was a strong easterly wind against us, we only gained thirty-five miles; camped seven miles below the Grand Rapids; continued rain throughout. The following day made a portage over the Rapids and camped on a small stony island ten miles above them. Showery. At this season the Rapids are seen to advantage, the river being low. The scenery at this season is likewise grand beyond description; the high mountains in the neighbourhood, which are for the most part covered with pines of several species, some of which grow to an enormous size, are all loaded with snow; the rainbow from the vapour of the agitated water, which rushes with furious rapidity over shattered rocks and through deep caverns producing an agreeable although at the same time a somewhat melancholy echo through the thick wooded valley; the reflections from the snow on the mountains, together with the vivid green of the gigantic pines, form a contrast of rural grandeur that can scarcely be surpassed.

Thursday, 23rd.—Having a strong westerly wind, we proceeded on our journey at daylight under sail and reached the lower part of The Dalles at dusk 6 miles below the Great Falls; camped in a small cove, under a shelving rock. Fortunately the night was fine and pleasant, clear moonlight, which was the more agreeable as our tent could not be well pitched. As the natives had collected in greater numbers than we expected, and showed some disposition to be troublesome on not getting such a large present of tobacco as they wanted, we were under the necessity of watching the whole night. Having a few of my small wax-tapers still remaining, which I lay great value on, I wrote a short note to Mr. Murray at Glasgow, and laid in paper a few Musci which were collected the preceding day.

Friday, 24th.—After a tedious night, daybreak was to me particularly gratifying, as might be well guessed, being surrounded by at least 450 savages who, judging from appearances, were everything but amicable. As no one in the brigade could converse with them better than myself, little could be done by persuasion. However, finding two of the principal men who understood the Chenook tongue, with which I am partially acquainted, the little I had I found on this occasion very useful. We took a little breakfast on the rocks at The Dalles, four miles below the Great Falls, at seven o'clock. The day was very pleasant, with a clear sky. At five in the evening we made the portage over the Falls, where we found the Indians very troublesome. I learned from Mr. McLeod they had collected for the purpose of pillaging the boats, which we soon found to be the case. After they had the usual present of tobacco, they became desirous of our camping there for the night, no doubt expecting to effect their purpose. The first thing that was observed was their cunningly throwing water on the gun locks, and on the boats being ordered to be put in the water they refused to allow them. As Mr. McLeod was putting his hand on one of their shoulders to push him back, a fellow immediately pulled from his quiver a bow and a handful of arrows, and presented it at Mr. McLeod. As I was standing on the outside of the crowd I perceived it, and, as no time was to be lost, I instantly slipped the cover off my gun, which at the time was charged with buckshot, and presented it at him, and invited him to fire his arrow, and then I should certainly shoot him. Just at this time a chief of the Kyeuuse tribe and three of his young men, who are the terror of all other tribes west of the mountains and great friends of the white people, as they call them, stepped in and settled the matter in a few words without any further trouble. This very friendly Indian, who is the finest figure of a man that I have seen, standing nearly 6 feet 6 inches high, accompanied us a few miles up the river, where we camped for the night, after being remunerated by Mr. McLeod for his friendship—I being King George's Chief or the Grass Man, as I am called. I bored a hole in the only shilling I had, one which has been in my pocket since I left London, and, the septum of his nose being perforated, I suspended it to it with a brass wire. This was to him the great seal of friendship.

After smoking, he returned to the Indian village and promised that he would not allow us to be molested. Of course no sleep was had this night, and to keep myself awake I wrote a letter to Dr. Hooker. Heavy rain during the night. The following day, the 25th, at daylight we resumed our route ; sleet and rain, with a keen north wind. Being almost benumbed with cold, I preferred walking along the banks of the river, and, although my path in many places was very rugged, I camped forty miles above the Falls, much fatigued. During the night and the following morning I found my knee troublesome and very stiff.

26th to the 28th.—Clear, fine, warm weather, maximum heat 64°, minimum 50°. At three o'clock on the 28th arrived at Wallawallah establishment, where I was very friendly received by S. Black, Esq., the person in charge. The whole country from the Great Falls to this place is nearly destitute of timber. Dry gravelly and rocky soils, with extensive plains. The largest shrub to be seen on the plains is *Tigarea tridentata*,[1] which we invariably used as fuel in boiling our little kettle, also several very curious species of shrubby *Artemisia*, and other shrubs which to me were perfectly unknown ; and the whole herbage very different indeed from the vegetation on the coast. To the south-east, at the distance of

---

[1] *Purshia tridentata*, S. Wats. Bibl. Ind. N. Am. Bot. p. 309.

ninety miles, is seen a ridge of high snowy mountains which run from north-east to south-west and terminate near the ocean, about 300 miles south of the Columbia; this place will afford very likely most of the plants found in the chain of the Rocky Mountains. Mr. Black has very kindly made arrangements for my journey early in June, which will at least occupy 15 to 20 days. The course of the river from this place to the ocean is south-west, many places are very rapid, not more than 50 to 70 yards, which renders it very dangerous.

Early on Thursday morning the 30th proceeded on our route. As the whole country was an extensive plain, I walked on the north side the river till ten o'clock, when we stopped for breakfast, opposite to Lewis and Clarke's River, a stream of considerable magnitude, 100 to 150 yards wide at many parts and likewise rapid. Salmon, I learn, are caught in great abundance as far up as the Falls, and on some of its branches in the immediate vicinity of the Rocky Mountains, passing through a tract of country not less than 1500 miles. The day being fine and clear, I wrote a note to Mrs. Atkinson and resumed my walking in the cool of the evening, picking anything on my way. Camped on a low grassy island forty miles above the establishment. This part of the country is entirely destitute of timber; soil, light brown earth, sandy and gravelly on the banks of the rivers, and blown in some places into hills or mounds 50 to 60 feet high. In such places I observed several species of *Lupinus*, *Oenothera*, some very singular bulbous plants, with some shrubby species of *Artemisia* and other *Syngenesia*, and the beautiful *Tigarea*,[1] the vegetable of the greatest growth on the plains. Keen north wind.

Friday, 31st.—Country the same as yesterday; clear weather, fine sky in the evening.

April 1st.—Here it becomes mountainous, of white clay, with scarcely a vestige of herbage or verdure to be seen, except in the valleys. The river here is much broader than lower down and makes a great bend running due east-south, parallel with the coast, and south-east. Camped on the Priest Rapids at seven o'clock in the evening. The river here is narrow, divided into two channels, with a narrow dall[2] through the small rocky island in an oblique direction. The rocks are very rugged, of limestone, and this is considered one of the most dangerous parts of the whole river. During the time of making the portage of nine miles I wrote to my old companion, Mr. Scouler of Glasgow.

April 2nd to the 6th.—Without delay continued to pursue our journey, always a little before day, camping at dusk. Arrived at the establishments on the Okanagan River, one of the northern branches of the Columbia, at eight in the evening, where we were very cordially received by Mr. Annance, the person in charge. From the Priest Rapids to this place the banks are steep, high granite and sandstone rocks, and on the moist places and valleys a species of *Pinus* (*P. rubra* ?[3]), of immense size. On some parts the

---

[1] *Purshia tridentata*, S. Wats. Bibl. Ind. N. Am. Bot. p. 309.
[2] A word of French origin referring to the rocks through which the river flows, and causing rapids.—ED.
[3] *Picea rubra*, Kew Hand-List, Conif., ed. 2, p. 87.

snow lies 3 to 5 feet deep.   Here, the whole country being covered with snow, nothing could yet be done.

April 7th and 8th.—Out in search of the grouse of the plain, but unable to find any.   Saw only one small black partridge, the same as that sent home in 1825.

April 9th.—Early in the morning, in company with my companions, I resumed my route on horseback over a neck of land to meet the boats which had proceeded on the river, taking a circuitous bend round the mountain.   Our path was very rough, over broken stones, which were partly covered with snow and rendered the footing precarious.   On the height of land a very beautiful yellow lichen is found over the dead brush-wood, it affords a very durable beautiful yellow colour and is used by the natives in dyeing.   Snow 2 to 4 feet deep.   Met the boats at eleven o'clock of the same day, when, after taking breakfast, again went on by water conveyance, distance by land thirty miles.   Weather dry and pleasant with fine clear and beautiful sky in the evenings.   Camped twenty miles above where we joined the boats.

Monday and Tuesday, 10th, 11th.—Weather warm and very pleasant, maximum 69° minimum 55°.   Arrived at the junction of the Spokane River with the Columbia at sunset, where we found John Warren Dease, Esq., commandant in the interior, and a party of fourteen men, on their way to the Kettle Falls, ninety miles further up the Columbia.   I was by this gentleman received with extreme kindness and had every attention and kindness that could add to my comfort.   This is a brother of the gentleman now accompanying Captain Franklin on his two journeys to the Polar Sea.   Mr. Dease, to whom I was made known through the general notice sent by that agreeable gentleman Mr. McLoughlin, at Fort Vancouver, gave me greater hopes than ever of making a rich harvest. He will do all in his power to assist me.   This part of the Columbia is by far the most beautiful that I have seen : very varied, extensive plains, with groups of pine-trees, like an English lawn, with rising bluffs or little eminences covered with small brushwood, and rugged rocks covered with ferns, mosses, and lichens.

12th.—Employed drying part of my paper which was wet, and putting into dry the few small plants collected on the journey.   In the afternoon and evening wrote to Joseph Sabine, Esq., to Mr. Munro, and to my brother and copying my notes.   Pleasant weather.

13th, *Thursday.*—Busy copying the remainder of my notes, as Mr. McLeod is to leave early in the morning for his long trip to Hudson's Bay. I am particularly obliged also to this gentleman for his friendly attention. He has in the most careful manner taken my small tin box of seeds in his own private box and will hand it over to Mr. McTavish.   He has also taken my package of notes.   I met here Mr. John Work, with whom I was acquainted last year, and who sent me a few seeds from the interior last November, and furnished me with some valuable information about the plants and mountain sheep in this neighbourhood.   I find that the package of seeds marked 'Wormwood of the Voyageurs' is *Tigarea tridentata* ; that marked by myself as if with a query is a very fine species of *Crataegus* found only in the interior. I am in a few days to proceed to the Kettle Falls and will make such stays at the different posts as appear most advantageous to my views.

Collected the following plants during my journey up the river, and for the more easy method begin with

(1) *Salix* sp. ; male and female, a small scrubby tree, found near rivulets and moist ground in the mountain valleys.

(2) *Juniperus* sp. ; a low straggling tree, and on the bare dry rocks only a few feet high ; on the shore of rivers and summit of the hills ; is not seen within 500 miles of the ocean, and the nearer the mountains or in cool situations grows more luxuriantly ; probably it may prove *J. excelsa* [1] of Pursh.

(3) *Gramineae*, perennial ; on dry rocky soils, 6 inches to 8 high ; plentiful.

(4) *Gramineae*, perennial ; of the same genus as the preceding, only much smaller ; rarely more than 2 inches ; both are found together.

(5) *Gnaphalium* sp. ; perennial ; a small plant, scarcely ever more than an inch high ; found on all barren elevated ground ; this is not found near the coast.

(6) *Claytonia* sp. ; annual ; leaves linear ; 6 to 8 inches high ; flowers small, white ; very plentiful on rocky soils and very luxuriant near villages and old encampments.

Caryophylleae, perennial ; (47) of 1825 ; found abundant on all moist rocks throughout the country.

(7) *Ribes aureum* ; not seen lower down on the Columbia than the Great Falls, where almost on every rock and crevice its beautiful blossoms form a fine contrast with the shade of the dark blue granite ; on the margins of the little rivulets ; 6 to 10 feet high. It is seldom seen to bear fruit in rich soils and the little to be seen is small, sickly, and liable to be attacked by insects. On the bare channels of rivers where there is scarcely any soil, in rocky dry situations on the banks of rivers, it thrives less luxuriantly, but bears by far more abundantly and the fruit is of better quality—particularly so if there is a portion of lime in the soil or rock. The fruit is large, about the size of the common white currant, in thick close strings, with an exquisitely fine flavour. Nuttall observes that on the Missouri the black variety is the more abundant. Here it is rarely seen ; that most common is a deep amber or between that and sulphur-yellow. Perhaps it might be well to try it in a *very dry poor soil, with a little lime.*

(8) *Geranium* sp. ; annual ; leaves compoundly pinnate ; flowers small, azure-purple ; on the sides of rivers ; in sandy and gravelly soils ; plentiful.

(9) *Cruciferae*, annual ; flowers yellow ; an erect-growing plant, a foot to 18 inches high ; plentiful near villages and grows strong on rich ground ; this may probably agree with one found on the shores of the river last year at Fort Vancouver, which for the present from want of seed-vessels I do not know.

<hr>

[1] *Juniperus occidentalis*, Sargent, Silva N. Am. x. p. 87.

(10) *Cruciferae, Alyssum* (?) sp.; perennial; a low reclining plant; found on gravelly and sandy soils; very plentiful around the establishment at Wallawallah, 450 miles from the ocean.

(11) *Ranunculus* sp.; perennial; a fine low plant, scarcely exceeding 2 to 3 inches high, on the mountains near springs; in the valleys and low grounds, under the shade of pines, 6 to 8 inches; plentiful at the junction of the Spokane River with the Columbia.

(12) ——, annual; abundant near villages; a low stinking plant; flowers white.

(13) *Umbelliferae*, perennial; flowers yellow; a low plant, 6 inches to a foot high, on all gravelly soils.

(14) *Umbelliferae*; a small plant, about the same size as the preceding; flowers white, has no smell; the former like that of caraways; plentiful.

(15) *Umbelliferae*, perennial; a low spreading plant; petioles and peduncle white and rough; also plentiful; flowers white; anthers purple; a fine small plant.

(16) *Umbelliferae*, perennial; flowers yellow; 8 inches to a foot high; growing near rivulets among stones; this may later in the season grow to a much larger size, as it seems a strong plant.

(17) *Umbelliferae*, perennial; has some resemblance to No. 14; found in the same places; flowers white; anthers bright purple.

(18) *Diadelphia*? a small creeping plant, found only on the grounds around the Great Falls, where last year, in the end of June, I found it in seed. *Cheiranthus*, annual; found last year in seed; now in blossom on all sandy light soils.

(19) *Syngenesia*, annual; flowers white; leaves sessile, alternate, somewhat dentate; stem hirsute; a low plant, not more than 6 to 10 inches high; on rocky and gravelly hills; plentiful. *Arbutus Menziesii* in blossom; took from the same plants specimens in flower where I gathered perfect berries last September; a splendid shrub.

(20) *Ribes* sp.; in blossom; the same as *R.* 217 collected 1825; flowers faint white, fragrant; on all rocky soils, from the Great Falls to the junction of the Spokane River.

(21) *Lilium* sp.; *L. pudicum*[1] of Pursh; I find that No. 25 of 1825, which I mistook for it, will prove a still more interesting plant, perfectly distinct from the genus *Lilium*, the style being invariably three-cleft; the present, which agrees perfectly with his description, is found in great abundance from the Falls upwards, on all dry high soils; this highly ornamental plant I must try to preserve roots to send home; roots eaten, both raw and roasted on the embers, by the natives and are collected in July and dried in the sun for winter store.

(22) *Syngenesia*, perennial; not far removed from *Bellis*; a low plant; in all dry gravelly soils.

(23) *Diandria*, perennial; leaves orbicular-reniform, partially lobed, smooth on the under side, upper and peduncle minutely pubescent; flowers blue; a low plant, in shady woods among moss, in rich decayed

[1] *Fritillaria pudica*, Baker, in Journ. Linn. Soc. xiv. p. 267.

vegetable soil; I have only found it on the ground around the Grand Rapids.

(24) *Phlox* sp.; shrubby; leaves opposite, sessile, linear, under side smooth, upper somewhat pilous; calyx ciliated; this handsome plant may be what Pursh has given as *P. speciosa* [1]; most likely his description taken from an imperfect specimen; flowers white, tinged with pink or rose colour at expansion, and then assumes a deeper hue of a bluish-purple; a foot to 18 inches; growing in small patches or tufts on all dry light soils; very plentiful on the plains near the junction to Lewis and Clarke's River with the Columbia.

(25) *Cruciferae, Coronopus* [2] sp.; annual; flowers white; a foot to 18 inches high; plentiful along the banks of the Columbia from the Falls upwards.

(26) *Sisyrinchium* sp.; annual; 8 inches to a foot high; on the summit of the low hills; plentiful; a white-flowered variety is usually found with it.

(27) *Salix* sp.; male and female; a small tree, or more properly a shrub, frequenting mountain springs and rivulets.

(28) *Oxytropis Lambertii*; abundant in rocky, sandy situations; grows very luxuriant on limestone hills from the Great Falls upwards; at the mouth of the Spokane River a variety, or may prove a second species, is found of more slender growth; not so silky; the calyx longer and not swollen.

(29) *Trillium* sp.; perennial; flowers sessile; reddish-brown leaves, nearly orbicular; on the margins of mountain pools and rivers, among low brushwood; a fine species; plentiful at the junction of the Spokane River.

(30) *Cruciferae*, annual; flowers rose colour or pink; leaves sessile, linear, minutely dentate; whole plant scabrous; a plant a foot to $2\frac{1}{2}$ high, rarely branching; on the gravelly banks of rivers, under solitary pine-trees; plentiful.

(31) *Phlox* sp.; a small shrubby species, 6 to 10 inches high, in individual tufts or plants, on the shores of rivers and limestone hills; flowers white and faint rose colour; this very handsome plant, which comes near *P. setacea*,[3] is found in conjunction with the other species.

(32) *Astragalus* sp.; perennial; stemless; leaflets numerous, silky, flowers white; a low plant, 4 to 12 inches high; found among stones in the channel of the river; not so plentiful as *Oxytropis*.

Dodecatheon; probably the same as that collected in 1825, in the Upper Country; on the elevated plains, under the shade of pines, a fine white variety takes the place of this species; when both are seen together

---

* [1] *Phlox speciosa* of Pursh is rarely to be seen in perfect seed. The intense heat, which generally sets in in May, long ere they arrive at maturity, completely dries them up. I found with difficulty a few on the plains in June, and in July on the high mountains near the north and south branches of Lewis and Clarke's River, I found a larger quantity, being in a more temperate situation.—(*August 5th.*)

[2] *Senebiera*, Benth. and Hook. f. Gen. Pl. i. p. 87.

[3] *Phlox subulata*, A. Gray, Syn. Fl. N. Am. ii. I. p. 131.

---

* Footnote made by Douglas.—ED.

they impart a grace to the scanty verdure of American spring that can only be equalled by the European daisy or the common primrose.

(33) *Pulmonaria* sp. ; perennial ; leaves glaucous, sessile ; flowers blue ; a small plant, 4 to 8 inches high ; abundant on the plains and open woods.

From Sunday the 16th till Wednesday 19th, continued to make small trips in the country contiguous to the junction of the Spokane River ; and being so early in the spring, more for the purpose of viewing the soil and face of the country, with any bird or animal I might pick up.

On Wednesday at eleven o'clock, in company with Mr. Dease and his party of fourteen men and two boats, I left this place for the new intended establishment called Fort Colville, near the Kettle Falls, ninety miles further up the river. I am much indebted to this gentleman for the care he took in placing my paper and other articles in a safe place in the boat, and for the kindness he showed myself by inviting me to a seat in his own boat. The whole distance is very mountainous and rugged, the nearer the Rocky Mountains more so, and more thickly wooded, of three species of *Pinus.* One *P. resinosa* (?) ; *P.* not unlike *taxifolia* [1] found near the coast, but by no means attains such a size ; *P. Larix* [2] is found in abundance in the mountain valleys, much larger than any I have seen on the other side of the continent or even read of. I measured some 30 feet in circumference, and several that were blown down by the late storms 144 feet long ; wood clean and perfectly straight. On the plains and valleys there is a thick sward of grass, and interspersed among the detached rocks are several species of shrubs which at this season I cannot ascertain ; the greater part of the hills covered with snow. Warm during the day, keen frosts at night ; maximum 65°, minimum 28°. Camped on the channel of the river as no place could be found more suitable, having made twenty-seven miles ; river very rapid. At 4 A.M. on Thursday raised camp and proceeded on our route prosperously throughout, having gained forty miles ; the whole distance I walked on foot except being crossed three times, as I could not pass by steep rocks. On Friday at daylight continued our journey, and as we had gained a very rapid place where a portage had to be made, we took breakfast a little earlier than usual, being nine o'clock. This rapid, which nearly equals the Grand Rapids, 150 miles from the ocean, having no name, I called it Thomson's Rapids after the first person who ever descended the whole chain of the river from its source to the ocean. About ten o'clock it began to rain heavily and continued so until four in the afternoon. Arrived at the Falls at six in the evening, thoroughly drenched to the skin, and gladly walked over the portage three-quarters of a mile to a small circular plain surrounded by high hills on all sides, where the new establishment is to be. After our tents were pitched we had a comfortable supper of salmon-trout, and dried buffalo-meat served up to us by the man who started the day before us with a band of horses. Although my plants were covered with a double oilcloth, I found it inadequate to keep them dry, and, lest any should be injured, such as were wet I put in dry paper,

---

[1] *Pseudotsuga Douglasii*, Mast. in Journ. R. Hort. Soc. xiv. p. 245.
[2] *Larix occidentalis*, Mast. loc. cit. p. 218.

and placed under some pieces of bark near the fire for the night.   The Fall is a perpendicular pitch of 24 feet across the whole northern arm of the river, through which the principal part of the water runs; the south branch is dashed over shattered rocks, making a semicircular curve similar to that on the other channel.   They meet a few yards below the Falls, where the whole body of the water is again dashed over a small cascade of 8 feet high. The river is only 43 yards wide, where it leaves the cascade in snowy flakes or foam, for the space of 120 yards.   The whole face of the country is exceedingly picturesque, and in many places the scene is grand with every variety of appearance that can be called beautiful.   The channels are divided by a small oval rocky island with a few stunted trees on it.

Sunday, April 23rd.—Heavy rain and snow on the hills.   Changed the paper on my plants.

Monday, 24th.—Clear and cold.   Made a walk to one of the hills to the south ; observed the three species of *Pentstemon*, shrubby, and a great profusion of dried capsules.   One very strong, perennial, and a third some-what smaller.   In the course of the day killed three partridges of the same species as the only male sent to England last October.   It is a very rare bird near the coast ; in the whole course of the season only two birds, both males, came under my notice.   As I observed before, it is found on the hills among the rocks and is very seldom to be seen on the plains.   It is by no means shy ; when raised they fly a few yards and will either light on a rock or on a pine, where they can be easily killed.   At this season they are not seen in great numbers together.   The number of young or the colour of the eggs I have not yet learned ; as I had spare time, of course they were carefully preserved.

Tuesday and Wednesday, April 25th and 26th.—Clear and cold.   Made a short walk on the banks of the river ; collected the following plants :

(34), *Acer* sp. ;   flowers green, young shoots red ;   a small shrub, 4 to 14 feet high ; in low woods, near springs or moist ground.

(35) *Betula* sp. ;   a tree sometimes attaining the height of 20 to 30 feet and 9 inches to a foot diameter on the margin of mountain springs and rivulets, where it is found most abundantly ; seldom more than 6 or 8 feet high ; plentiful.

(36) *Trillium* sp.; flowers sessile, brownish-red ; leaves ovate-orbicular ; a fine species, inhabiting low moist peaty soils among *Betula* and *Salix* ; this I take to be *T. ovatum* of Pursh.

Observed *Berberis Aquifolium* in greater abundance than in the lower country ; plants much smaller, and a greater profusion of blossoms, growing among shattered rocks where there is scarcely any earth.

Collinsia of Nuttall is also plentiful, but likewise smaller.

Killed a male curlew and, having a spare hour in the evening, I preserved the skin.   They are plentiful on the dry plains, and on being raised perch on trees.   This bird seems to differ materially from the European, which is almost always seen in the vicinity of morasses or moist ground.   Egg, which I found in the nest, about the size of the common partridge, of a light brown with blue spots, the small end more pointed.

27*th*, 28*th*.— Showers of rain ; raw and cold ; snow on the hills. Continued my walks in quest of plants :

(37) *Erythronium grandiflorum* of Pursh ; this exceedingly beautiful plant came under my notice fifteen or sixteen days ago, but being not then in blossom I took it for *Fritillaria* ; abundant over all the undulating country, under the shade of solitary pines, in light dry soils ; it has a most splendid effect in conjunction with *Dodecatheon* and a small species of *Pulmonaria* ; omit not to procure seed and roots of such a desirable plant.

(38) *Claytonia lanceolata*[1] (Pursh) ; abundant in all open pine-woods, in light soils ; its small roots are eaten by the natives, both in a raw state and cooked by roasting in the embers ; when raw it is bitter and in every shape an insipid root ; this species is different from most others of the genus, as it is seldom seen luxuriant in rich soils such as near old villages or encampments.

Saturday 29*th*, *Sunday* 30*th*, *Monday*, *May* 1*st*.— Weather changeable, showery ; maximum 56°, minimum 28°. As usual, always on my walks and generally adding one or two to the numbers.

(39) *Pinus Larix*[2] ; abundant in the valleys ; I have already observed that its size is much greater than any on the other side of the continent.

(40) *Shepherdia* (Nuttall), *Hippophaë* (Pursh) ; leaves ovate-lanceolate, deciduous, covered with rusty scales on the under side and stellately silky on the upper ; stamens eight, situated between the calyx and the eight glands ; an upright slender shrub, 5 to 10 feet high, inhabiting the subalpine grounds under partially shady woods ; I am for the present unable to find the female.[3]

(41) *Diandria, Monogynia* ; perennial ; calyx tripartite ; capsule two-celled ; seeds unknown ; radical leaves cordate, serrate, glabrous, on long petioles, cauline sessile or somewhat amplexicaul, with small ovate bractea at each flower ; filaments red ; a plant 1 to 2½ feet high, flowering in a spike ; found abundantly on the plains among grass and several species of *Artemisia* and other *Syngenesia* not yet in blossom.

(42) *Claytonia alsinoides*[4] (Pursh) ; flowers small, white ; a small plant ; found on all rich soils near rivers.

Tuesday, *May* 2*nd*.— Rainy throughout the day, and a heavy fall of snow on the hills.

Wednesday and Thursday, May 3*rd and* 4*th*.— Made a walk on the southern banks of the river on the 3rd and in the opposite direction on the 4th. Observed and collected as follows : two species of *Rosa*, one strong upright bush with red wood, destitute of spines ; one small prickly on rocky places, neither of the leaves expanded. A luxuriant evergreen shrub, with ovate, serrate, and smooth leaves, probably *Clethra*, abundant

[1] *Claytonia caroliniana* var. *sessilifolia*, S. Wats. Bibl. N. Am. Bot. p. 117.
[2] *Larix occidentalis*, Mast. in Journ. R. Hort. Soc. xiv. p. 218.
* [3] *Shepherdia* of Nuttall. I have with much care, but in vain, endeavoured to find this curious species in perfect fruit, although the males and females are usually seen together.—(*August* 15*th*.)
[4] *Claytonia sibirica*, S. Wats. Bibl. Ind. N. Am. Bot. p. 119.

* Footnote made by Douglas.—ED.

in the woods, from the great profusion of mutilated capsules. This must be a plant worthy of strict attention. A small bulbous plant just peeping through the ground at the point of rocks to the south, six miles back from the river. To-day found abundance of *Hippophaë* or *Shepherdia* (female) which escaped my notice the preceding.

(42 [*bis*]) *Fragaria* sp. ; petioles and under side of the leaf pubescent ; upper smooth ; flowers large, white, fragrant ; abundant on all open dry and rocky soils ; this strongly resembles one found over all the lower country, but lest there should be even any difference as a variety of fruit, I lay some in.[1]

(43) *Viola* sp. ; leaves cordate, nearly entire, smooth ; flowers blue ; a small stemless plant, 2 to 4 inches high, plentiful on low open moist ground.

(44) *Viola* sp., perennial ; flowers white ; a variety of the preceding ; found in the same place.

(45) *Tetrandria, Monogynia* ; calyx quadripartite-ovate ; petals four, minute, narrower than the calyx, fringed and revolute, scarlet ; anthers yellow ; style half the length of the filaments ; leaves opposite, ovate-lanceolate, serrate, smooth ; this little evergreen shrub, I think, came under my notice on the mountains at the Grand Rapids last year in an imperfect state, which I took to be *Vaccinium* ; abundant in the mountain valleys, under the shade of *Betula* and *Populus*.[2]

(46) *Umbelliferae*, perennial ; flowers purple ; one of the strongest of the tribe found in the upper country ; the tender shoots are eaten by the natives ; very plentiful in all rocky situations and sandy soils near rivers.

(47) *Vaccinium* sp. ; flowers pink colour ; a small plant, 4 to 6 inches high, in open pine-woods.[3] Maximum heat 57° ; minimum 36° ; wind north.

Friday, 5th.—Made a walk to the hills on the opposite side of the river and found the following :

(48) *Viola* sp. ; perennial ; leaves ovate-lanceolate, smooth ; flowers yellow ; seldom exceeding 6 or 8 inches high ; abundant on open ground.

Saturday, May 6th.—Rainy ; made an excursion to the opposite side of the river. Killed one small female pheasant ; found usually on the plains, it is a shy bird and difficult to be had. The plumage is not so fine as I should have wished ; very thin on the breast, probably hatching its eggs. I am to preserve the skin if I am fortunate enough to prepare it nicely. Two species of *Prunus*, not yet in blossom, one with large ovate leaves, and one with smaller, narrower, and serrated leaves. A fine species of *Pentstemon* just coming to blossom, with lanceolate-

---

* [1] *Fragaria*, small, flavour good, the same as that found on the coast. In the interior not so abundant ; but, as might be expected, better from the dryness of the climate. —(I had this by report.)

* [2] No fruit could I find of this curious plant.—(*August 20th.*)

* [3] *Vaccinium :* fruit abundant, small, globular colour, light brown ; with an agreeable acid.—(*July 19th.*)

---

* Footnotes made by Douglas.—ED.

denticulate leaves; flowers appear to be red. Returned at dusk, drenched in wet. Collected the following:

(49) *Lupinus* sp.; perennial; leaflets five to nine, equally silky on both sides; seldom more than one to three leaves on the stem; flowers pale blue, upper lip in the centre white; a small plant, 10 to 16 inches high, in open woods and banks of rivers; plentiful; this has some resemblance to one collected last year.

(50) *Gymnocaulis uniflora* [1] of Nuttall; on moist shady places near the Kettle Falls, on the Columbia. In the same place I picked up *Pterospora andromedea*, measuring 4¼ feet high.

(51) *Senecio* sp., perennial; leaves lanceolate-serrate, cauline, somewhat amplexicaul; flowers white; a strong plant, 2 to 3 feet high; on the plains; abundant.

(52) *Atragene* sp.; flowers blue; margins of rivulets, among thick brushwood; plentiful.

(53) *Carex* sp.; female, a small plant, 4 to 6 inches high; on elevated grounds; plentiful.

Collected also species of *Musci*, one *Hypnum*, one *Polytrichum*, one *Bartramia*, and *Bryum*.

Sunday, *7th.*—Employed laying in what were collected yesterday, changing the paper of others, drying my clothes, &c. Pleasant. Maximum heat 61°; minimum 47°.

Monday, *May 8th.*—Weather cold and raw, cloudy, with showers of hail. Made a short walk to a neighbouring meadow, but found nothing different from before.

Tuesday, *9th.*—Left the Kettle Falls on the Columbia River at 10 A.M., with two horses, one carrying my provisions, which consisted of buffalo dried meat, a little tea and sugar, my blanket and paper; the other for carrying me over the bad places of the way. I had for my guides two young men, sons of a Mr. Jaques Raphael Finlay, a Canadian Sauteur, who is at present residing in the abandoned establishment of Spokane, in which direction I was going. Mr. Finlay being a man of extensive information as to the appearance of the country, animals, and so on, Mr. Dease kindly gave me a note to him requesting that he would show me anything that he deemed curious in the way of plants, &c. Took my departure in a northerly direction over the mountains, towards the Spokane River, distant about 100 or 110 miles. As my path, running along the skirt of the mountains, was at this season very bad and scarcely passable, the numerous mountain rivulets being so much swollen by the melting of the snow, and the meadows being overflowed or so soft that the horses could not pass, obliged me to make a more circuitous route than if it had been later in the season. Camped under a large pine on a rising bluff in the centre of a large plain at four o'clock, having made about twenty-seven miles. After hobbling the horses, took a walk around my camp, where I found

(54) *Pentandria monogynia*, perennial; calyx five-cleft, obtuse; corolla five, narrower than the calyx; stigma bifid; flowers white; half the inside of the corolla covered with strong yellow hairs and purple veins; peduncle

[1] *Aphyllon uniflorum*, Ind. Kew. fasc. ii. p. 1076.

solitary, one flower; leaves lanceolate, smooth, entire; in low wet ground; this is not far removed from *Menyanthes*.

(55) *Vicia* sp.; perennial; flowers large, purple; a very showy plant; plentiful under the shade of solitary pines and outskirts of the woods. Observed a species of *Ribes* very different from any yet in my possession, but being not in blossom left it for the present until my return. The scenery picturesque in the extreme.

Wednesday, 10*th.*—Rose at daylight and had my horses saddled, and being desirous of making the most of my time I took no breakfast further than a little dried meat and a drink of water, and proceeded on my journey at five o'clock. At twelve noon reached a small rapid river called Barrière River by my guides, which took up an hour in crossing. As there were no Indians near the place, we had to choose either making a raft or to swim. As the latter was the easier method, and all of us good water-men, we unsaddled the horses and drove them in. They all went over well except the last, which entangled itself by the hind legs among some brushwood and struggled much for a considerable time; fortunately the wood gave way and he reached the shore much better than I had any reason to expect. I made two trips on my back, one with my paper and pen, the other with my blanket and clothes—holding my property above water in my hands. My guides made three trips each with the saddles and provisions. Breadth of river 30 yards; heat of the water 40°. During this time there was a very heavy shower of hail, and being nearly half-an-hour in the water I was so much benumbed with cold that I was under the necessity of kindling a fire. After handing my guides a pipe of tobacco and making ourselves comfortably warm, I continued my route through a delightful undulating country till three o'clock, when I began to ascend a second ridge of mountains, which I crossed and camped at dusk at their base in a thick woody valley near a small stream of water on the dry rocky ground. The small beautiful species of *Phlox* which I found some time since on the Columbia gave the whole open places a fine effect. Flower changeable, white, blue, and fine pink colours.

(56) *Dioecia* (?); annual; calyx none; corolla three to four cleft; filaments minute; anthers sessile on the centre of each petal; no female flower; a most singular-jointed, leafless, succulent parasite on a small species of pine, belonging to the second section of Pursh; different from any yet in my possession, but I regret that it could not be found in perfection; like *Viscum* it does not survive the death of its supporter; abundant on all the slender twigs, and particularly so where the pines are in a light sandy dry soil; I took it first for a species of lichen, and passed it as such, thinking it already collected.

(57) belongs to the same genus and may prove only a variety of the same species, this being a much stronger plant of a darker brown colour, and is uniformly found on a different pine, which I take to be *Pinus resinosa*; not in blossom; in every other respect agrees with the preceding. In crossing the mountain just mentioned I killed seven black partridges, the same as the one preserved two weeks since. In the ovary of the females thirteen to seventeen eggs; found one nest with seven eggs about the size of common pigeon's egg, bright brownish-dun colour,

with large and small red spots. Blew one egg as a specimen and cooked the others; together with the partridges and buffalo meat I had a comfortable supper.

Thursday, 11*th*.—Heavy rain during the night, which roused me long ere day. In the twilight of the morn I raised camp, the weather assuming a more inviting appearance. At seven in the morning gained the summit of the last range of hills between the two rivers, and had one of the most sublime views I ever beheld As I approached the banks of the Spokane River the soil became more barren, except small belts of low ground in the valleys—near the mountain rills. Reached the old establishment at Spokane at eleven o'clock, where I was very kindly received by Mr. Finlay.   He regretted exceedingly that he had not a single morsel of food to offer me.   He and his family were living for the last six weeks on the roots of *Phalangium Quamash* [1] (called by the natives all over the country *Camass*) and a species of black lichen which grows on the pines. The manner of preparing it is as follows : It is gathered from the trees and all the small dead twigs taken out of it, and then immersed in water until it becomes perfectly flexible, and afterwards placed on a heap of heated stones with a layer of grass or leaves between it and the stones to prevent its being burned ; then covered over with the same material and a thin covering of earth and allowed to remain until cooked, which generally takes a night.   Then before it cools it is compressed into thin cakes and is fit for use. This process is similar to the preparing of *Phalangium*. A cake of this sort and a small basin of water was all he had to offer me.   By the kindness of Mr. Dease, I had ample provisions for fourteen days, with a good stock of game in the saddle-bags which I killed on my way, and this enabled me to share the half of my stock with him ; such fare as I had, although very palatable, cannot be considered fine living, but was to him the best meal he had enjoyed for some time. As the principal object of my journey was to get my firelock arranged by him, being the only person within the space of eight hundred miles who could do it, and being an item of the utmost consequence to have done soon, I lost no time in informing him of my request. Unfortunately he did not speak the English language, and my very partial knowledge of French prevented me from obtaining information which I should have acquired.   In the afternoon I made a walk up the river and returned at dusk, when I found he had obligingly put my gun in good order, for which I presented him with a pound of tobacco, being the only thing I had to give.

Friday, 12*th*.—Immediately after breakfast, at six in the morning, in company with one of his sons, I made a short journey to the neighbouring hills. Collected as follows :

(58) *Syngenesia*, perennial ; leaves opposite, cordate, serrate, pubescent on the under side as also the peduncle and stem, upper side scabrous ; flowers yellow ; in open woods ; a handsome plant, a foot to 18 inches high.

(59) *Xylosteon ciliatum*, var. *album*, [2] of Pursh (?) ; in rocky dry soils ; appears not to be plentiful ; a low slender shrub ; faint white.

---

[1] *Camassia esculenta*, Baker, in Journ. Linn. Soc. xiii. p. 257.
[2] *Symphoricarpos racemosus*, fide Nutt.; see Torr. and Gray, Fl. N. Am. ii. p. 9.

(60) *Cynoglossum* (?) sp. ; perennial ; leaves alternate, pinnate ; stem leaves and calyx woolly ; flowers capitate, white ; anther yellow ; a fine little plant, 6 inches to a foot high ; on dry sandy hills ; plentiful.

(61) *Oxytropis* or *Astragalus missouriensis*, var. *alba*, of Nuttall (?) ; flowers faint white and yellow, carmine tinged with purple ; this very interesting small plant is found in great abundance on all dry gravelly soils and among rocks.

(62) *Ribes* sp. ; leaves three-lobed, smooth, serrate, covered on the back with small golden glands ; flowers erect, in a long spike, white, fragrant ; wood white ; an exceedingly handsome strong-growing plant, 4 to 10 feet high, with an aromatic scent like the common black currant ; in rocky places ; plentiful ; make every effort to procure seeds of this highly ornamental plant ; as I think it new. In conjunction with the above, *R. aureum* ; *R.* sp., No. 20 ; *R.* with small green flowers all collected ; also *Alyssum* sp., *Collomia linearis* of Nuttall, apparently a small variety of that collected last year. As they were finer specimens than any in my possession, I could not resist laying a few of each in, having a spare sheet of paper. Mr. Finlay tells me that *R. aureum* in that neighbourhood produces very fine large yellow fruit ; that he never saw it black or brown. The white one, red small solitary berries of a pleasant taste. This agrees with my note last year,[1] as I took specimens in blossom this spring from the same bush that I gathered fruit last year at the Great Falls on the Columbia ; the green flowering one, a small black gooseberry. No. 62 he never saw before. I requested he should dry me seeds of all the sorts as well as a fine species of *Allium*, the roots of which were brought from forty miles above, on the banks of the Spokane River. Root as large as a nut, very pleasant and mild.

Saturday, 13th.—As I thought of bending my steps again towards the Columbia, Mr. Finlay offered that one of his sons should escort me, which I accepted. Before parting with him I made inquiry about a sort of sheep found in this neighbourhood, about the same size as that described by Lewis and Clarke, but instead of wool it has short thick coarse hair of a brownish-grey, from which it gets the name of Mouton Gris of the voyageurs.[2] The horns of the male, of a dirty-white colour, form a volute,

---

* [1] *Ribes* ; peduncle very long, erect ; fruit prolific, large, round, smooth, black and glossy, juicy ; taste good, exactly like the common black currant, but more acid. This may (should it produce fruit in England) prove worthy of cultivation as a fruit in addition to its showy beautiful fragrant blossoms—(*July 20th*, 1826).

* [2] Mouton Gris of the voyageurs, or grey sheep. Although I have made every possible exertion in my power to put myself in possession of this interesting animal, I am still unable to procure any—and can say little of it, having never had an opportunity of seeing it alive.

It is by all the hunters said to agree with that discovered by Lewis and Clarke on the head-waters of the Missouri, near their pass in the Rocky Mountains, which is certainly erroneous. As I had an opportunity of seeing the whole animals collected during their expedition (now in Philadelphia Museum), I can without hesitation say that the Mouton Gris of the voyageurs is a perfectly distinct and still more interesting animal. In size and shape like the common sheep, sometimes 150 to 240 lb. avoirdupois, destitute of wool, hair short, coarse and thick, of a light brownish-grey ; flank, back part of the thigh, and on the hips of a faint dirty-white. Horns of the male very large, curved backwards, forming a perfect volute, the point inclining inwards ;

---

* Footnotes made by Douglas.—ED.

sometimes weighing 18 to 24 lb. The horns of the female are bent backwards and curved outwards at the point, about 10 inches or a foot long. The flesh is fine. I offered a small compensation to the sons to procure me skins of male and female, at the same time showing them what way they should be prepared. He assured me that in all probability he would be able to find them about August, as he was going on a hunting trip to the higher grounds contiguous to the Rocky Mountains. Close to the old establishment an Indian burying-ground is to be seen, certainly one of the most curious spectacles I have seen in the country. All the property of the dead, consisting of war implements, garments, gambling articles, in fact everything. Even the favourite horse of the departed is shot with his bow and arrow, and his skin with the hoofs and skull hung over the remains of deceased owner. On trees around the ground small bundles are to be seen tied up in the same manner as they tie provisions when travelling. I could not learn if this was as food or as a sacrifice to some of their deities. The body is placed in the grave in a sitting position, with the knees touching the chin and the arms folded across the chest. It is very difficult to get any information on this point, for nothing seems to hurt their feelings more than even mentioning the name of a departed friend. Left Spokane at 8 A.M. with one guide; went on the same track that I came. As I saw nothing different on my way from what I had previously observed, my stoppages were fewer; gained Barrière River a little before dusk, which I crossed in the same way as I did a few days before; heavy rain all the afternoon. Camped a few yards from the banks of the river, in the shade of some pines.

Sunday, 14th.—Very rainy during the whole night; although tolerably well sheltered and had a large fire to sit at, yet I felt cold, my blanket and

---

frequent instances have been known of the horns so overgrown that they prevented the animal from feeding, the skulls being found with the horns grown in the jaws, which may naturally lead one to suppose the animal died of starvation. Horns of a dirty yellowish-white colour; tail short, hairy like a goat. The horns of the female are small in comparison to the male, 5 to 10 inches long, slightly bent backwards and inclining outwards at the points; differs in no way from the male in colour; said to have one, sometimes two, kids at a time. The flesh said to be fine and equal to the domesticated sheep, the lean of a more brown colour, beautifully marbled with fat of a delicate white colour. Inhabits the highest mountains, and seldom seen abundantly except on those whose summits are covered with eternal snow. On the portage between the waters of the Athabasca and Columbia Rivers, in the Rocky Mountains, and in the country between Peace and Smoky Rivers, they are seen in large herds; on the Columbia, near the Lakes, and on the mountains of McGillivray's and Flathead Rivers, and that tract of country between the Columbia and the higher parts of Fraser's River to the north, in fewer number, and in the south towards the waters of Rio del Norte near the Spanish possessions. Their voice is precisely the same as the sheep, *Ma-aa-aa.* They are shy, and when discovered immediately ascend the mountains and endeavour to place themselves on the most inaccessible places of the rocks, and will, on gaining such situations, composedly stand and steadfastly look down on you, where by an expert rifleman they are oftentimes brought from the height of several hundred feet. Lewis and Clarke's sheep is (if I recollect aright) of a dirty-white with hair and wool of coarse texture, and stands under the name of *Ovis communis* beside the antelope or *Mouton blanche* of the voyageurs, which is seen on the highest peaks of the mountains.

To Mr. Sabine, *May 20th.*

August 27th: Purchased a pair of male horns of the above from an Indian, for which I gave him three charges of ammunition and a few crumbs of tobacco. They are small, and in my opinion the animal could not be more than two and a half or three years old—as some of the larger and older ones weigh 18 to 24 lb.—On the Great Falls of the Columbia.—D. D.

clothing being wet. As I could not sleep I rose at two o'clock and with some difficulty dried my blanket and a spare shirt, in which I placed my paper containing the few plants collected. Afterwards boiled my small kettle and made some tea. Felt a severe pain between my shoulders, which I thought might arise from the cold in swimming and lying in wet clothes. Therefore, as I had no medicine to take, I set out a little before 4 A.M. on foot, driving the horses before me, thinking that perspiring would remove it, which it partly did. On arriving at my first night's encampment at midday, I stopped a short time to look for the currant in perfection which I saw on my way out just coming to blossom, and fortunately found it in a fine state.

(63) *Ribes* sp. ; leaves equally three-lobed, serrate, both sides covered with strong viscous, glandulous hairs; peduncle and petiole equally so; corolla faint green on the outside, yellow inside, in a thick close raceme ; wood white, resembling that of white raspberry ; leaves very fragrant, exactly like the scent of *Pelargonium odoratissimum* ; this desirable plant will no doubt prove *R. viscosissimum* of Pursh [1]; all the seed of this must be taken great care of, as well as the other species ; in rocky places ; I have as yet only seen it in this one place. Reached the Kettle Falls on the Columbia in the evening, and although I have not obtained a great number of plants, yet with the repairing of my gun and the few plants collected, I must say I felt satisfied.

Monday and Tuesday, 15*th*, 16*th*.—On the morning after my arrival the pain between my shoulders returned and became so bad, as also a severe headache, that I was under the necessity of keeping my bed. As I was feverish and likely to become worse, I took some salts and then a few grains of Dover's powder, which relieved me greatly. I regretted it the less as the weather was so rainy and boisterous, with thunder, that I could have done but little good although in good health.

Wednesday, 17*th*.—As the weather was still unsteady, with showers, I was afraid to venture out lest I should have a relapse. Therefore I employed myself turning and changing the paper of what were latest collected. Cold and raw, wind northerly.

Thursday, 18*th*.—Collected the following near my residence at Kettle Falls :

(63 [*bis*]) *Juniperus* sp. (?) ; leaves opposite, glaucous on the upper side ; this fine plant, of which I could only find the female, is seen on rocky grounds near springs ; 3 to 4 feet high, somewhat procumbent ; however I am inclined to think it attains a larger size ; rare.

(64) *Geum* sp.; perennial ; flowers faint yellow ; leaves radical, pinnate, ciliate ; stem and peduncle hairy ; a fine small plant, 12 to 30 inches high ; plentiful on the plains, on the banks of rivers, and under the shade of pines.

* [1] *Ribes viscosissimum* of Pursh; fruit long, oblong, hairy ; hairs viscid; skin thin, dry, and divested of fleshy substance ; seeds very prolific, small; taste musty and very disagreeable, causes vomiting. Will never be of any use as a fruit, but a great addition to the ornamental garden.—(*July 27th*, 1826.)

---

* Footnote made by Douglas.—ED.

(65) *Arabis* sp.; annual; on moist mountain rocks and banks of rivers; flowers white.

(66) *Ranunculus* sp.; perennial; flowers yellow; abundant in all low moist meadows.

(67) *Juncus* sp.; perennial; in the like places; abundant.

Friday, 19th.—Made a journey at daylight to a neighbouring hill, and returned at dusk. In the course of the day found *Ribes* (63) in great abundance and very luxuriant, 6 to 10 feet high, spreading and somewhat reclining, in rocky situations. Also abundance of *Ribes* (62) at the foot of the hill in low damp ground among *Populus* and *Alnus* with *Betula*, stronger than what I saw the other day.

(64 [*bis*]) *Ribes* section *Grossularia*; leaves equally five-lobed, serrate, smooth; petiole slightly hairy; peduncle long, slender, more hairy than the petiole; flowers small, spreading, yellow, tinged with red in the inside; germ thickly beset with short glandulous hairs; wood whitish, covered with small straight spines; abundant; this resembles one which I found last year in fruit on the summit of the hills at the Grand Rapids; the only specimen I had unfortunately was lost.

(65 [*bis*]) *Viola* sp.; perennial; leaves smooth, cordate, round, and somewhat reniform; flowers small, yellow, with dark purple veins; under the shade of pines, in rich vegetable soil.

(66 [*bis*]) *Prunus* sp.; flowers white, in a cluster; leaves lanceolate, obtuse, smooth, minutely serrate; fragrant, with a bitter astringent taste [1]; abundant on rocky grounds, near river and mountain springs; 6 to 12 feet high.

(67 [*bis*]) *Saxifraga nivalis*; on the summit of the high mountains; plentiful, in conjunction with the following:

(68) *Saxifraga* sp.; perennial; small, 4 to 8 inches high; flowers white.

Saturday, 20th.—Clear, warm, and pleasant, wind southerly. Early in the morning made a journey on the south side of the river up as far as twenty miles, and returned at dusk; country the same as at Kettle Falls. Collected the following:

(69) *Pentandria, Monogynia,* allied to *Myosotis*; perennial; anthers short, surrounding the orifice; flowers blue; radical leaves linear-oblong; cauline linear, alternate, sessile, or somewhat amplexicaul; leaves and stem hirsute; a foot to 2½ high, on dry gravelly soils.

(70) *Anemone* sp.; perennial; radical leaves digitate; upper surface smooth, under (as well as the stalk) covered with a soft thick pubescence; petiole of the floral leaves, short; one to three flowered, white; abundant in the same places as the preceding plant.

Sunday, May 21st.—Being a little fatigued with walking so much the preceding days, I was unable to go out; I tied up one bundle of dry plants and made a small box to place them in against damp and insects. Cloudy in the morning, warm afternoon.

---

[1] Prunus; fruit small, scarlet, astringent, and is in every point an insipid fruit.—(*July 15th.*)

---

* Footnote made by Douglas.—ED.

Monday, 22nd.—Crossed the Columbia to Dease River, one of its northern branches, which I ascended ten miles in a canoe with two Indians and a Mr. Kittson who was going to examine it, being never entered before by Europeans. I walked along the banks while they came up with the canoe ; found the stream so very rapid that to get higher was fruitless, so we just left ourselves time to return in the evening. Supposed to take its waters from the Rocky Mountains. Found in blossom what I supposed to be *Pentstemon* some weeks ago.

(71) *Didynamia, Angiospermia ;* calyx five-leaved, bibracteate ; corolla bilabiate, lower [lip] plaited and bearded ; bark slightly fringed ; anthers lanuginous ; sterile filament shorter ; four to seven hairs on the upper-side near the middle ; seeds angular ; stem fruticose, reclining ; leaves opposite, sessile, lanceolate, smooth, slightly denticulate, floral, broader and more obtuse, somewhat pubescent ; upper part of the stem, peduncle, and calyx equally so ; this plant common in some points to *Pentstemon frutescens ;* if it is, the anthers and sterile filament must have been overlooked ; seeds being not membraneously margined makes it differ from that genus ; it may be an anomaly ; abundant on rocky and gravelly soils on the upper parts of the Columbia and its branches.

(72) *Polemonium* (?), perennial ; calyx five-cleft ; corolla partially rotate ; tube short, closed at the base with valves ; stigma trifid, cap not in perfection ; leaves pinnate, slightly pubescent ; calyx somewhat glutinous ; flowers bright blue ; a delicate little plant, 10 to 20 inches high ; on rocky situations and light soils ; in great abundance on the banks of this river, the only place it has come under my notice.

Tuesday, May 23rd.—As usual went out in search of any plant that might appear different from that already collected ; found the following :

(73) *Umbelliferae,* perennial ; root large, fusiform, tastes somewhat like a parsnip ; radical leaves entire (sinuated when in rich damp soils), lanceo-late, smooth, three-nerved, floral pinnate, amplexicaul ; flowers sulphur-yellow colour ; plentiful in all low swampy grounds ; the roots are gathered by the natives and boiled or roasted as an article of food (taste insipid) ; called by them Missouii.[1]

(74) *Umbelliferae,* perennial ; leaves ternate, cordate, serrate, smooth ; flowers yellow ; like the former, found on low wet ground and margin of mountain springs.

(75) *Heuchera* sp. ; perennial ; scape pilous, leaves lobed, acute, upper side smooth, under nerved and slightly pubescent, dentate, and mucronate ; flowers white ; this does not differ much from *H. caulescens*[2] of Pursh, may probably prove to be it ; abundant on the subalpine hills in rocky places, usually in partially shady spots ; a fine plant. In the course of the day found abundantly *Collomia linearis* of Nuttall on the channel of the Columbia and its creeks, in sandy soils ; less luxuriant than in the lower country. Observed a large plant belonging to *Malvaceae,* in open woods, not yet in blossom, also a small species of *Artemisia,* in rocky places, with pinnate, revolute leaves, woolly

---

[1] This word is almost illegible and may be ' Missouri.'—ED.
[2] *Heuchera villosa,* S. Wats. Bibl. Ind. N. Am. Bot. p. 326.

underneath, and a plant in wet ground which from the dry capsules I take to be a fine species of *Helonias* ; leaves entire, smooth ; dies down to the ground in the winter. Pleasant, with showers of rain throughout the day.

Wednesday, 24th.—Showery, with a south-west wind. After turning some plants and taking others out of the presser, went out in quest of more. Returned in the evening with the following :

(76) *Brodiaea* (?) sp. ; perennial ; corolla six-partite, irregular, three partly covering three, the inner undulate, shorter than the outer ; stamens same length as the pistil, outer a little shorter ; flowers fine blue ; root a round, solid bulb ; two to three long grassy leaves ; this certainly differs from *B.* collected last year ; may on examination prove a different genus ; abundant in light sandy soils, under the shade of solitary pines.

(77) *Epilobium* sp. ; perennial ; leaves lanceolate, alternate, smooth, entire ; flowers fine red ; calyx of a darker hue ; a beautiful little plant, 6 inches to a foot high ; at the junction of Barrière River with the Columbia.

(78) *Crataegus* sp. ; the only one of the genus I have seen in the interior ; on the edges of rivers and creeks ; a low spreading shrub.

(79) *Allium* sp. ; perennial ; flowers purple ; on the banks of rivers ; this plant is the only vegetable that I have to use in my food ; I get it generally stewed down in a little dried buffalo-meat or game.

(80) *Populus* sp. ; a large tree, plentiful on the banks of rivers.

(81) *Populus* sp. ; a tree 20 to 40 feet high, 6 inches to 18 in diameter ; bark smooth and white ; margin of mountain lakes and springs.

(82) *Ribes* (section *Grossularia*) ; flowers small green ; leaves round, lobed, serrate, smooth ; plentiful in rocky situations.

Thursday, 25th.—Warm and pleasant. In the course of my walk along the river, at the Falls, killed two specimens of swallows, but all too much destroyed by the shot to preserve. The male of the one, white on the belly and side of the head ; back and upper part of the head green ; rump bright purple ; point of the tail and wings light brown. The female differs little from the male ; head brownish-green ; rump of a fainter hue ; a small bird, not larger than the English wren. The second, sex unknown ; belly light brown or amber ; back a dark glossy-purple ; point of the wings and tail brown. The tail of this one longer than the body, at the extremity only two feathers ; a little larger than the former. Killed a fine large male black partridge, which being in good order I shall skin in the morning to complete two pair of that fine species. Found the following plants :

(83) *Pentandria, Monogynia* (allied to *Echium*) ; calyx five leaves, linear ; corolla five-cleft, obtuse ; filaments longer than the corolla, hairy ; style bifid, purple-blue ; leaves alternate, linear, revolute, hirsute ; stem somewhat less hirsute than the leaves ; annual ; a fine plant, 8 inches to 18 high, rarely branching ; on light hilly soils ; abundant.

(84) *Collomia* sp. ; annual ; stem round, smooth ; lower leaves opposite, sessile, linear ; floral leaves and upper part of the stem slightly

pubescent; petals toothed; a more straggling plant than *linearis*; less viscid; flowers larger, faint pink colour; found in conjunction with it in all sandy soils on the banks of rivers.

(85) *Myosotis* sp.; perennial; leaves alternate, sessile, linear, hirsute; stem equally so; only one plant I could find for the present; flowers small, white.

Friday, 26th.—At daylight went on a trip to the hills south of my residence. Very warm, thermometer 86° at noon. I felt so much fatigued and overpowered with the heat that I sat down under the shade of *Thuya occidentalis* in a valley near a small spring, where I fell asleep and did not wake till 4 P.M. As I was then twenty miles from home, I would have taken up my lodgings for the night; but as Mr. Dease would feel anxious about me I hastened home, which took me six hours, the road being mountainous and rugged. Mr. D. had just given orders to send some Indians in search of me in the morning, thinking something had befallen me, when I made my appearance, and on informing him of my delay he laughed heartily. Killed a female curlew and a small male pheasant; the latter too much destroyed for preserving.

(86) *Pentandria, Monogynia*; calyx five-partite, acute; corolla five-partite, obtuse; leaves verticillate, linear; a slender annual plant, 3 to 9 inches high; on gravelly soils; rare. Observed *Pinus balsamea*[1] plentiful on the mountains.

Saturday and Sunday, 27th and 28th.—Weather cold, raw, and unpleasant; high winds, with showers of hail and rain. On Sunday made a short turn up the south bank of the Columbia, but found nothing different from what I have already seen. In the course of the day I observed *Rubus* (sp. 109 of 1825) in flower, the shrubs much smaller than in the lower country. Another species just coming into blossom, which I think is not in the collection. Also a second evergreen species, and gathered some better specimens of *Saxifraga* than were collected some days ago.

Monday and Tuesday, 29th and 30th.—

(87) *Prunus* sp.; flowers white, in a long raceme; fragrant, resembling *Crataegus*; leaves ovate, smooth, serrate; a small tree, and in dry light soils a low shrub, frequenting rocky situations on the subalpine hills; plentiful.

(88) *Gnaphalium* sp.; perennial; flowers small, white, in a corymb; leaves alternate, linear-lanceolate; a low plant, 6 inches to 18 high; on all hilly, light gravelly soils.

(89) *Umbelliferae*, perennial; flowers white; leaves ternate, three-lobed, smooth, dentate; in low moist ground, near springs and rivulets, in shady places; rare.

(90) *Lupinus* sp.; perennial; leaves digitate; leaflets seven to thirteen, silky on both sides; flowers beautiful blue; this does not appear to differ much from one found last year so abundant around Fort Vancouver; abundant in all light soils.

(91) *Gramineae*, perennial; on low dry soil.

[1] *Abies balsamea*, Mast. in Journ. R. Hort. Soc. xiv. p. 189.

(92) *Gramineae*, perennial ; in low valleys ; a strong grass, 3 to 4 feet high ; plentiful. *Poa* sp. ; perennial ; a tall species, plentiful on hilly situations among rocks.

(93) *Carex* sp. ; on low damp soils ; a fine specimen ; plentiful. Observed the same species of *Onosmodium* with white flowers that is found so abundant at the Great Falls and on most of the sandy islands below ; laid in six specimens of the beautiful *Pentstemon*, being much better ones than those already collected.

Wednesday, 31st.—Collected near the Falls :

(94) *Rubus* sp. ; leaves three-lobed, serrate ; shoots hispid ; flowers small, white ; on rocks and dry gravelly soils near rivers.

(95) *Icosandria, Monogynia* ; flowers white ; leaves cordate, partially three-lobed, dentate ; 4 to 6 feet high ; on rocky situations ; plentiful. Rainy throughout.

Thursday, June 1st.—Made a walk to the hills to the south and found only two which I had not seen before :

(96) *Tetrandria, Monogynia* ; calyx four-partite, linear ; petals four, acute, faint blue ; leaves opposite, sessile, linear ; a low perennial plant, a foot to 18 inches ; plentiful on the meadows.

(97) *Syngenesia*, perennial ; in fissures of rocks ; leaves pinnate ; flowers brownish-yellow ; has a strong scent like mint ; this plant I saw some days ago, but not in blossom. Warm and pleasant.

Friday to Sunday, 2nd to the 4th.—Employed packing three bundles of dry plants, being all that are in a state fit to be sent to the coast. Two pair of mountain or rock grouse, one pair of curlews, and one small female pheasant. Changing the paper of some in a half-dry state and arranging my affairs for a journey to the plains below. On Sunday wrote a letter to my brother. I am to-morrow to leave this place by a boat at daybreak.

June 5th.—Rose at half-past two o'clock and had all my articles given over in charge to Mr. Dease. My tent struck and breakfast before five, when I took my leave, in company with Mr. William Kittson, of the wild romantic scenery of Kettle Falls. We went on horseback two miles from the new establishment, where the boats had been laid up, and embarked at seven precisely. The river by the melting of the snow is much swollen, fully twelve or fourteen feet where it is six hundred yards wide ; on getting into the current the boats passed along like an arrow from the bow. Half-an-hour took us to Thomson's Rapids, where, as I observed on my ascent, the water is dashed over shattered rocks, producing an awful agitation of the water from side to side. On being visited by Mr. Pierre L'Etang, the guide, he observed the water was in fine order for jumping the Rapid, as he termed it. Good as it appeared to him, I confessed my timidity prevented me from remaining in the boat. Although I am no coward in the water and have stood unmoved, indeed with pleasure, at the agitation of the ocean raging in the greatest pitch, yet to descend such a place I can never do unless necessity calls for it. Therefore Mr. Kittson and I walked along the rocks. No language can convey an adequate idea of the dexterity shown by the Canadian boatmen : they pass through rapids, whirlpools, and narrow channels where, by the strength of such an immense body of water forcing its way, it is risen in the middle to a perfect convexity. In such places, where you think the next moment you are to be dashed to pieces against the steep rocks, they approach and pass with an indescribable coolness, leaving it behind cheering themselves with an exulting boat-song. Reached the junction of the Spokane River twenty minutes after three o'clock of the same day, having in the short space of eight hours made a distance of ninety miles, which may give some idea of the current. As Mr. K. made a stay of an hour, securing some boats from the sun, I took a walk around our old April encampment, when I added several items to the collection. Camped at dusk opposite the Cinqpoil River, forty miles further down on the south side of the river. I was much obliged by Mr. K. kindly putting to shore for anything that attracted my attention.

Tuesday, 6th.—Embarked a few minutes before three and continued our route : passed the Little Dalles at eleven, where I walked over the rocks and, as usual, gleaned something for my collection. At one, arrived at Okanagan establishment, where I found my old friend Mr. John Work, William Conolly, Esq., a Mr. P. C. Pambrun, and a James Douglas, with a party of men from Western Caledonia, and a Mr. Francis Ermatinger from Thomson's River, a brother of the young man who accompanied me in the spring, all on their way to Fort Vancouver. I shall ever feel no small degree of pleasure on thinking of the kindness I had from these people, which is naturally doubly esteemed in this distant uninhabited country. I must mention in particular the genuine and unaffected friendliness of Mr. Conolly, who instantly begged that I would consider myself as an old acquaintance. Mr. Work, whom I have so often spoken of, kindly preserved for me a large female grouse and a male black rock-grouse, both very well done, with a few eggs of the former ; but as my time was so much taken up collecting plants and changing the paper of those gathered on my way down, I had no time to make a box to place them in, so I left them until the autumn with Mr. Ermatinger.[1] Made a turn round the rocks west of Okanagan River, and again made some more additions.

* [1] From October till April they are seen abundantly and are easily killed by the Indians with the bow on the banks of the Columbia from the junction of the Spokane River to the Wallawallah River below, being the boundary of the plains, a chain of three hundred miles. Are seen in large flocks at the Priest Rapids, about half-way between the two places. I am informed by several of the persons in authority they also inhabit the grounds around Lewis and Clarke's River or Snake River, Kooskooskee River, and their numerous branches. It would appear that the great humidity of the atmosphere, which is so sensibly felt on and near the coast and is in some degree the case even two hundred miles from it, is prejudicial to them, and the cold on the east the nearer they approach the dividing range of mountains, equally so. Between the parallels of Lat. 43° to 48° N., Long. 116° to 119° West.—*June 10th.*

August 21st.—Raised two large flocks near the Grand Coulee and had to content myself with a sight of them, having no gun. Perhaps there might be ten or twelve broods in each flock, say 100 or 120 birds. I could readily go within twenty yards. The large grouse of the plains is by no means a rare bird, although at this season seldom seen near the banks of the river, as they retire back to the elevated dry grounds for the purpose of hatching. The male in particular is a very noble bird, measuring 2 feet to 3½ inches in length, including the tail, and weighing 5 to 8 lb. avoirdupois. The flesh is very dry, and white like that of the ptarmigan, but tender, and affords perhaps as little nutriment as any known animal substance. In the breast of the male

* Footnotes made by Douglas.—ED.

Wednesday, 7th.—At eight this morning with a brigade of five boats I left for Wallawallah, at the junction of Lewis and Clarke's River with the Columbia, which I intend to make my headquarters for six or eight weeks. Passed the Stony Islands, a place in the river about half a mile in length, exceedingly rugged and dangerous. At four and shortly afterwards, camped on the south side of the river, earlier than usual, two of the boats having been broken. This circumstance gave me a few hours among the rocks, which I spent to great advantage. Killed a large rattlesnake, 3 feet long, on the edge of the river among some stones. Thermometer at noon 92° in the shade. At night the heavens appeared an entire sheet of lightning, until midnight, without thunder or rain.

Thursday, 8th.—As usual started at daylight and took breakfast below the Priest Rapids, on some very fine fresh salmon and buffalo tongue. As Mr. Conolly was very desirous of reaching the next establishment this night, no time could be lost. He informed me that his stay at that place would be very short, therefore in the afternoon I wrote to Joseph Sabine, Esq. ; but as a strong south wind with heavy rain began about five o'clock, obliged me to leave my letter in a half-finished state. Arrived at Wallawallah, where I was kindly received by Mr. Black. Having had very little sleep since I left Kettle Falls, I thought of indulging six or seven hours at least, so I laid myself down early on the floor of the Indian Hall, but was very shortly afterwards roused from my slumber by an indescribable herd of fleas, and had to sleep out among the bushes ; the annoyance of two species of ants, one very large, black, ¾ of an inch long, and a small red one, rendered it worse, so this night I did not sleep and gladly hailed the returning day. As soon as I could see to make a pen I finished the following letter to Mr. Sabine :—

Junction of Lewis and Clarke's River with the Columbia :
June 9th, 1826.

DEAR SIR,—As an unexpected opportunity has occurred of communicating with the coast, I willingly embrace it, at the same time as I have it in my power of sending the whole of my gleanings up to this time, amounting to upwards of a hundred distinct specimens not in the collection

are two remarkable spots destitute of plumage, of a bright brownish-yellow colour, serving him as a crop. In size the female is smaller and less varied in plumage. Are not shy ; will sit till within 30 or 40 yards and sometimes ten or twelve shots may be fired at the same flock without them rising. Their flight is swift but steady ; when they first leave the ground, they produce a burring sound with the wings, like the common pheasant in England, and alternately with the wings give three or four flaps and then float along for some distance, gradually falling while floating, and rising when flapping, and so unwieldy do they appear at first that they look as if wounded. They are very careless in the nests, give themselves little trouble : a few straws of dead grass is all, placed under the branches of *Tigarea tridentata* [1] or *Artemisia* among the sand. It is only in the country (which is particularly dry, either sand or gravel) where these shrubs grow they are found, and feed on the buds and tender leaves of both. Has ten to seventeen eggs ; egg small in proportion to the bird (being only a little larger than the common English partridge), in form and colour the same, only this has a few brownish small spots or freckles at the thick end.

A river is said to exist called by the American hunters Prairie Hen River, thought to fall into Rio del Norte in the Spanish possessions ; if this is the same bird, so called by Lewis and Clarke, the range of country it inhabits must be very great.

---

[1] *Purshia tridentata*, S. Wats. Bibl. Ind. N. Am. Bot. p. 309.

of 1825. I accordingly do so. Among which are six species of *Ribes*; two I think will prove new : *R. viscosissimum* of Pursh, which is surpassed by few of the genus ; his description will require some amendment. Four species, also interesting, but not so showy as the preceding. The others, *R. aureum* and one belonging to the section *Grossularia*, with green flowers. A few days after I wrote you from the junction of the Spokane River with the Columbia, on the 12th of April, across the Continent, I left that place for the Kettle Falls, ninety miles higher up the river, which I made my headquarters until the 5th of this month. This time was taken up making journeys in the directions that seemed best calculated to afford the richest harvest, and although I did not reap so abundantly as I expected, I do not consider my time altogether thrown away, as many are new or most imperfectly known species. About the 25th of June I shall make a journey to a ridge of snowy mountains, about 150 miles distant from this place in a southerly direction, which will occupy fifteen or sixteen days. After securing the result of this trip, I am to make a voyage up Lewis and Clarke's River as far as the Forks, where I will make a stay of ten or twelve days, more or less, as appears necessary, when I will return overland in a north-easterly direction to my spring encampment at Kettle Falls. Shortly afterwards I will accompany a Mr. Work, who goes on a trading excursion through the country contiguous to the Rocky Mountains, not far remote from the pass of Lewis and Clarke. After that period I shall gradually retrace my steps over the places I have visited, or may yet visit, and in all probability will reach the ocean about November. The difficulty of carrying as many of the different things collected as would appear desirable is very great : many times on my journeys I am under the necessity of restricting myself to a small number, to give place for a smaller proportion of all. I have been fortunate enough to procure two pair of a very handsome species of Rock Grouse, found only on mountainous grounds. A male of this species I killed last year on the coast, where they are very rare. None of this species is found east of the mountains, so it may prove new. A pair of curlews, having singular habits from most of the tribe, frequents dry soils, and roosts on trees. A female small pheasant is all I have been able to add to the collection that are ready for sending. They are packed in a small box, with three bundles of plants. Having so much to do, it is entirely out of my power to send at this time a copy of my journal, a circumstance I regret exceedingly. I have five splendid species of *Pentstemon*, only one of them described, which is *P coeruleus* of Nuttall. Abundance of *Tigarea tridentata*[1] in flower and fruit, several species of *Rubus, Lupinus,* two species of *Prunus,* all different from those found on the coast last year. I am now in the finest place for the large fruit and hope shortly to procure some. It will at all times afford me much gratification to mention the kindness and assistance I have from the persons in authority. As the ship will sail for England before I can reach the coast, I will not have it in my power to write you again. Thank God, I enjoy most excellent health. There is nothing in the world could afford me a greater degree of pleasure than hearing from you. With sincerity I hope

---

[1] *Purshia tridentata,* S. Wats. Bibl. Ind. N. Am. Bot. p. 309.

it may come to pass in the course of the autumn.—I am, Dear Sir, your most faithful and humble servant, D. Douglas.

To Joseph Sabine, Esq.

Afterwards wrote to Mr. McLoughlin and all my kind friends at Fort Vancouver, respecting my articles going by the ship. Mr. Work obligingly undertook to see my box and articles landed safe, and Mr. Conolly handsomely presented me with 12 feet of tobacco (better than 2 lb.) to assist me in my travels, during their absence at the sea. Being as it were the currency of the country and particularly scarce, I esteemed it invaluable, as it will enable me to have guides and other services performed more willingly. On their departure at noon, after changing the paper of my plants, I now proceeded to extract from my note-book the following :

(98) *Phlox* sp. ; stem shrubby, pubescent ; leaves opposite, revolute, the base dilated, smooth ; calyx glaucous ; flowers white and rose colour ; very large and fragrant, in partially shady woods at the junction of the Spokane River ; this may only prove a variety of *Phlox speciosa*, but is nevertheless a desirable plant, much stronger, larger flower, and more profuse.

(99) *Pentstemon* sp. ; stem smooth, white, rarely exceeding a foot high ; leaves opposite, sessile, serrate, lanceolate ; flowers small, white ; sterile filament naked, short ; inside of the calyx pubescent ; a beautiful interesting species, has some affinity to *P. albidus* of Nuttall ; plentiful on rocky situations, near the Cinqpoil River and on the rocks at the Stony Island on the Columbia.

(100) *Pentandria, Monogynia*, annual ; calyx deeply five-cleft, woolly, segments linear, with a bractea about the size of the segments ; corolla funnel-form, very long, spreading ; stamens same length as the corolla ; stigma three-cleft ; flowers bright scarlet ; leaves alternate, pinnate, somewhat succulent ; stem smooth, branching, 2 to 3 feet high ; a most beautiful plant, with a great profusion of flowers ; plentiful from Kettle Falls for the distance of three hundred miles further down, on dry light soils.

(101) *Allium* sp. ; perennial ; leaves plain ; flowers faint purple and white ; a low plant, 6 to 10 inches high ; abundant on a low sandy point at the junction of the Spokane River.

(102) *Linum Lewisii*,[1] perennial ; on all dry elevated open places plentiful, and in individual plants under the shade of solitary pines.

(103) *Chrysanthemum* sp. ; leaves obovate (stem leaves, bidentate), woolly on the under side ; this has some affinity to one found on the coast last year ; plentiful in all gravelly soils. Observed *Phlox Sabinii*[2] in great abundance in blossom, and although I secured plenty of it last year, I could not help laying in a dozen more ; also *Viburnum* sp., with white flowers and berries, so plentiful near the coast.

(104) *Silene* (?) perennial ; flowers white ; leaves opposite, linear-lanceolate, entire, smooth ; a slender plant, seldom more than a foot in height ; in gravelly soils ; rare.

[1] *Linum perenne*, S. Wats. Bibl. Ind. N. Am. Bot. p. 146.
[2] *Phlox speciosa* var. *Sabini*, A. Gray, Syn. Fl. N. Am. ii. i. p. 134.

(105) *Myosotis* sp.; perennial; leaves alternate, sessile, linear-lanceolate, hispid; stem very hispid; flowers white; stamens enclosed within the yellow orifice; this very handsome plant, which may prove *M. hirsutissima* [*sic*], is found plentifully all over the plains; a foot to 18 inches high.

(106) *Salvia* sp.; shrubby; leaves obovate; petioles short, soft, or somewhat fleshy (under the microscope thickly covered with minute yellow glands); flowers verticillate, large, bright blue, surrounded by orbicular-ciliated bracteae; a splendid shrub, with a powerful and not disagreeable scent; 2 to 4 feet high; plentiful on all hilly rocky situations; this would be a most valuable addition.

(107) *Phlox* sp.; suffruticose; leaves subulate; young shoots slightly pubescent; segments of the calyx subulate, same length as the tube of the corolla; flowers spreading, faint yellow and white; this handsome but seemingly very delicate little plant will form also a good addition to that type of splendid plants; an upright shrub, a foot to 15 inches high; rare; on rocks of the Stony Islands and Little Dalles.

(108) *Pentstemon* sp.; perennial; leaves glabrous, opposite, linear-lanceolate, entire; flowers very large, bright azure-blue, without a tinge of purple, rose, or any other colour; sterile filament, short (naked); inside of the corolla, smooth; this resembles *P. coeruleus* of Nuttall; climate may make a slight difference (sterile filament, he observes, is bearded and the corolla nearly equally five-cleft, which is not the case with the present); 3 to 4 feet high; seen abundantly in individual plants between Spokane River and Okanagan River, in light soils.

(109) *Malva* sp.; perennial; leaves partially five-lobed, scabrous; stem and calyx woolly or tomentose; flowers small, bright scarlet; may prove *M. coccinea*[1] of Nuttall; found in great abundance over the whole plains and makes the same contrast among other things that the common *Papaver* does in the fields of Europe.

(110) *Pentandria, Monogynia*, annual; calyx five-leaved, linear; corolla tubulous; petiole obtuse, spreading; style bifid; capsule one-celled; seeds numerous; small oblong radical leaves, sinuate, partly succulent, cauline entire; stem and calyx beset with glandulous hairs; flowers blue; abundant on the plains; a low plant, a foot high.

(111) *Pentstemon* sp.; perennial; leaves amplexicaul, cordate-acute, dentate, pubescent; likewise the stem; calyx more so; flowers bright blue and purple, beard of the sterile filament brown; plentiful on rocks near Okanagan; a foot to 18 inches high.

(112) *Santolina*, perennial; leaves linear-lanceolate, alternate, tomentose; flowers yellow; a low plant, 6 to 8 inches high; plentiful on dry rocks.

(113) *Santolina* (?), perennial; leaves linear, white, alternate, woolly; flowers white; 10 to 15 inches high; plentiful on the plains, in sandy soil.

(114) *Cruciferae*, annual; flowers white, in a thick close spike; stamens double the length of the corolla; radical leaves ovate-lanceolate, sinuate, or deeply dentate, glabrous, cauline nearly entire; a strong-growing plant, 4 to 5 feet high; abundant on the plains, in all soils.

---

[1] *Malvastrum coccineum*, S. Wats. Bibl. Ind. N. Am. Bot. p. 137.

(115) *Pentandria, Monogynia* ; annual ; corolla five-cleft, undulate ; anthers sessile, enclosed within the orifice ; pistil one-third the length of the tube ; tube long ; flowers white, in an umbel surrounded by five lanceolate bracteae ; very fragrant during the night, leaves opposite, ovate-lanceolate, entire ; stem brown and glutinous ; a low but wide-spreading plant, a foot to 18 inches high ; this or a second species came under my notice last year at the Great Falls ; in dry sandy soils, a little below the Stony Islands, on the south side of the Columbia.

(116) *Decandria, Trigynia,* perennial ; calyx five-leaved ; leaves linear, acute ; corolla five petals ; petals linear, obtuse, double the breadth of the calyx ; leaves opposite, subulate, dilated at the base ; stem jointed ; flowers white ; 4 to 6 inches high ; in the sandy plains ; plentiful.

(117) *Eriogonum* sp. ; suffruticose ; leaves oblong-ovate, tomentose under, smooth above ; flowers small, bright yellow ; this may prove *E. sericeum*[1] of Nuttall ; a low plant, 1½ to 2 feet high ; on the Stony Islands and a few miles below it ; abundant ; a fine plant.

(118) *Pentandria, Monogynia* ; calyx five-leaved ; corolla five-cleft, round ; style bifid, pubescent at the base, filaments exserted ; leaves pinnate, pubescent ; stem creeping, hirsute ; root creeping ; flowers purple ; a straggling perennial plant, frequenting crevices of rocks ; grows very strong on moist rocks.

(119) *Cruciferae,* perennial ; stem smooth, branching ; leaves alternate, sessile, linear, smooth ; flowers yellow ; plentiful at the Priest Rapids.

(120) *Eriogonum* sp. ; flowers yellow ; may prove a variety of *E.* found on the coast last year (not rare) at the Stony Islands, the only place it came under my notice.

Thermometer in the shade 90° ; minimum 63°. Having had little sleep for the last five nights, I felt somewhat fatigued. I went to bed earlier than usual, and shortly after dusk an Indian arrived from Fort Vancouver with news of the arrival of a ship in the river, and brought me letters and a small parcel of newspapers. I grasped the parcel eagerly and tore it open, turning over my letters ; at last I found one in Mr. Goode's handwriting ; all was right. I thought one was from Mr. Sabine, but on opening I found it from Mr. G. Immediately opened my note from Mr. McLoughlin, who informed me he had kept what he conceived to be letters from the Society until the people would return, not deeming it prudent to risk them by the Indian. I had one from Mr. William Booth, which with Mr. G.'s is all that have as yet come to hand. Never in my life did I feel in such a state ; an uneasy, melancholy, but pleasing sensation stole on my mind, with an inordinate longing for the remaining part, and although I did not hear directly from my friends, I now for once in my life enjoy and relish the luxury of hearing from England. I had letters from all my kind friends on the coast, full of expressive hopes that my labours might be amply rewarded. It is a circumstance worthy of notice that I should write to England in the morning and receive letters on the same day, for in this uninhabited distant land the post calls but seldom. The express for the coast with my letters had only left me six hours when the Indian arrived.

[1] *Eriogonum flavum,* Benth. in DC. Prod. xiv. p. 8.

Saturday, June 10*th.*—Morning cloudy ; warm last night. I pored over my two letters till after midnight, when I lay down on my mat. It is needless to say, although I had not slept twelve hours for the five preceding nights, this one I passed without as much as closing my eyes. About noon I felt very unwell and did not deem it prudent to make much exertion during the intensive heat of the day. Therefore I employed myself mending some of my shoes and walked along the river in the afternoon, where I collected as follows :

(121) *Lupinus* sp. ; annual ; small, hairy ; flowers blue and white ; legume two-seeded ; 4 to 8 inches high, on the open barren plains ; this is no doubt *L. pusillus* of Nuttall.

(122) *Diadelphia*, perennial ; legume double-curved, many-seeded, seeds small ; leaves pinnate, ovate, obtuse ; stem and leaves hispid ; flowers purple and white ; a most beautiful low-spreading plant, 6 inches to a foot high ; plentiful on dry gravelly soils.

(123) *Melilotus* (?), perennial ; flowers white and blue ; this plant I saw last year near the Great Falls on the Columbia, but in an imperfect state ; abundant over all the plains.

Sunday, 11*th.*—Heavy rain during the night ; warm and clear. Turned and arranged some plants.

Monday, 12*th.*—Warm. Collected the following :

(124) *Polydelphia*, annual ; calyx five-leaved, lanceolate ; petals obtuse, five ; stem white, smooth ; leaves amplexicaul, very rough ; capsule sessile, long, one-celled ; seeds small and numerous ; flowers yellow (if I recollect, a magnificent plant of the same genus I found at the Great Falls last year in blossom, but could obtain no perfect seeds of it) ; the present is seen in little patches among *Tigarea*[1] and some species of *Artemisia* on the dry sandy plain ; not very abundant.

(125) *Hexandria, Monogynia*, annual ; calyx four-partite ; corolla four petals ; petals lanceolate-obtuse, two small (narrower) ; leaves digitate, glaucous ; leaflet lanceolate, three to five ; flowers terminal stamens exserted ; flowers yellow ; a plant 1 to 2½ feet high, frequenting dry soils ; not unhandsome but has a very fœtid burned-like scent. Found abundance of *Clarkia pulchella* on the rising ground and laid in a few specimens.

(126) *Cruciferae*, annual ; this plant may prove only a variety of No. 114, as there appears to be little difference except the present has purple flowers and the leaves narrower and more sinuated, which situation may affect ; on dry rocks ; a showy plant, flowering in a thick close spike ; 4 to 10 feet high. At the junction of the Wallawallah River with the Columbia, I found on the high grounds a species of *Lupinus* which I have little doubt is *Lupinus villosus*, variable flowers, sometimes pale purple, approaching to a dingy-yellow. Collected a few seeds of *Phlox speciosa* ; this seems not to produce seeds in abundance ; probably the sudden intensity of the summer heat is too great.

Tuesday, 13*th.*—High winds and heavy rains during the night ; I had the greater part of my paper and all my clothing wet, not a dry stitch to put

[1] *Purshia*, Benth. and Hook. f. Gen. Pl., i. p. 617.

on. Employed exposing my paper to the sun and turning my plants until twelve o'clock, when I went out and gleaned the following :

(127) *Pentandria, Monogynia*, perennial ; a second species of the genus (115), also very fragrant during the night; plant glaucous; stem smooth, white, purple at the joints, brittle, and somewhat succulent; leaves ovate-obtuse, entire ; root creeping and produces shoots like the genus *Salix* and *Populus* when exposed above ground ; a low but wide-spreading plant; plentiful on the plains, near the junction of Lewis and Clarke's River with the Columbia ; would be a valuable addition to the garden ; this I think is the plant I saw last year at the Falls and had not time to look at at the moment (that found this year at the Stony Island a second species).

(128) *Pentstemon* sp.; perennial; very smooth and glaucous; radical leaves lanceolate, with cauline amplexicaul, cordate-acute ; corolla nearly equally five-cleft, fine blue and rarely with any other colour (sometimes assuming a tinge of purple on the tube after a few days' expansion) ; sterile filament shorter than the perfect, bearded on the upper side, hairs light brown ; peduncles, many flowered ; a most beautiful species, a foot to 3 feet high ; abundant on the banks of Wallawallah River, the only place I have as yet seen it; perfectly distinct from *P*. 108 ; both will, I think, prove new.

(129) *Myosotis* (?) sp.; perennial ; leaves alternate, nearly sessile, lanceolate-obtuse, entire, glaucous ; stem glaucous and somewhat succulent ; flowers white, dark in the centre ; rare ; on the banks of the Wallawallah River, on moist ground.

(130) *Oenothera* sp.; annual; leaves small, glabrous, sessile, alternate, linear ; capsule long, slender, twisted; flowers faint rose colour ; a minute plant, never exceeding 6 or 8 inches high ; on the plains ; abundant in dry sandy soils.

(131) *Eriogonum* sp.; perennial ; leaves linear-lanceolate, villous ; flowers white ; abundant on the plains.

(132) *Rosa* sp.; seems to be the only species; on the banks of rivers on the plains ; flowers large.

Wednesday, 14th.—Collected as follows :

(133) *Clematis* sp.; flowers white; on shady places, among rocks ; not plentiful.

(134) *Erigeron* (?) sp.; pubescent; flowers white ; 8 inches to a foot high ; on rocky places ; rare.

(135) *Orobanche* sp.; annual (not a parasite) ; abundant on the dry sand-scorched plains.

(136) *Diadelphia*; perennial; stem low, branching ; leaves pinnate ; leaflets linear, pubescent; legume many-seeded ; flowers blue and purple; on the plains; abundant.

(137) Allied to *Rumex*, perennial; leaves glaucous, linear-lanceolate, entire ; wings of the capsule bright scarlet; root creeping ; a strong coarse plant; abundant on the plains at the junction of Lewis and Clarke's River.

(138) *Gramineae*, perennial; a tall strong grass ; on the plains, with the three following :

(139) *Gramineae*, perennial;

(140) *Gramineae*, perennial;

(141) *Gramineae*, perennial.

(142) *Gramineae*, perennial; a fine tall grass.

(143) *Quercus* sp.; perennial; small, 8 inches to a foot high; also on the plains, but rare. Very high south-west wind, producing such a boisterous swell on the river that the canoes could not go to fish; no salmon caught for the last three days. Killed three small *Arctomys*, two males and a female, 9 inches long, tail 1¼ inch long; on the back, light grey with numerous small white spots; belly same colour, but destitute of the spots; sides, legs, and lower part of the neck, of a whitish colour; nose and round the anus, light brown; iris fine blue purple-black. The female differs in no respect from the male, except in size is a little smaller. Burrows in the sands under brushwood and lives on the leaves and fruit of *Tigarea tridentata* [1] and the leaves of several species of *Artemisia*, both found on the plains. Having nothing to eat, they with some boiled horseflesh formed my supper. I found the former rancid or rather of a musty taste, probably from the bitter strong-scented plants on which it feeds; called by the Wallawallah and Kyuuse Indians *Limia*.

Thursday, 15th.—Very windy, with showers of rain; at 4 A.M. started on a walk to some rocky grounds on the opposite side of the Columbia; after taking a little breakfast, the same as my supper last night, I proceeded. Long before twelve o'clock I felt fatigued, as I could not get so much as a mouthful of water. My eyes began to trouble me much, the wind blowing the sand, and the sun's reflection from it is of great detriment to me. Returned at four in the afternoon, fatigued. My eyes so inflamed and painful that I can scarcely see distinctly an object ten yards distant. Gathered a few more seeds of *Phlox speciosa*.

(144) *Myosotis* sp.; perennial; stem hirsute; calyx more so; leaves sessile, lanceolate, entire, smooth above, slightly silky below; flowers white, with five yellow prominent nectaries surrounding the orifice; a fine low plant, 8 to 14 inches high; on the banks of the Columbia, a few miles below the junction of Lewis and Clarke's River; rare.

(145) *Syngenesia* sp.; perennial; smooth-branching and twiggy; leaves long, sessile, linear, entire, smooth; flowers fine pink colour; root and branches yield a viscid, milky, bitter juice; a foot to 18 inches high; on the banks of rivers; not very plentiful.

(146) *Opuntia ferox* of Nuttall; abundant on all dry rocky and sandy soils and many times gives me great trouble picking the spines from my feet; flowers yellow; style and stigma purple; filament and anthers yellow.

Friday, June 16th.—Wind more moderate, warm and pleasant; making preparations for a journey to the south on the Blue Mountains and Grande Ronde, distant about 140 miles. Sent to the Indian camp to inform my guide to be ready to-morrow at sunrise. Last night I was much annoyed by a herd of rats, which devoured every particle of seed I

[1] *Purshia tridentata*, S. Wats. Bibl. Ind. N. Am. Bot. p. 309.

had collected, cut a bundle of dry plants almost right through, carried off my razor and soap-brush. One, as he was in the act of depriving me of my inkstand, which I had just been using before I lay down and was lying close to my pillow, I lifted my gun (which is my night companion as well as day, and lies generally alongside of me, the muzzle to my feet) and gave him the contents. I found it a very strange species, body 10 inches long, tail 7; hairy belly nearly white, back light brown, point of the hairs darker; ears very large, ¾ of an inch long; whiskers jet black, 3 inches; long nose, pointed. In the hurry to recover my inkstand and the great desire that I had of securing him, I did not take time to change the shot for a smaller sort. Rose early, soon after daylight, and watched; I had not sat up above half an hour when a second came. I handed it a lighter shot and did not destroy the skin. In every respect like the male, only about an inch shorter, the head smaller, body as thick; was not with young. I am informed they are found in great abundance in the Rocky Mountains, particularly to the north of the Peace and McKenzie Rivers, where in the winter they do much injury to everything that comes in the way.

(147) *Syngenesia* sp.; perennial; leaves opposite, ovate, three-nerved, smooth above, slightly pubescent under; flowers sessile, yellow, sweet-scented; calyx five-leaved; a singular plant, 8 inches to 18 inches; abundant on the plains, in light dry sandy soils.

(148) *Diadelphia*, perennial; stem smooth; leaves pinnate, nine to fifteen, smooth, ovate-lanceolate; flowers white; peduncle pubescent; root creeping; a foot to 1½ high; on gravelly banks; not plentiful.

Saturday, 17th.—My guide did not arrive from the camp until 8 A.M., and as I was uncertain if he would come that day, the horses were not brought in from the meadow, nor my provisions put up. Considerable time was taken up explaining to him the nature of my journey, which was done in the following way : I told Mr. Black in English my intended route, who translated it to his Canadian interpreter, and this person communicated it to the Indian in the Kyuuse language, to which tribe he belongs. As a proof of the fickle disposition and keenness of bargain making in these people, he made without delay strict inquiry what he should get for his trouble. This being soon settled, then came the smaller list of present wants, beginning, as his family had been starving for the last two months, and he going just at the commencement of the salmon season, by asking Mr. Black to allow them something to eat should they call, which was promised. Afterwards a pair of shoes, and, as his leggings were much worn, leather to make new ones was necessary ; a scalping knife, a small piece of tobacco, and a strip of red coarse cloth to make an ornamental cap. This occupied two hours and was sealed by volumes of smoke from a large stone pipe. Mr. B. kindly offered to send a boy twelve years of age, the son of the interpreter, who speaks the language fluently, with me, which I gladly accepted. As he spoke a little French, I would be the better able to make known my wants to my guide. I had provided for me three excellent horses for carrying my paper, blanket, and provisions, which was equally divided, and as I choose to walk except on bad places of the road or crossing the creeks, I placed a little more on my horse. Mr. B. had put up for me a supply of pemmican, a little biscuit, sugar, and tea, and I was amply supplied for ten days with dry salmon for the guides ; with what we would kill I might consider myself comfortable. Towards midday I took my leave in a south-easterly direction, along the banks of Wallawallah River, which I crossed a short distance from its junction with the Columbia. Proceeded slowly along on the south side of the river, always making halts collecting anything different from what I had seen. As water is very scarce in crossing the plains, and the day being far spent I camped on the edge of a small spring among some birch.

Sunday, 18th.—I was dreadfully annoyed with mosquitoes during the night and was roused at two in the morning by a heavy shower of rain which lasted till half-past four. I had the horses brought in and started at five, and continued my journey. This morning I found great relief, the atmosphere being cool and the sand prevented from blowing. This part of the country to the north is an entire level plain of gravel and sand, destitute of timber, not even a shrub exceeding 4 feet in height, except a few low straggling birch and willows on the sides of rivulets or springs. On the south the ground is undulating, also bare of wood, and on some places veins of decomposed brown granite are to be seen, and in three places volcanic rock. Passed the southern branc of the Wallawallah River at eight ; on the eastern shores, which are steep, of gravel and lime in a recent formation, abundance of *Ribes aureum* is seen, with perfectly ripe fruit. The fruit is of exquisite flavour, much finer than any I am acquainted with, both yellow and black, but the former more abundant. The racemes are not large but very numerous, the berries about the size of the common currant and equally prolific. It might be well to try some in very dry light gravelly or shingle soil, as I observe it is never seen with fruit where the soil is rich, and but with a scanty crop in moist situations. Continued my route in a southerly direction. As I approached the high ground I found the air gradually becoming cool. Camped at the foot of the range at 5 P.M. As I had not tasted anything to-day, I made myself a basin of tea, on which with a little dry salmon and my other provisions I made a comfortable supper, and then took a turn round my encampment. Observed several interesting plants which I had not seen before, but left them until my return. To save time in the morning I made a little additional tea in the evening and left it in the kettle over night.

Monday, 19th.—After breakfast, shortly after sunrise, I began my ascent in an easterly direction on foot, the guides going on before with the horses ; the ascent was gradual for a short distance (about fifteen miles), where the path in many places became rugged and difficult to pass over. Reached the middle of the mountain where the wood begins, where I camped at mid-afternoon much fatigued. I had walked not less than thirty-five or forty miles. I found the timber to consist of the same species that are found prevalent on the mountainous ground, and among the brushwood two species of *Ribes*, *R. viscosissimum* and another already collected, the former very plentiful. Also a species of *Lonicera* or *Xylosteon*, which I gathered sparingly on the banks of the Spokane River six weeks since in this high altitude, not yet in blossom.

Tuesday, 20th.—As usual started at daylight with a view of reaching the height of land against dusk. The further I went the more difficult I found my undertaking. At midday I made a short stop, where I passed the first snow and collected several plants. Immediately after eating a little dried salmon and a mouthful of water from a chilly crystal spring, I continued my route until 4 P.M., where the horses were stopped by deep wreaths of eternal snow, about 1500 feet below the extreme height of the range. As my object was, if possible, to reach the low alluvial grounds on the opposite side, where I had great expectations, my disappointment may be imagined. However, in the meantime I selected my camp under a projecting rock, saw the horses hobbled, and as it appeared to me my guide seemed some-what alarmed, I thought it prudent to give him a little time to cool or change his opinion. Therefore I set out on foot with my gun and a small quantity of paper under my arm to gain the summit, leaving them to take care of the horses and camp. In the lower parts I found it exceedingly fatiguing walking on the soft snow, having no snowshoes, but on reaching within a few hundred feet of the top, where there was a hard crust of frost, I without the least difficulty placed my foot on the highest peak of those untrodden regions where never European was before me. The height must be great—7000 to 7500 from the platform of the mountain, and on the least calculation 9000 above the level of the sea. (Thermometer at 5 P.M., 26° Fahr.) Two days before the maximum heat at the foot of the mountains was 92°, and this day I have every reason to think was equally warm. The view of the surrounding country is extensive and grand. I had not been there above three-quarters of an hour when the upper part of the mountain was suddenly enveloped in dense black cloud; then commenced a most dreadful storm of thunder, lightning, hail, and wind. I never beheld anything that could equal the lightning. Sometimes it would appear in massy sheets, as if the heavens were in a blaze; at others, in vivid zigzag flashes at short intervals with the thunder resounding through the valleys below, and before the echo of the former peal died away the succeeding was begun, so that it was impressed on my mind as if only one. The wind was whistling through the low stunted dead pines accompanied by the merciless cutting hail. As my situation was not a desirable one for spending the night, and it was creeping on me, I hastily bent my steps to my camp below, which I providentially reached at eight o'clock, just in the twilight, the storm raging still without the least appearance of abating. The horses were so alarmed that I found it necessary to tie them to some trees close to the camp. As no fire could be kept in, my supper was of the same quality as my breakfast; and as all my clothes were wet, and having nothing to change, I stripped and rolled myself in my blanket and went soundly to sleep shortly afterwards. Precisely at twelve I was so benumbed with cold that on endeavouring to get up I found my knees refused to do their office. I scoured them well with a rough towel, and as the storm was over made a cheering fire. I could not resist the temptation of making a little tea, which I found restored me greatly (thermometer 26°). If I have any zeal, for once and the first time it began to cool. Hung my clothes up to dry and lay down and slept until three o'clock.

Wednesday, 21st.—I found the spirits of my guide and interpreter greatly damped, and as I had not abandoned the idea of gaining the other side by a more circuitous and less elevated route, a little more to the south than my present situation, I found their fear increasing; and, as I had every reason to suppose, the young rascally boy told the Indian the reverse of what I wished him to do. At last the Indian told me that I might be able to go, but as I had a river to cross (the Utalla), which at this season was very large, I would require to swim or make a raft; and as his nation was at war with the Snake tribe and we going on the confines of their lands, in all likelihood they would steal our horses and perhaps kill us. As it would be very improper to force him to go with me, and impossible for me to go alone, I was reluctantly obliged for the present to give up the idea of crossing. Therefore I lost no time in gleaning on this side. Camped half-way between my Monday's and Tuesday's encampments.

Thursday, Friday, and Saturday, 22nd, 23rd, 24th.—I continued my labours, shifting and moving about in the most suitable places for my pursuit, until I had all that appeared peculiar to that district of the hills, when I bent my steps towards the Columbia again, which I reached in the afternoon of Saturday, being eight days on my journey, during which time I saw not a human being except my guides. I found the boy had told the Indian not to go with me as it was dangerous. From what I have seen in these mountains it is my intention, as soon as the plants are dry that are collected, to start for another eight days in a different direction.

Sunday, 25th.—In the early part of the day I placed in dry paper some of the plants, and under the presser the more recently gathered ones, and then put up some seeds which I had hung up to dry ten days since. I come now to take from my note-book the following collected on my journey:

(149) *Paeonia* sp.; perennial; root large and jointed, partly creeping; stem glaucous, red; leaves alternate, compoundly lobed, smooth, and glaucous; flowers small petals same length as the stamens, centre and the outside dark purple, on the edge and inside bright yellow; a low plant, 6 inches to a foot high; in great abundance, in clumps, among low bushes on the sunny side of the mountains, flowering in perfection on the confines of perpetual snow; lower down it is seen in feeble enervated plants, and in the more temperate regions completely disappears; this valuable addition will I trust be an acquisition to the garden; if in my power, seeds of it must be had.

(150) *Anemone* sp.; perennial; flowers one, drooping, dark purple; stem leaves, and outside of the petals, pubescent; a low plant, 9 to 15 inches high; seen in conjunction with the preceding; also in abundance.

(151) *Lupinus argenteus*, of Pursh; perennial; only seen on the sub-alpine or undulating grounds, in dry gravelly soils; seldom exceeding 14 or 18 inches high; a very fine plant; found abundantly with the small annual species so common on the alluvial plains of the Columbia, growing together; by some means or other procure seeds of all these.

(152) *Lupinus* sp.; perennial; *Turneri* or *aureus* (D. D.); leaves digitate, five to ten; leaflets lanceolate, densely covered with silky white hairs; stem nearly smooth; flowers verticillate, very large, bright golden-

yellow; calyx thickly covered with rigid hairs; this extremely handsome plant, certainly the finest of the genus (not even excepting *L. nootkatensis*), is found on the mountainous grounds covering the whole face of the country, and at the distance of a few miles appears much like *Spartium scoparium*[1] on the wild and beautiful heaths of England; 18 inches to 2½ feet high; often producing a spike of perfect blossoms a foot to 15 inches, as the specimens in the collection will show; I beg the former name will be adopted after Mr. Turner.

(153) *Cerastium* sp.; perennial; leaves opposite, ovate-lanceolate, entire, smooth; stem spreading; flowers white; plentiful on the mountains, in shady places.

(154) *Salix* sp. (male); a low shrub; on the hills; rare; I regret the female of this species could not be found.

(155) *Pedicularis* sp.; annual; flowers white; a low plant; on moist ground on the summit of the hills.

(156) *Hexandria, Monogynia*, perennial; calyx three-leaved, acute; corolla three petals, obtuse, the inside covered with strong purple and white hairs; stigma trifid; capsule angular, immature; flowers white; stem low; seldom more than one leaf sheathing, linear-lanceolate; a small bulbous plant, 4 to 8 inches high; plentiful on the highest peaks of the mountains, flowering delightfully among the snow.

(157) *Mitella*, perennial; will probably prove *diphylla*; in the close shady pine-woods, in rich decayed vegetable soils.

(158) *Cerastium* sp.; perennial; leaves opposite, sessile, linear, or subulate, slightly pubescent; flowers white; probably *C. alpinum*; on rocks on the hills, in small tufts.

(159) *Fumaria* sp.; perennial; root bulbous or granulous; flowers white; leaves glaucous; a low delicate plant; on high ground, growing out of the crevices of rocks and near decayed wood; plentiful.

(160) *Draba* sp.; perennial (?), biennial (?); leaves orbicular, ovate, cauline, alternate, amplexicaul, glaucous; flowers white; a low spreading plant, frequenting springs and moist rocks; on the summit of the mountains; abundant.

(161) *Viola* sp.; perennial; leaves orbicular, smooth, nearly entire; flowers small, bright yellow; abundant in all the upland and mountain woods.

(162) *Saxifraga* sp.; perennial; leaves orbicular, dentate, glaucous; flowers white; 10 to 15 inches high; on moist rocks, flowering among the snow.

(163) *Valeriana* (?), perennial; leaves opposite, pinnate; flowers appear to be purple; not quite in flower; rare.

(164) *Pentstemon* sp.; perennial; corolla nearly equally five-cleft, striated with dark purple veins; sterile filament thickly covered with white hairs; leaves lanceolate, dentate, serrulate, upper ones nearly entire, smooth; flowers purple-rose; a small but very fine plant; abundant on the channels of mountain rivulets and rocky situations; does not seem to have been observed by Nuttall or Pursh.

[1] *Cytisus scoparius*, Hook. f. Student's Fl. Brit. Isl., ed. 3, p. 92.

(165) *Pentstemon* sp.; perennial; flowers large, purple-blue; sterile filament same length as the fruitful, and less hairy; anthers lanuginous; leaves lanceolate-ovate, serrate, smooth; a splendid plant, 2 to 3 feet high; in large thick tufts in rocky places and channels of rivulets; this may prove a species of *Chelone*, but no seeds of the last could be found and this not yet ripe.

(166) *Diadelphia*, perennial; flowers yellow; a very strong plant; on the channels of rivers and springs; rare; in flower and seed.

(167) *Helianthus* sp.; perennial; leaves alternate, sessile, lanceolate, rough; stem hispid; flowers yellow; 2 to 2½ feet high; on mountainous grounds, in dry soil.

(168) *Amentaceae*; leaves linear, sessile, somewhat succulent; a low spreading, scrubby shrub; abundant on the plains and on rocky places near rivers.

(169) *Astragalus* sp.; perennial; stem strong and somewhat creeping; flowers white; capsule slender, smooth; leaves slightly pubescent; very plentiful, in small tufts, on the plains near the skirts of the mountains, in all dry light soils; 2 feet high.

(170) *Dioecia*, perennial; leaves ternate, glaucous, nearly entire; on the hills; rare.

(171) *Anemone* sp.; flowers white; a low plant; on the upland and mountain woods, on dead wood, and all shady soils; abundant; perhaps *A. nemorosa*.

(172) *Sedum* sp.; perennial; flowers yellow; abundant on moist rocks, dry channels of rivers, particularly in hilly situations.

(173) *Syngenesia* sp.; perennial; perhaps not far removed from *Santolina*; on the plains, in light dry soils; rare. I find that (120) *Eriogonum* will prove a distinct and constant variety or species, as it is uniformly seen with yellow blossoms, much larger than any other of the genus; abundant on decayed granite rocks or dry channels of rivers, growing luxuriantly with scarcely a particle of earth to support it.

(174) *Lupinaster macrocephalus* [1] (Pursh); this beautiful plant is seen in a short, thick, green luxuriant sward on the summit of the Blue Mountains on the confines of the snow; it is not seen until upwards of 3000 feet from the platform of the mountain; seeds of this desirable plant must be found.

(175) ——— ? perennial; flowers yellow; leaves lanceolate; in moist ground, near springs; I had not time to look for this plant in a living state.

(176) *Astragalus* sp.; perennial; small, creeping; leaves smooth; capsule short, curved, few-seeded; on the plains, in sandy soils; rare.

(177) *Myosotis* sp.; annual; on the sides of rivers, not plentiful, and in the mountain valleys.

(178) *Syngenesia* sp.; annual; flowers white, three, four, five petals, generally three-petal, three-toothed, centre half the size of the outer dentures; leaves linear, sessile; a low spreading showy little plant, frequenting rocky soils.

(179) *Malva* sp.; perennial; flowers large, fine rose colour; leaves

[1] *Trifolium megacephalum*, S. Wats. Bibl. Ind. N. Am. Bot. p. 264.

three to five lobed, dentate, rough ; a strong growing plant, 4 to 6 feet high ; margins of pools and rivers ; abundant.

(180) *Senecio* sp. ; perennial ; leaves lanceolate, serrate, smooth ; flowers yellow ; 5 to 6 feet high ; frequenting the same places as the former.

Monday, 26th.—As I had still time to spare and being somewhat disappointed on my trip last week, I thought of making a second journey to the same mountains in a different direction. On my guide being told last night to be ready at daybreak, he instantly began to complain that the fatigue of the former journey had weakened him so much that he did not think he would be able to proceed. To-day he did not make his appearance till 6 o'clock A.M., when he refused to go. He was certainly a little broken down, but I saw nothing that would seriously injure him ; and I was on the eve of giving him a little corporal chastisement to teach him that I was not to pay him and not to have his services. He lost no time in making his escape. I learned that the Young Wasp, the interpreter's son, who accompanied me, had told him I was a great medicine-man, which is always understood as a necromancer, or being possessed of or conversant with evil spirits, and had the power of doing great wonders; and should he go with me, if he did not do as I wanted, though very likely I did not kill him, he might depend I would turn him into a grisly bear to run and live in the mountains, and he should never see his wife again, which of course acted powerfully on him. The boy was paid by his father according to the merit of his services. I was detained till ten o'clock, when Mr. Black procured me another guide, whom I took the more willingly as he was no smoker and at the time such a knave, that no one would dare to steal from him ; and it is worthy of notice that there has hardly ever been an instance of dishonesty known when trust was placed in them by depositing property in their hands. Proceeded along the north banks of the Wallawallah River, and then took to the north branch, intending to touch on the range of mountains a little more to the north from where I was last week. As my present journey is through the same sort of country and climate as the preceding, I found until I reached the mountains, which occupied better than two days, but few plants that I had not seen before. I was more fortunate than before, as the weather continued dry but oppressively warm. Maximum heat 84° to 98° ; minimum 64°, and on the mountains 31°.

Wednesday and Thursday, 28th, 29th.—Continued my journey in the mountains. Reached the snow on the evening of 28th ; camped at dusk. As I saw that little different could be gleaned from what I had collected before, I did not think it proper to make my stay long, particularly as it was my wish to have a few days' collecting and drying seeds found on the plains, against the coming of the people from the ocean, who are daily expected. Therefore as I had got a few more species for the collection, with a few papers of seeds, I left the mountains at mid-afternoon and gained the Columbia on Saturday at nine in the morning, much worn down and suffering great pain from violent inflammation of the eyes. To read or write I cannot, but in the morning without pain. Collected as follows :

(175 [*bis*]) *Allium* sp. ; perennial; flowers white ; a small low plant, 4 to 5 inches high ; on the mountains near the snow ; plentiful.

(176 [*bis*]) *Allium* sp. ; perennial; leaves flat, smooth, and glaucous; flowers large, purple, fragrant; bulb very mild ; a foot to 15 inches high ; on the subalpine hills, in dry soils ; abundant.

(177 [*bis*]) *Veratrum* (?) sp. ; leaves amplexicaul, ovate, acute, plaited, pubescent below ; flowers white ; a strong plant, 5 to 8 feet high ; on the undulating grounds, banks of rivers and outskirts of woods ; plentiful.

(178 [*bis*]) *Tetrandria, Monogynia*, biennial; calyx four-partite; corolla four segments, ovate, spreading, with a long gland in the centre of each covered with purple hairs; flowers faint with purple spots, verticillate, numerous ; leaves verticillate, sessile (generally four), lanceolate, entire, smooth ; a strong plant, 4 to 9 feet high ; abundant on the low hills ; may prove to be *Frasera*, which for want of seeds cannot be ascertained.

(179 [*bis*]) *Oenothera* sp.; biennial ; leaves alternate, sessile, ciliated, slightly pubescent above ; flowers very small, faint purple; stem pubescent ; 4 to 5 feet high ; near rivers ; abundant.

(180 [*bis*]) *Lupinus* sp. ; perennial; leaves digitate, five to nine ; leaflets lanceolate, silky, more so below; stem slightly pubescent; flowers small, white ; a foot to 18 inches high ; this fine plant, which is seen constantly with white flowers, and invariably alone, without any others of the genus, I first saw at Okanagan on the Columbia, and took it at that time only for a variety, and if it proves constant, it will be an addition ; in dry soils, at the foot of the mountains ; abundant. Collected fine specimens of *Tigarea* [1] in perfect seed.

(181) *Diadelphia*, perennial; flowers bright red ; calyx deeply five-partite, with a solitary bractea larger than the segment; corolla five-petal ; stamens five, seldom more than three fertile; pistil exserted ; capsule one-celled, one-seeded ; seed small, orbicular; leaves pinnate, three, five, seven, glabrous, punctate ; a singular low but branching plant, 18 inches to 2½ feet high; on the plains at the foot of the mountains, in small clumps, in light poor soils ; rather rare.

(182) *Asclepias* sp.; perennial ; flowers whitish-yellow ; leaves nearly sessile, ovate, oblong, slightly pubescent beneath ; a strong plant, frequenting low points of land ; near rivers and lakes, common.

(183) *Chenopodium* sp.; perennial; leaves nearly sessile, slightly dentate, maculate, spots red ; near Indian villages; common.

(184) *Geum* sp. ; perennial ; flowers small, yellow ; on the mountains; abundant.

(185) *Bartsia* sp.; perennial; leaves alternate, sessile, entire, linear-lanceolate ; nerves on the back scabrous ; flowers of a deep glossy-scarlet; plant 2 to 3 feet high, rarely branching ; on the subalpine mountains; plentiful ; a distinct species from that found near the coast last year.

(186) *Delphinium* sp.; perennial ; leaves smooth, multipartite ; flowers bright azure-blue, on a long spike ; margins of rivers and lakes, in deep alluvial soils.

[1] *Purshia*, Benth. and Hook. f. Gen. Pl. i. p. 617.

(187) *Cruciferae*, annual; stem smooth and glossy; 6 to 12 feet high; never more than one stem from the same plant; leaves alternate, smooth, linear, acute; flowers appear to be faint red and white, not yet in full blossom; on the plains, in dry soils; not abundant.

(188) *Bromus* sp.; perennial; in low meadows; plentiful on the Wallawallah River.

(189) *Spiraea* sp.; partly shrubby; leaves orbicular, smooth, serrate; flowers white; on the low hills and channels of rivulets; plentiful.

(190) *Cypripedium* sp.; on high grounds, among low brushwood.

(191) *Serapias* sp.; perennial; in low wet points of land among grass; flowers yellow, with purple veins.

(192) *Orchis* sp.; flowers small, white, fragrant; in the same place as the former; also plentiful.

(193) *Rhus* sp.; flowers dingy-white and yellow; near springs on the hills; abundant; a foot to 18 inches high.

(194) *Trifolium* sp.; perennial; stem and petiole pubescent; leaflets very long, underside less pubescent than the stem, and minutely ciliated; flowers very large and pure white; a splendid species and it appears not to be included in the flora; on the low hills; plentiful.

Tuesday, July 4th.—Employed since my arrival drying the plants collected, gleaning and putting up seeds; weather changeable.

(195) *Malva* sp.; perennial; stem reclining; leaves reniform, orbicular, scabrous below, pubescent above; flowers fine yellow; in low sandy soils, near springs; not abundant.

Wednesday, 5th.—Made a short excursion on the high grounds south of Wallawallah River, where I collected a few seeds and one species of *Oenothera.* On my return in the evening I found Messrs. McDonald and MacKay on their way to the sea (the same persons I accompanied last year in August), a few days' march on the Multnomah River. As they are to proceed by water as soon as the boats are in order, and have offered to take anything I have to send to the coast, I shall without delay pack up all I have ready for sending.

Thursday 6th to Saturday 8th.—Employed making a box, airing plants and seeds, and packing the same. Very warm. Thunder in the evenings.

Sunday, July 9th, 1826.—At the junction of Lewis and Clarke's River. Wrote as follows to Jos. Sabine, Esq. :—

DEAR SIR,—This day month I wrote you from this place, and at that time I stated how my time would be taken up during the summer. I have a few days ago arrived from a fatiguing journey on the Blue Mountains, spoken of in my last letter, and have been very successful. I have found on those alpine snowy regions a most beautiful species of *Paeonia, Lupinaster macrocephalus,*[1] a splendid species of *Trifolium* equally fine, *Lupinus argenteus* of Pursh, and another species by far the finest of the tribe, not even excepting *L. nootkatensis*; has a spike a foot to 20 inches of full blossom of a deep golden-yellow; I have placed Mr. Turner's name behind it; the plant 4 to 6 feet high. One species of *Pentstemon* different from any spoken of

---

[1] *Trifolium megacephalum*, S. Wats. Bibl. Ind. N. Am. Bot. p. 264.

before, with an assemblage of smaller plants. I have been continually on my feet, scarcely three nights in one encampment. As I have accidentally met with a Mr. McDonald on his return from a hunting excursion in the south, and the same person I accompanied on a few days' march last August on the Multnomah River, he has kindly offered to take the result of my labours for the last month, which I willingly accept. The collection consists of three bundles of dry plants (ninety-seven distinct species), forty-five papers of seeds, three *Arctomys*, and one curious rat, which I hope you will receive safe. On the evening of the 9th of June, I received a letter from Mr. Goode, and one from Mr. Booth by an Indian from the coast. Mr. McLoughlin did not deem it safe to send what he conceived your letter, and very properly kept it until his people should return from the ocean. I am just going down the river, two days' march, in hopes of receiving yours, for my patience is completely exhausted; and as I am now ready to bend my steps again to the upper country, I cannot well go without hearing from you. I have not a moment to spare, as the people are starting. May I ask Mr. Goode to make a note to Mr. Atkinson saying I am well, who will inform my brother, as I have no time to write him ? I can only think of Mr. Turner, Lindley, and Munro. It is impossible I can write them a single line.—I am, Dear Sir, your obedient servant, D. DOUGLAS.

To JOSEPH SABINE, Esq.

P.S.—July 11, on the Great Falls of the Columbia at sunrise : I arrived here last last night and have secured your communication. To say I am happy would only convey but a faint idea of the rapture I enjoy in hearing from you. I have no time ; am just setting out again for the upper country and am glad to think I may reach the coast before the ship sails for England.—D. D.

Embarked at 10 A.M. and proceeded with a speed not less than twelve miles an hour, the river being at its greatest height. About two a strong wind from the west set in, which greatly retarded our progress and obliged us to put on shore an hour before sun-down. As no salmon could be caught, there being too great a swell for the canoes to go out, and having nothing to eat, a horse was killed, part of the flesh of which and a mouthful of water I made my supper. Very warm during the day, 97° Fahr. in the shade ; thunder and vivid sheet-lightning during the night. Having no tent, was dreadfully annoyed by mosquitoes.

Monday, July 10th.—After a cheerless night and but little refreshed, resumed my route at daybreak, the morning being calm and pleasant ; put ashore at Day's River, a southern branch of the Columbia, at eleven o'clock to breakfast, which occupied an hour. As there was enough boiled last night no fire was necessary except for smoking. This part of the river being very rapid, gained already about fifty or fifty-five miles. While eating my food an Indian who was standing alongside of me managed to steal my knife, which was tied to my jacket by a string, and being the only one used for all purposes I was loth to part with it. I offered a reward of a little tobacco for its recovery, without effect. At last I commenced a search and found it secreted under the belt of one of the knaves. When detected he claimed the premium, but as he did not give it on the first application, I paid him, and paid him so well, with my fists that he will, I daresay, not forget the *Man of Grass* for some days to come. As the wind blew up the river very strong after midday, I had to put to shore twelve miles above the Great Falls, not being able to get round a rocky island at this part of the river. Remained until 4 P.M. when the wind abated, and then proceeded. On my gaining the Great Dalles six miles below the Great Falls at seven o'clock, I observed smoke rising among the rocks; thinking it to be Indians fishing, I walked to the lower end in quest of salmon, but instead of Indians I was delighted beyond measure to find it the camp of the brigade from the sea. I cannot describe the feeling which seizes me even on seeing a person again, although I am but partially acquainted with them. After travelling in the society of savages for days together and can but speak a few words of their language, assuredly the face of a Christian although strange speaks friendship. It was the more agreeable to me as I am previously acquainted with all the persons in authority; and I should be destitute of every feeling of gratitude if I did not mention the kindness and hospitality shown by all. One caused water to be brought me to wash, while another was handing me a clean shirt, and a third employing himself cooking my supper. My old friends Mr. Work and Mr. Archb. McDonald handed me my letters, which were grasped greedily and eagerly broken open. Received one from Joseph Sabine, Esq., and Mr. Munro one, both gratifying; one from Mr. Atkinson, and one from my brother, with a note from Mr. McLoughlin and some other friends at the sea. There is a sensation felt on receiving news after such a long silence, and in such a remote corner of the globe more easily felt than described. I am not ashamed to say (although it might be thought weakness by some) I rose from my mat four different times during the night to read my letters; in fact, before morning I might say I had them by heart—my eyes never closed.

Tuesday, 11*th*, *to Saturday*, 15*th July.*—At daylight made a note to Mr. Sabine mentioning the receival of his letter, and proceeded with the boats up the river at sunrise. I commenced walking and continued to do so throughout, gathering seeds and any plant, starting in the morning and walking to camp with the people in the evening; found nothing particular. Weather warm and dry generally, with thunder in the evening. Arrived at Wallawallah at noon on Saturday.

Sunday, 16*th.*—Gathered some seeds, which I packed in a small chest with some note-books and other things, to be taken to Kettle Falls by water. Thermometer in the shade 96°.

Monday, 17*th.*—In company with Mr. Work and McDonald started on a journey by water with a party of twenty-eight men for the fork of Lewis and Clarke's River, about 150 miles from the Columbia, and as their marches would be short, I hope to put myself in possession of most of the plants found along the banks of that river. Camped fifteen miles up that river.

Tuesday 18*th to Monday* 24*th.*—This river is of considerable magnitude, some places 250 to 300 yards broad, in many places very deep and rapid:

at least four times as much water as there is in the Thames ; the general course east. Twenty-five miles from its junction with the Columbia from an undulating dry, barren country it changes to high rugged mountains, and not a blade of grass to be seen except in the valleys or near the springs, where a little withstands the intense heat. Rose always at day-break and camped about 10 A.M., and rested till 3 or 4 P.M., the heat being too great for any exertion in the middle of the day, when we made generally fifteen or twenty miles further in the cool of the evening. Some idea may be formed of fatigue experienced on this voyage, when the thermometer frequently stood from 98° to 106° of Fahr. in the shade, destitute of a screen from the scorching sun. The only thing I might say that renders it superior to the deserts of Arabia is abundance of good water enjoyed in inland voyages. That excepted, there is but little difference. Salmon are caught in the river and in some of its branches near the Rocky Moun-tains, but by no means so plentiful as in the Columbia nor of such good quality. We obtained occasionally a few of them from the Indians, but their extreme indolence prevents them from catching barely what serves themselves. Our general fare was horse-flesh cooked by boiling, and sometimes roasted on the point of a spike before the fire. I learn that the wants of the natives are simple and they require but little to support life in original simplicity. From the oppressive heat I found great relief by bathing morning and evening, and although it causes weakness in some degree, I have some doubt if I had not I should not have been able to continue my trip. On Monday arrived at the branches of the river at dusk, where was a camp of three different nations, upwards of five hundred men able to bear arms. One called Pelusbpa, one the Pierced nose (Chawhaptan) and Chamuiemuch. The chiefs or principal men came and stayed with us till bed-time and presented some favourite horses.

Tuesday 25th till Monday 31st July.—As I understood from my companions that their stay would be for a few days, I was desirous of making a trip to the mountains, distant about sixty miles, the same ridge I visited last month further to the south-east ; but as they had not yet made any arrangement with the natives, it was not thought prudent to go from the camp, so I was guided by their advice. On Wednesday a conference was held and ended favourably and with great splendour by dancing, singing, haranguing, and smoking. All were dressed in their best garments, and on the whole presented a fine spectacle and certainly a new one to me. On Thursday at daylight—Mr. McDonald having been so kind as to send one of his men (Cock de Lard) with me, more as a companion than guide, for he was as much a stranger as myself—we set out in a south-easterly direction, the country undulating and very barren. In the course of the day passed only two springs, and as I was uncertain if more were near and the day far spent, I camped at four o'clock in the afternoon. Found only one species of *Pentstemon* and a few seeds. Very warm. On Friday reached the mountains at nine o'clock and took my breakfast (dry salmon and water) among some very large trees of *Thuya occidentalis*, the spot pointed out to me by the Indians where Lewis and Clarke built their canoes, on their way to the ocean, twenty-one years ago. I left my man to take care of the horses at the foot of the mountain, while I ascended to see if it afforded anything different from what I had seen before. Reached the highest peak of the first range at 2 P.M., on the top of which is a very remarkable spring, a circle 11 feet in diameter, the water rising from 9 inches to 3½ feet above the surface, lowering and rising at intervals, in sudden gushes ; the stream that flows from it is 15 feet broad, and 2½ deep, of course running with great force as its fall is 1½ foot in 10 and it disappears at the foot of the hill in a small marsh. I could find no bottom to the spring at the depth of 60 feet. Surrounding the spring there is a thicket of a species of *Ribes* belonging to section *Grossulariae*, 12 to 15 feet high, with fine delicate fruit of a very superior flavour and large, nearly as large as a musket-ball. This fine species I have not seen before ; should it prove new, I hope it may be called *R. Munroi*, as I have called the spring Munro's Fountain ; at the same time how delighted he would feel to see such in the garden. Found in seed, nearly ripe, *Ribe viscosissimum*, and, lest I should not meet with it in a better state, gathered a quantity of it. Found a few seeds of *Paeonia*, but not so ripe as I should have wished, with a small species of *Vaccinium* and a few seeds of *Xylosteum* which I saw in blossom on the mountain near Spokane in May. I joined the man and horses at six o'clock and set out for my encampment of last night. On arriving and looking for something to eat, I found that only salmon for one day had been put in by the man in mistake, and both having a good appetite we mutually agreed to make for the camp. Set out at dusk (Cock de Lard undertaking to be guide) before the moon rose, at least before it became visible ; he took us out of the way about ten miles. Arrived at the camp at sunrise, when I threw myself down in one of the tents to sleep. I had not been asleep more than two hours when I was hurriedly aroused to take on myself the profession of a soldier, a misunder-standing having taken place between the interpreter and one of the chiefs ; the latter accusing the former of not translating faithfully, words became high till at last the poor man of language had a handful of his long jet hair torn out by the roots. On the Indian being reproved, he went off in a fit of rage and summoned his followers, amounting to seventy-three men. All arrived and came to our camp with their guns cocked and every bow strung. As every one of our party had done all in his power that it should be mutually and amicably adjusted and been refused, every one seemed more careless for the result than another. We (thirty-one of us) stood to our arms and demanded if war was wanted ; it was answered ' No, we want only the interpreter killed, and as he was no chief there could be but little ill done.' They were told that whatever person we had in our party, whether chief or not, or if it was only an Indian under our protection, should they attempt to kill or disturb him in the least, certainly they would know we had been already in war. The coolness that seemed to be the prominent feature in our countenance had the desired effect of cooling their desire for war and made them glad to ask for peace, which on our part was as willingly granted them. Many speeches were made on the

[1] *Lonicera*, Benth. and Hook. f. Gen. Pl. ii. p. 5.

occasion, and, if it may be allowed to judge from gesture and the language of nature, many of them possess qualifications that would be no disgrace to a modern orator. Although there is much repetition in their harangues, delivered with much vehemence and intense feeling, they are uniformly natural and are certainly calculated either to tie the knot of affection and sympathy, or rouse the mind to discord and war. I have observed speakers and hearers so overcome that they sobbed and cried aloud, and the proceedings delayed until they recovered. This affair was concluded in the usual way—exchange of presents. Although friendship had again been restored, it would have been imprudent to have gone from the camp; therefore I employed myself putting in order those collected and airing some seeds. On Sunday at midday we rose camp and pitched on the northern shore of the north branch.

Monday, July 31st, to Friday, August 4th.—Early in the morning I had the plants and seeds which were collected carefully secured in one of the saddle-bags. Parted with Mr. McDonald, who descended the river; and Mr. Work with two men and myself took our departure overland in a north-easterly course to Kettle Falls on the Columbia. On gaining the top of the hills near the river we found the road good in many places and in others very bad, with badger and rabbit holes, which at this season are covered with grass, rendering them more dangerous. Made about forty-five miles. Camped at mid-afternoon on a low piece of ground where there was water; passed only one spring in the course of the day. The whole country destitute of timber; light, dry, gravelly soil, with a scanty sward of grass. Gathered seeds of two species of *Astragalus*. Thermometer 97°; heavy dews during the night; minimum heat 53°; rain towards daybreak. Tuesday and Wednesday started early in the morning and went briskly on till eleven, when we halted to breakfast and rest during the heat of day. I opened my saddle-bags and exposed my seeds, lest they should have been wet or damp with the rain last night, and then made a circuit on the high grounds to the south, leaving my watch with Mr. Work to know when to start, and as they would pass near my track would pick me up. Remaining longer than expected I went back and found him and the two men fast asleep; packed up again and proceeded till dusk, when we camped under a solitary poplar on the margin of a stagnant pond full of *Nuphar luteum, advena*, and one species of *Potamogeton*; water very bad. Warm during the day, a fanning wind during the night, which prevented us from being troubled with mosquitoes. Gathered some seeds of *Mimulus albus* [*sic*] and laid in specimens, and put up a paper of seeds of a small species of the same genus. On Wednesday came to the high grounds on the south side of the Spokane River, which is intersected with belts of wood. We intended to stop at a small rivulet in the second one, but our guide (one of the men) missed the road and made too far to the east, where we could get no good water until three o'clock, when we made a stop to take breakfast near a small lake with fine cool water. Made a stay of two hours and proceeded from this lake due north through woods and plains until we came to the banks of the Spokane River, two miles below the falls of that river; the fall is a perpendicular pitch of 10 feet across the whole breadth of the river. Proceeded down the south side of the river seven miles, and as night stole in on us, and having to pass some very high perpendicular rocks, we chose to camp rather than go to the old establishment so late at night. Gathered some seeds of a species of *Lilium* growing in open or partially shady woods. I am sorry that no seeds of *Lilium pudicum*[1] can be found perfect, and, as I could not get any, dug up some roots—perhaps they may keep to England. Warm; thermometer 99° maximum, 56° minimum. Heavy dew.

Thursday and Friday, 3rd, 4th.—At nine o'clock in the morning crossed the Spokane River to the old establishment on the south side, where we found old Mr. Finlay, who gave us abundance of fine fresh salmon from his barrier, placed in a small branch of the main river. After breakfast, and having the horses crossed, left that place at noon for the Columbia. An hour's ride from that place passed the Indian camp on the north side of the river, where they were employed fishing. Their barrier, which is made of willows and placed across the whole channel in an oblique direction, in order that the current which is very rapid will have less effect on it, has a small square of 35 yards enclosed on all sides with funnels of basket-work (just made in the same manner as all traps in England), and placed on the under side, through which the salmon passes and finds himself secure in the barrier. When the spearing commences, the funnels are closed with a little brushwood. Seventeen hundred were taken this day, now two o'clock; how many may be in the snare I know not, but not once out of twelve will they miss bringing a fish to the surface on the barb. The spear is pointed with bone and laced tight to a pointed piece of wood a foot long and at pleasure locks on the staff and comes out of the socket when the fish is struck; it is fastened to the staff by a cord. Fifteen hundred and sometimes two thousand are taken in the course of the day. Camped a little before sundown not far from one of my encampments in May on my way to Spokane. Gathered a few seeds of a strong species of currant with white blossoms. Cold during the night.

On Friday as usual started at daybreak. About seven o'clock I went off the way to gather some more seeds of the *Ribes* I saw last night. As I found them but scanty it took up a considerable time picking a sufficiency. Overtook my companions ten miles beyond Barrière River, which I had to swim in spring. Rode across it to-day, halted at noon to breakfast, and proceeded on at two o'clock. On crossing Cedar River, a small rapid stream nine miles from the Columbia, my horse on gaining the opposite shore, which is steep and slippery, threw back his head and struck me in the face and I was plunged head foremost in the river. I fortunately received no further injury than a good ducking and got wet what seeds I had collected during the day, which were in my pockets and knapsack with my note-book. Arrived at Kettle Falls at 7 P.M., having been absent two months, and was cordially and hospitably entertained by Mr. Dease.

August 5th and 6th.—In addition to a good many seeds, collected the following plants on my journey:

(196) *Helonias* sp.; perennial; root somewhat bulbous; leaves long,

---

[1] *Fritillaria pudica*, Baker, in Journ. Linn. Soc. xiv. p. 267.

grassy ; flowers white, in a long spike ; abundant in low valleys between Spokane and Kettle Falls.

(197) *Mimulus* sp. ; annual ; leaves alternate, nearly sessile, slightly dentate ; stem pubescent ; flowers small, yellow ; on moist rocks and mountainous grounds ; by no means so plentiful as *M. luteus*.

(198) *Gentiana* sp. ; perennial ; leaves opposite, lanceolate, smooth, entire ; flowers fine blue ; in the mountain valleys in small tufts ; abundant.

(199) *Pedicularis* sp. ; annual ; flowers light yellow, slender, and rarely branching ; in wet meadows ; abundant.

(200) *Potentilla* sp. ; shrubby ; leaves somewhat pinnate, hirsute ; flowers yellow ; a low plant, 2 to 3 feet high, frequenting margins of mountain lakes and rivulets ; abundant ; perhaps may prove *P. fruticosa*.

(201) *Astragalus* sp. ; perennial ; smooth ; flowers bright blue ; a strong plant, 2 to 3 feet high ; on moist ground.

(202) *Silene* sp. ; perennial ; leaves opposite, sessile, lanceolate ; flowers white ; plant glutinous ; a low plant, a foot to 18 inches high ; in small tufts on rocky and gravelly soils near Spokane.

(203) *Didynamia* sp. ; perennial ; leaves opposite, slightly dentate ; flowers purple-rose ; margins of streams ; abundant.

(204) *Syngenesia* sp. ; perennial ; on rocks in open woods near Spokane.

(205) *Trifolium* sp. ; annual ; flowers faint red ; on wet soils in the valleys ; not plentiful.

(206) *Orchidaceae*, perennial ; aquatic ; leafless ; flowers yellow ; a singular plant, in a small pond, growing around the leaves of *Nymphaea* ; rare.

(207) *Trifolium* sp. ; perennial ; leaves very long, minutely serrate ; flowers very large, bright red ; a strong plant, 2 to 3 feet high ; this fine plant, which I consider a new species, will I have no doubt be a profitable fodder in cattle-feeding.

(208) *Astragalus* sp. ; perennial ; in open woods in damp soils ; abundant ; flowers white.

(209) *Eriogonum* sp. ; perennial ; leaves oblong ; flowers sulphur-yellow ; abundant in all dry, hilly, sandy soils.

(210) *Oenothera* sp. ; perennial ; plant low and branching ; leaves ovate, slightly dentate ; flowers terminal, white, fragrant ; a low plant, 6 inches to a foot high ; on the banks of Lewis and Clarke's River.

(211) *Artemisia* sp. ; flowers dirty-yellow ; abundant on the banks of rivers.

(212) *Syngenesia* sp. ; perennial ; flowers light purple-blue ; on the channels of rivers ; abundant.

(213) *Trifolium* sp. ; perennial ; flowers red and white ; a low spreading plant ; plentiful in low meadows.

(214) *Melissa* (?) sp. ; perennial ; flowers faint purple ; on the mountains at the junction of Lewis and Clarke's River ; 2 to 4 feet high ; in all rocky soils.

(215) *Oenothera* sp.; annual; may prove the same as 210, but has red flowers; found in the same places.

(216) *Syngenesia* sp.; flowers yellow; a stunted low shrub, 2 to 3 feet high; on the plains; abundant.

(217) —— a low shrubby tree, 1 to 3 feet high, with a white bark; the wood very hard; this I saw last year a low stunted plant, and took it to be *Maclura*; capsule small, one-seeded.

(218) *Lupinus* sp.; perennial; hirsute; flowers blue; legume three to four seeded; on the plains, in small patches; rare.

(219) *Chenopodium* sp.; annual; near Indian camp and all rich ground; abundant.

(220) *Ribes* sp.; leaves partly five-lobed, slightly pubescent; fruit large, round, smooth, bright red, with a fine flavour; a very strong plant, 8 to 12 feet high; round Monro's Fountain; on the mountains at the branches of Lewis and Clarke's River.

(221) *Aconitum* sp.; in shady woods in rich soils; flowers blue; 4 to 6 feet high.

(222) *Syngenesia* sp.; perennial; flowers dark purple; on the mountains.

(223) *Frasera* (?) sp.; biennial; leaves verticillate, ovate-lanceolate, smooth, entire; flowers white; on the mountains; rare; 2 to 4 feet high.

(224) *Hexandria, Monogynia*; a fine bulb; may prove different from that found last year at the Great Falls of the Columbia; on the mountains of Lewis and Clarke's River.

(225) *Pentstemon* sp.; perennial; leaves sessile, somewhat amplexicaule, ovate, nearly entire, pubescent; whole plant glutinous; flowers large, fine purple-blue; this fine plant will I think be a great addition to that fine genus; on the hills near the branches of Lewis and Clarke's River; rare.

(226) *Eriogonum* sp.; tomentose; flowers white; on all dry gravelly soils; abundant.

(227) *Clethra* sp.; flowers white; leaves ovate, serrate; a fine strong evergreen shrub, 4 to 6 feet high; abundant in mountain woods.

August 7th, Monday, to Tuesday the 15th.—Continued collecting seeds, drying, and packing. The boats came on Monday, but unfortunately the small box containing seeds, plants, and sundries which I left at the junction of Lewis and Clarke's River to be brought in one of them was forgotten, and I regret it the more as it contains forty-six papers of seeds to be sent across the Rocky Mountains this autumn, which of course I will take with me in the spring. By the 15th found seeds of most of the plants around Kettle Falls; except four species of *Pinus*; a note of which I have made to Mr. Work, to gather them for me in October. I regret that no perfect seeds of *Lilium pudicum*[1] nor *Claytonia lanceolata*[2] could be found, being both scorched up shortly after flowering. Dug a few roots of the former, which may keep; not one of the latter could I find. I had a journey projected to

---

[1] *Fritillaria pudica*, Baker, in Journ. Linn. Soc. xiv. p. 267.
[2] *Claytonia caroliniana* var. *sessilifolia*, S. Wats. Bibl. Ind. N. Am. Bot. p. 117.

the upper parts of Flathead or Clarke's River in company with Mr. Work, as he is under the necessity of making a new arrangement and altering his course in consequence of some misunderstanding (or war party) among the different tribes inhabiting that country. I do not think it would be of much moment to go on his route, as the country is exactly the same as that in this neighbourhood.

Wednesday, 16*th.*—As I learned from Mr. McLoughlin that the vessel would not sail for England until the 1st of September, and as the bulk of my plants are at the sea, and having a collection of seeds amounting to 120 of this year's gleanings, and at the same time, as I also learn that this may be the last ship for some years going direct for England, makes me very desirous of sending all that is possible by her. Should Mr. Simpson (the Governor) not arrive in a few days, I must endeavour to reach the ocean by some means or other, taking with me my collection of seeds to be sent home. To-day packed in a small box five quires of paper, three brown and two cartridge, sixty-nine papers of seeds—a portion of those collected on my journey on Lewis and Clarke's River and since my arrival here— with two linen shirts to be sent across the Rocky Mountains; where I will find it early in June at Fort Edmonton. Very warm. Mr. Dease spoke to the *Little Wolf,* a chief of the Okanagan tribe, about guiding me to that place. The river, which is still high with the cascades, rapids, dalles, and whirlpools, renders it impossible to go by water without six or eight men in a canoe; and indeed there are none at this place sufficiently large at present.

17*th.*—Packed one bundle of dry plants in my trunk among my little stock of clothing, consisting of one shirt, one pair of stockings and a night-cap, and a pair of old mitts, with an Indian bag of curious workmanship, made of Indian hemp, *Apocynum* sp., *Helonias tenax,*[1] and eagles' quills, used for carrying roots and other articles in. A party of twenty-one men and two females belong to the Cootanie tribe, whose lands lay on the shores of the small lake called Cootanie Lake, the source of the Columbia, and that small neck of land at the head-waters of McGillivray's River. An old quarrel of nine years' standing existing between them and the tribes on the Columbia lakes, sixty miles above this place, who are here at present at the salmon fishing at the Falls, gave Mr. Dease and every other person much uneasiness. The parties met stark naked in our camp, painted, some red, black, white, and yellow, with their bows strung, and such as had muskets and ammunition were charged. War-caps of calumet-eagle feathers were the only particle of dress they had on. As one was in the act of letting the arrow from his bow, aiming at a chief of the other party, Mr. Dease fortunately brought him a blow on the nose which stunned him. The arrow grazed the skin and passed along the rib opposite the heart without doing much injury. The whole day was spent in clamour and haranguing, and as we were not too sure what might be the result, we were prepared for the worst. Mr. D. proposed that they should make peace to-morrow, and that it would be much better they should go to each other's lands as friends than butchering each other like dogs. His advice they said they should

---

[1] *Xerophyllum asphodeloides* var. Baker, in Journ. Linn. Soc. xvii. p. 467.

follow; that they would come early in the morning. The Wolf, being one of the principals on one side, told me he cannot go to-morrow, as the peace is to be made, which could not be well done without his presence.

Friday, 18*th*.—Bustle and uproar; towards evening peace was signed and sealed by an exchange of presents; and as there is to be a great feast on the occasion, the Wolf is uncertain when he can be spared from his office. As my time is very short, Mr. D. spoke to an Indian who is in the habit of attaching himself to the establishment and going on journeys with his people, to go with me—to which he agreed at once. So I will start to-morrow early.

Saturday, 19*th*.—Detained till eight o'clock, although Mr. Dease had sent out early for the horses, but being far off could not be caught sooner. I had put up for my journey some dried meat of buffalo flesh, a little sugar and tea, and a small tin pot. My gun being left, out of order, in the lower country, Mr. Work kindly gave me the loan of a double-barrelled rifle-pistol; and going alone as I am, it is perhaps much better to have as little as possible of tempting articles about me. I left this delightful place highly gratified, having made a tolerable addition to my collection and received every kindness from the hospitable people which they had in their power to show. Being short of clothing, Mr. Dease gave me a pair of leather trousers made of deerskin and a few pairs of shoes, which in my present state were very acceptable. He provided me with three of his best horses, one for my guide, one for carrying my little articles, and one for myself. The only thing in the way of clothing except what was on my back, was one shirt and one blanket, and in this shape I set out for Okanagan, distant 250 miles north-west of this place. It was my intention to have gone by water, but was dissuaded by Mr. D., that part of the river at this season being high and from numerous cascades and rapids perhaps dangerous. Proceeded along the south banks of the Columbia, intending to cut the angle between the Columbia and Spokane Rivers. My path very mountainous and rugged, in many places covered with timber of the same sorts as are commonly seen abundantly over all the country Nothing occurred to-day, my guide (to whom I cannot speak a single syllable) seems to conduct himself very well. Camped at sundown near a small spring, surrounded by a thicket of birch and willow, at the foot of a high conical hill ten miles from the Spokane River.

Sunday, 20*th*.—Shortly after two o'clock I had my horses saddled, and, the ground being very uneven and stony, drove them before me; and the moon shining delightfully clear, I found it by far more pleasant travelling during the night than the day. Arrived on the Spokane River nine miles from the Columbia, where there was a large number of Indian lodges, being a fishing ground. After making a short stay and presenting them with a little tobacco, four of them accompanied me two miles further down the river, where they assisted me in crossing the horses and carried myself and all my property across in their canoe. At ten o'clock left the woody country and began my course through a trackless barren plain, not a vestige of green herbage to be seen, soil gravel and sand. About one o'clock I halted, to rest the horses and take some breakfast, opposite the

Grand Rapids—having already made nearly fifty miles—and made a small pan of tea, which I let stand till it cooled and settled, and then sucked the water of the leaves. In the interval I gleaned a few seeds, bathed in the rapid, which recruited me greatly, and again in the cool of the evening resumed my route, course west. Towards dusk came to a small pool of stagnant water, very bad, and having nothing to qualify it I was urged to continue till eleven o'clock, when I came to a small spring, but without a single twig for fuel. I made an effort to boil my little pan with dry grass, a large species of *Triticum*, but was unable to succeed. Being an old encampment I fortunately found some horse-droppings, by the aid of which and the grass mentioned I managed to make some tea, when shortly afterwards I laid myself down to sleep on the grass. I have a tent, but generally am so much fatigued that the labour of pitching it is too great. Here it could not be done for want of wood, and tent-poles cannot be carried.

Monday, 21*st*.—To-day I overslept myself; started at four o'clock. The country same as yesterday; at eight passed what is called by the voyageurs the Grand Coulee, a most singular channel and at one time must have been the channel of the Columbia. Some places from eight to nine miles broad; parts perfectly level and places with all the appearances of falls of very extraordinary height and cascades. The perpendicular rocks in the middle, which bear evident vestiges of islands, and those on the sides in many places are 1500 to 1800 feet high. The rock is volcanic and in some places small fragments of vitrified lava are to be seen. As I am situated, I can carry only pieces the size of nuts. The whole chain of this wonderful specimen of Nature is about 200 miles, communicating with the present bed of the Columbia at the Stony Islands, making a circular curve $1\frac{1}{2}°$ further south, and of course longer than the present chain. The same plants peculiar to the rocky shores of the Columbia are to be seen here, and in an intermediate spot near the north side a very large spring is to be seen which forms a small lake. I stayed to refresh the horses, there being a fine thick sward of grass on its banks. The water was very cold, of a bitterish disagreeable taste like sulphur. My horses would not drink it, although they had had no water since last night. At noon continued my route and all along till dusk. The whole country covered with shattered stones, and I would advise those who derive pleasure from macadamised roads to come here, and I pledge myself they will find it done by Nature. Coming to a low gravelly point where there were some small pools of water with its surfaces covered with *Lemna*, or duck weed, and shaded by long grass, one of the horses, eager to obtain water, fell in head foremost. My guide and myself made every effort to extricate it, but were too weak. As I was just putting some powder in the pan of my pistol to put an end to the poor animal's misery, the Indian, having had some skin pulled off his right hand by the cord, through a fit of ill-nature struck the poor creature on the nose a tremendous blow with his foot, on which the horse reared up to defend himself and placed his fore-feet on the bank, which was steep, when the Indian immediately caught him by the bridle and I pricked him in the flank with my pen-knife, and not being accustomed to such treatment, with much exertion he wrestled himself from his supposed grave. The water was so bad that it was impossible for me to use it, and as I was more thirsty than hungry I passed the night without anything whatever.

Tuesday, *22nd*.—Last night being very warm, with the whole firmament in a blaze of sheet lightning, and parched to a cinder, I passed a few miserable hours of rest but no sleep, and as usual set out before day ; and, my road being less mountainous, with little exertion I found myself on the Columbia at midday opposite the establishment. Seeing an old man spearing salmon I had the horses watered and hobbled, and crossed in a small canoe with my guide. Here I found my old friends Messrs. McDonald and Ermatinger, who received me with every kindness. After washing and having a clean shirt handed me, I sat down to a comfortable dinner. I was glad to find the small box, which I thought might have been overlooked in one of the portages, brought to this place by the former gentleman and left until I should pass. As my time was of great consequence I communicated to them my wish, and immediately they purchased a small canoe for me, and hired for me two Indians to go with me to the junction of Lewis and Clarke's River. In the meantime I wrote to Mr. Dease by my old guide, who behaved himself in every way worthy of trust and is to make a stay of two or three days to rest, and I then put up a few seeds and changed some plants collected on the journey. As I felt somewhat wearied I went early to bed ; the doors being left open by reason of the heat, and the windows, which are made of parchments being by no means close, gave the mosquitoes free access, I was under the necessity of abandoning the house at midnight and took myself to a sort of gallery over the door or gate, where I slept soundly. Before leaving this place early in the morning after breakfast, I had a little tea and sugar offered me, which I thankfully accepted, and a small tin pot made in the form of a shaving pot, the only cooking-utensil. They regretted the only provision they had that would do for carrying was dried salmon, but as I still had, through the goodness of Mr. D., my last host, a little dried meal, in that respect I was not so bad. Left at seven o'clock. In passing a long rapid (two and a half miles) about ten miles below the house, I took the precaution to take out my paper, seeds, and blanket, and was walking along the shore with them while the Indians ran the canoe down. When in the middle of the rapid a heavy surge broke over them and swept every article out of it except the dry meat, which being weighty by chance was wedged in the canoe, it being very narrow. The loss of the tea and sugar with the pot was a great one in my present situation, but I considered myself happy, having saved my papers and seeds. As I have said something of the river on my ascent, I need only observe my encampments. Camped at the mouth of the Piskahoas River at dusk. To-day, although I paddled all day—at least served as steersman—I did not feel fatigued but my hands were much blistered.

Thursday, *24th*.—By eight o'clock gained the Stony Islands, an extremely dangerous part of the river where the channels are very narrow, not more than 20 to 30 feet broad. As my guides were little acquainted with this part of the river, I hired an Indian of the place to pilot my canoe and after landing her safe below, I paid with a few crumbs of tobacco and a smoke from my own pipe. As I had nothing to cook I ate some crumbs of dried meat and salmon, and when I wanted to smoke kindled my pipe with my lens, so I was not under the necessity of making a stay to kindle a fire. Reached the top of the Priest Rapids at six o'clock, and although late I undertook to run the canoe down, making my old guide (they were father and son) carry my little parcels, he being tired. Night stole in on me too soon, and I was obliged to camp on the north side of the river in the middle of the rapid ; four and a half miles.

Friday, 25th.—I could not leave my encampment before daylight, having still four and a half miles of very bad water. I had left by land an hour before the canoe, and, after waiting nearly an hour at the foot of the rapid, as my guides did not make their appearance I became alarmed for their safety and returned, when I discovered them about a mile and a half above where I halted, comfortably seated in a small cove treating some of their friends to a smoke with some tobacco I had given them the preceding evening. As I had now a fine sheet of water without any rapids, but a very powerful current, I went rapidly on, like the day before scarcely out of my canoe, and arrived at Wallawallah, the establishment near Lewis and Clarke's River, at sundown. I felt so much reduced that I was too weak to eat, and after informing Mr. Black of my going to the sea and asking him to procure me two guides to carry me to the Great Falls in the morning, I laid myself down to rest on a heap of firewood, to be free from mosquitoes.

Saturday, 26th.—Wrote to Mr. McDonald by the old guide and gave him ten charges of ammunition and a little tobacco to buy his food on the way home, and after obtaining a larger canoe from Mr. Black in lieu of my present one, and two guides, I took my leave for the sea at six o'clock. At the foot of a rapid twenty-five miles below I purchased a fresh salmon, the half of which I roasted on a stick for breakfast and reserved the other half for next day, lest I should not get anything. As I knew all the bad places of the river, I went on all night drifting before the stream, taking the steering in turn, and as I had to pass a camp of Indians who are noted pillagers, made me anxious to pass during the night, which I accomplished.

Sunday, 27th.—Precisely at noon I reached the Great Falls, and finding my canoe too heavy to carry over the rocks I left it and hired one to carry me to the Dalles six miles below. Here I purchased a pair of horns of a male grey sheep of the voyageurs, for which I paid three balls and powder to fire them. The Indian had the skin dressed, forming a sort of shirt, but refused it me unless I should give him mine in return, which at present I cannot spare. On the Dalles were at least from five hundred to seven hundred persons. I learned that the chief Pawquanawaka, who would have been my last guide to the sea, was not at home ; but as I am now in my own province again, and understand the language tolerably well, I had no difficulty in procuring two, and was glad to find one who was well known to me. While he and his companion brought the canoe down the Dalles, after being refreshed with a few nuts and whortleberries

I proceeded over a point of land fifteen miles, taking an Indian to assist me in carrying my things. The canoe did not appear till an hour after dark. In the evening a large party of seventy-three men came to smoke with me, and all seemed to behave decently till I discovered that my tobacco-box was off. I had hung up my jacket and vest to dry, being drenched in the canoe descending the Dalles. As soon as I discovered this I perched myself on a rock, and in their own tongue I gave then a furious reprimand, calling them all the low names used to each other among themselves. I told them they saw me only one *blanket man* but I was more than that, I was the *grass man*, and was not afraid. I could not recover it. After all the quarrel, I slept here unmolested.

Monday, 28th.—Detained by a strong west wind till eight o'clock, when it became more moderate and I proceeded. Made but little progress. Camped fifteen miles above the Grand Rapids.

Tuesday, 29th.—As the wind increased with the day, I could not venture out in the stream, and even near the shore the waves were so high that I had to carry all my property on my back along the high shelving rocks, leaving the Indians to bring down the canoe. Arrived at the village on the Grand Rapids at three and repaired to the house of Chumtalia, the chief, and my old guide last year, where I had some salmon and whortleberries laid before me on a mat. I made a hearty meal and then spoke of procuring a large canoe and Indians to take me to the sea. He offered to go himself, but as he was busily employed curing salmon I was loth to accept his services, and took in preference his brother and nephew, with a fine large canoe, and proceeded down the lower end of the rapid in the evening. Camped on a low bank of sand in the channel where there was no herbage, so of course was not annoyed with insects. Long before daylight I was under way lest I should be detained with wind, which for the last three days rose with the sun. Passed Point Vancouver at sunrise. I had the gratification of landing safe at Fort Vancouver at midday, after traversing nearly eight hundred miles of the Columbia Valley in twelve days and unattended by a single person, my Indian guides excepted. My old friends here gave every attention a wayworn wanderer is entitled to. On their discovering me plodding up the low plain from the river to the house alone, unpleasant thoughts struck them. As the river was never seen higher than it has been this year, and of course caution is requisite in descending, they apprehended I was the only survivor. I confess astonishment came over me to meet a people from whom I had had more kindness a thousand times than I could ever have expected, look so strange on me ; but as soon as I dispelled the cloud of melancholy that sat on every brow I had that unaffected welcome so characteristic among people so far from home. I had a shirt, a pair of leather trousers, an old straw hat, neither shoe nor stocking nor handkerchief of any description, and perhaps from my careworn visage had some appearance of escaping from the gates of death. In the list of the little society I have here is a Peter Skeene Ogden, Esq. (brother of the Solicitor-General for the Canadas), a man of much information and seemingly a very friendly-disposed person ; also Capt. Davidson, of the ship

Dryade. I am glad to find that all my collections arrived safe here, except a male curlew and a female partridge. As I had no time to lose (to-morrow being the day appointed for the clearing of the ship), I unpacked and repacked seeds brought with me and those sent before from the interior.

August 31*st.*—Early in the morning, by Mr. McLoughlin's kindness in furnishing me with boxes, I had shortly after noon done all and wrote to Joseph Sabine, Esq. ; and should I have time in the evening I shall make a note to Mr. Munro and Mr. Atkinson. This I did, for although somewhat broken down by fatigue and in a state little qualified for giving my friends a year's news, I could not allow such an opportunity to pass if within the bounds of possibility. Wrote to Mr. Munro, Atkinson, Goode, Booth, and my brother. Another to the many restless nights I have experienced.

Friday, September 1*st.*—In the morning I saw my chests placed in one of the boats which were going with cargo to the ship. It was my intention to have seen them on board the vessel, but the captain arriving, to whom I gave a note regarding their situation in the ship, rendered it unnecessary ; and as I stood in need of a little ease I returned up the plain, having bid adieu to my countrymen about to visit England.

Saturday, 2*nd, to Friday,* 15*th.*—Weather warm and cloudy, with heavy dews at night. Employed myself gleaning a few seeds of choice plants collected last year : *Ribes sanguineum, Gaultheria Shallon, Acer macrophyllum, Berberis Aquifolium, Acer circinnatum* ; laid in specimens of *Pinus taxifolia* [1] with fine cones, and collected a few sections of the various woods, gums, and barks of the different timbers that compose the forest in this neighbourhood. On Thursday, I consulted with Mr. McLoughlin about my proposed journey to that country south of the Columbia and the Multnomah and towards the Umpqua River. I had spoken to Jean Baptist Mackay, one of the hunters, who goes sometimes through the country contiguous to the latter river ; but as he had been here in July, fully six weeks sooner than expectation, and was gone before I arrived, perhaps I might have difficulty in overtaking him. Mr. McLoughlin informed me that a party in a few days would be despatched for that quarter under the superintendence of A. R. McLeod, Esq., a gentleman who has given me much civility, and that there would be nothing to prevent my accompanying him. I could not allow this favourable opportunity to escape, and expressed my thanks to the persons in confidence for their assistance. Mr. McLeod left on Friday to go by land to Mackay's abandoned establishment on the Multnomah, fifty-six miles from its junction with the Columbia, where he is to remain until joined by the remainder of the party, which will leave in a few days.

Saturday, 16*th, to Tuesday,* 19*th.*—Employed making preparations for my march. As my gun has entirely failed me, I am under the necessity of purchasing a new one, which only costs £2. Being a new country and no knowledge whatever south of the Umpqua, each has to confine himself to as little encumbrance as possible ; and as nearly the whole must be land carriage, this increases the difficulty. Packed six quires of paper and other

[1] *Pseudotsuga Douglasii*, Mast. in Journ. R. Hort. Soc. xiv. p. 245.

little articles for my business, and provided myself with a small copper kettle and a few trifles, with a little tobacco for presents and to pay my way on my return. Of personal property (except what will be on me), one strong linen and one flannel shirt; and as heavy rains may be expected, being near the coast, I will indulge myself with two blankets and my tent. On Monday Mr. McLoughlin kindly sent by land, to await me on the Multnomah, one of his finest and most powerful horses for carrying my baggage or riding, as he may be required, which is of great service.

Wednesday, 20th.—At midday left Fort Vancouver in company with Mr. Manson, one of the persons in authority, and a party of twelve men and one boat, with their hunting implements. Camped five miles up the Multnomah, on the east bank of the river, at dusk.

21st.—Proceeded on our journey at sunrise and camped on the rocks on the Falls, among some drift-wood. Warm during the day, with heavy dews at night.

Friday, 22nd.—The boat being injured in hauling it up the rocks, two hours were spent in gumming; gathered a few seeds of different plants, among which *Pentstemon Richardsonii*, perfectly ripe. Proceeded at seven o'clock and put ashore seven miles above to breakfast, which occupied an hour. Reached Mr. McLeod's encampment at four o'clock, and after having our tents pitched sat down to supper at dusk. Pleasant, with a beautiful sky in the evening.

23rd to 25th.—Detained longer at our encampment than expectation in consequence of several horses missing, having strayed in the woods. Gathered a few seeds: two species of *Rosa, Viburnum* with large oval black berries, one species of *Caprifolium,*[1] and a *Ribes* like *R. sanguineum,* and, lest it should prove different, took a few seeds and specimens. Weather warm and pleasant, heavy dews at night.

Tuesday, 26th.—Detained still at our encampment, as five of the horses still strayed. Pleasant; dried some seeds and laid in a few specimens.

27th.—Morning spent collecting the horses, and before all were ready for starting it was near noon. Took our course due west, towards the coast; passed a small creek, two and a half miles on, and camped shortly after two o'clock. Country undulating; soil rich, light, with beautiful solitary oaks and pines interspersed through it, and must have a fine effect, but being all burned and not a single blade of grass except on the margins of rivulets to be seen. This obliged us to camp earlier than we would have otherwise done. As we had no fresh meat, each took his piece and went out. I raised two small deer and fired at one without effect. Mr. McLeod is not yet arrived at the camp and most probably will bring something home. Marched to-day five miles.

Thursday, 28th.—Mr. McLeod returned shortly after dusk last night and brought with him one of the Indian guides from the coast south of the country inhabited by the Killimuks. All unfortunate in the chase, and although nine small deer were seen in a group, yet by their keeping in the thickets near the small stream a few miles from our encampment, prevented the hunter from approaching them. Morning pleasant but chilly,

---

[1] *Lonicera*, Benth. and Hook. f. Gen. Pl. ii. p. 5.

with heavy dew. Thermometer 41°. Started at eight o'clock, keeping a south-west course. Passed two small streams. About noon, in a small hummock of *Corylus* and *Pteris aquilina*,[1] started four deer, one of which was killed by one of the hunters with his rifle at two hundred yards' distance. The ball entered the left shoulder and passed through the neck on the opposite side, yet she ran three hundred yards before she fell. Camped on the south side of Yamhill River, a small stream about twenty-five yards wide ; channel for the greater part mud and sand. Two hundred yards below where we forded are fine cascades 7 feet high. Country much the same as yesterday ; fine rich soil ; oaks more abundant, and pines scarcer and more diminutive in growth. Spread out the seeds that were not perfectly dry and some paper that got wet, and then took a turn up the river a few miles from our encampment. Picked up a species of *Donia* in flower and seed and a small annual plant closely allied to *Phlox*, both in rich light dry loam in open woods. Hunters out in search of deer and not yet come home. I expect a fine fall, as seventeen shots were heard in various directions in the woods.

Friday, 29th.—Morning dull and cloudy. Only one young deer was killed last night, by Mr. McLeod ; the mother he wounded, but being in the dusk of the evening she could not be found. At eight one of the hunters returned with a very large fine doe. Started at nine and kept a south-west course, and camped not far from the point of a low hill at three o'clock. Heavy rain for the remainder of day and the greater part of the night. Country not different from yesterday. Travelled about thirteen miles. Nothing new came under my notice.

Saturday, 30th.—Cloudy until noon, after-part of the day clear and fine with a fanning westerly wind. In the morning dried some of my things which got wet the preceding day. Started at nine and continued our route in a southerly direction, on the opposite side of the hill from where we were yesterday. Most parts of the country burned ; only on little patches in the valleys and on the flats near the low hills that verdure is to be seen. Some of the natives tell me it is done for the purpose of urging the deer to frequent certain parts, to feed, which they leave unburned, and of course they are easily killed. Others say that it is done in order that they might the better find wild honey and grasshoppers, which both serve as articles of winter food. I walked along the low hills but found nothing different. Saw four large bucks. Reached the camp at four o'clock, cleaned my gun, had supper and went out with a hunting party. One female and her kid were killed by the Indian hunter. Mr. McLeod passed a ball through the right shoulder of one and afterwards had two other shots fired with effect, but not mortal. Notwithstanding, although to appearance seriously injured, she made her escape. Returned shortly after dark.

Sunday, October 1st.—Heavy dew during the night ; clear and pleasant during the day, with a refreshing westerly wind. Started at the usual hour and continued our route. Had to make a circuitous turn east of south, south, and south-west to avoid two deep ravines that were impass-

---

[1] *Pteridium aquilinum*, Christensen. Ind. Fil. p. 591.

able for the horses. Walked the greater part of the day, but found nothing new. On my way observed some trees of *Arbutus laurifolia*,[1] 15 inches to 2 feet in diameter, 30 to 45 feet high, much larger than any that I saw last year on the Columbia; fruit nearly ripe; soil deep rich black loam, near springs, and on a gravelly bottom. Passed at noon some Indians digging the roots of *Phalangium Quamash*[2] in one of the low plains. Bulbs much larger than any I have seen, except those on Lewis and Clarke's River. Camped at four on the banks of a small stream which falls into the Multnomah three miles to the east. In the like journeys Sunday is known only by the people changing their linen, and such of them as can read in the evenings peruse religious tracts, &c., whose tenets are agreeable to the Church of Rome. In the dusk I walked out with my gun. I had not gone more than half a mile from the camp when I observed a very large wasp-nest, which had been attached to a tree, lying on the plain where the ground was perfectly bare and the herbage burned, taken there by the bears. At the same time John Kennedy, one of the hunters, was out after deer and saw a very large male grizzly bear enter a small hummock of low brushwood two hundred yards from me. Being too dark, we thought it prudent to leave him unmolested; perhaps we may stand a chance of seeing him on our way. Marched eighteen miles.

Monday, 2nd.—Morning, heavy dew and chilly; clear and fine during the day; sheet lightning in the evening. Early in the morning a small doe was killed near our encampment and was placed on one of the horses. As we passed on the track, I went in search of the large bear seen last night but could see nothing of him. At noon passed two deep gullies which gave much trouble, the banks being thickly covered with brushwood, willow, dogwood, and low alder. Course nearly due south, inclining to the west. Country same as yesterday, rich, but not yet a vestige of green herbage; all burned except in the deep ravines. Covered with *Pteris aquilina*,[3] *Solidago*, and a strong species of *Carduus*. On the elevated grounds where the soil is a deep rich loam, 3 to 7 feet thick on a clay bottom, some of the oaks measure 18 to 24 feet in circumference, but rarely exceeding 30 feet of trunk in height. On the less fertile places, on a gravelly dry bottom, where the trees are scrubby and small, a curious species of *Viscum*, with ovate leaves, is found abundantly. I recollect Dr. Hooker asking me if I ever saw it on oak in Scotland, which I never did. As no place could be found suitable for fodder for the horses, we had to travel till four o'clock, when we camped at a low point of land near a woody rivulet. Marched twenty-one miles. My feet to-night are very painful and my toes cut with the burned stumps of a strong species of *Arundo* and *Spiraea tomentosa*.

Tuesday, 3rd.—Last night, as we were nearly out of provisions I walked out with my gun, in company with Mr. McLeod, a few miles from the camp. Saw only two deer, but being too dark did not fire. Shortly before starting

[1] A. Gray, Syn. Fl. N. Am. ii. i. p. 27, says may be *Prunus caroliniana*, but is indeterminable.

[2] *Camassia esculenta*, Baker, in Journ. Linn. Soc. xiii. p. 257.

[3] *Pteridium aquilinum*, Christensen, Ind. Fil. p. 591.

at nine o'clock our Indian hunter returned with one, having been out all night. Morning raw and chilly, heavy showers during the day, which obliged us to camp at one o'clock, having gained only nine miles. Observed a small hawk of a light mottled grey, neck and head bright azure. The Large Buzzard, so common on the shores of the Columbia, is also plentiful here ; saw nine in one flock. Gathered seed of a species of *Gentiana*, leaves ovate, acute, entire ; 2 feet high, on moist black soil ; near springs.

Wednesday, 4th.—The morning being cloudy and overcast, we did not start so soon. As it cleared up about ten, the horses were saddled and we proceeded on our route in a southerly direction. Passed in the course of the day three small streams, which all fall into the Multnomah ten miles below this place. As no place could be found fit for camping we were obliged to go on until five o'clock, when we put up on the south side of a muddy stream, banks covered with *Fraxinus*. No deer killed this day, although several were seen. Nothing particular occurred. Marched twenty-four miles ; somewhat fatigued.

Thursday, 5th.—After a scanty breakfast proceeded at nine o'clock in a south course. Country more hilly. At one o'clock passed on the left, about twenty-five or thirty miles distant, Mount Jefferson, of Lewis and Clarke, covered with snow as low down as the summit of the lower mountains by which it is surrounded. About twenty miles to the east of it, two mountains of greater altitude are to be seen, also covered with snow, in an unknown tract of country called by the natives who inhabit it 'Clamite.' On the low hills observed *Pinus resinosa*, of very large dimensions, 4 to 6 feet in diameter, 90 to 130 feet long. Cones not perfectly ripe. (Secure specimens on your return.) Killed a very large grey squirrel, 2 feet long from the point of the tail to the snout. Saw a curious variety of the ground or striped, and also the flying, but could secure neither. Camped on the side of a low woody stream in the centre of a small plain—which, like the whole of the country I have passed through, is burned. One of the hunters killed a small doe shortly after leaving our encampment, which will provide an unexpected supper. Day very fine, sky clear, with a strong north-west wind. Thermometer at noon 74° in the shade. Marched nineteen miles.

Friday, 6th.—Heavy dew during the night ; clear and fine at noon, with a refreshing northerly breeze. Two of the hunters went out early in the morning and killed three small deer, which were very acceptable, as all we had killed before was cooked for breakfast. Started at nine o'clock keeping our usual southerly course. Gathered a few seeds of what I take to be a very fine species of *Argemone*, but as it was not in a perfect state, and having no book, I cannot say. At noon we were joined by Jean Baptist Mackay and two Iroquois hunters on their way to the Umpqua River. Mackay informs me he had five days ago sent one of his people on to collect the cones I spoke of to him last spring, lest the season should be past before he would be there, he not knowing of our projected journey. The day being very warm and the horses much fatigued, we were obliged to camp earlier than usual, lest we should not get in time to a feeding-place. Near our camp there is a long narrow lake, the margins covered with

Scirpus, Typha angustifolia, and abundance of *Nymphaea advena.*[1] Country the same as yesterday ; on the outskirts of the woods the *Pyrus* so common on the shores of the Columbia is plentiful here. Marched sixteen miles. As I walked nearly the whole of the last three days, my feet are very sore from the burned stumps of the low brushwood and strong grasses.

Saturday, 7th.—Morning very pleasant, clear sunshine and warm during the day throughout. In consequence of one of the Indians belonging to the party having lost his way the night before, and his horse unable to overtake us, we had to wait for him till twelve o'clock. I bargained with Baptist Mackay for a skin of a very large female grizzly bear which he had killed seven days before. I gave him an old small blanket and a little tobacco. This was to make myself an under-robe to lie on, as I found it cold, from the dew, lying on the grass. Mackay is to endeavour to kill me male and female, so that I might have it in my power to measure them if not to skin them. John Kennedy had this morning gone out hunting two hours before day, and about ten o'clock was attacked by a large male grizzly bear. He was within a few yards of him before he was discovered, and as he saw that it was impossible to outrun him he fired his rifle without effect and instantly sprang up a small oak-tree which happened to be near him. The bear caught with one paw under the right arm and the left on his back. Very fortunately his clothing was not strong, or he must have perished. His blanket, coat, and trousers were almost torn to pieces. This species of bear cannot climb trees. A party went out in search of him but could not fall in with him. Country more hilly than that I passed through on the former days, and not so rich ; limestone is seen in abundance in the channels of several of the small rivulets. Soil poorer than before and equally dry. On the summit of the low hills, in dry parched soil on clayey or rocky bottoms, a species of *Quercus* is found in abundance, 2 to 3 feet in diameter but seldom more than 30 to 40 feet of trunk. This may prove *Quercus tinctoria.*[2] Camped in a low woody valley at three o'clock, having marched seven miles.

Sunday, 8th.—Morning cool and pleasant ; day clear and warm. Thermometer in the shade 82° ; much sheet lightning in the evening, wind westerly. Started at nine o'clock, keeping as usual a southerly course. Country more hilly and less fertile. *Quercus* and *Fraxinus* abound in the valleys and *Pinus* on the mountains—*Pinus taxifolia,*[3] but of diminutive stature compared with what are seen on the more fertile spots of the Columbia and Multnomah Rivers. No plants came to-day under my notice. Camped on the side of a low hill under the shade of some oaks ; marched five hours ; gained thirteen miles. No deer killed and had the last fragments cooked for supper, which gave us all but a scanty meal. Shortly after dusk one of the hunters fortunately killed one small deer, which will serve us for breakfast. We are just living from hand to mouth. All the hunters observe that the animals are very scarce and those shy in consequence of the country being burned.

---

[1] *Nuphar advena,* S. Wats. Bibl. Ind. N. Am. Bot. p. 37.
[2] *Quercus velutina,* Sargent, Silva N. Am. viii. p. 137.
[3] *Pseudotsuga Douglasii,* Mast. in Journ. R. Hort. Soc. xiv. p. 245.

Monday, *9th.*—Morning cloudy ; drizzly rain. As we expected wet weather, did not start till noon. Hunters out and fell in with a small herd of elk ; but being in the close and almost impenetrable thickets, only one could be secured, which fell after receiving eleven shots. At this season the males are very lean and tough eating ; weighed about 500 lb. Horns very large, 33 inches between the tips, with five prongs on each, all inclining forward ; the two largest, 11 inches long, running parallel with the nose and reaching nearly to the nostrils ; body of a uniform brown, with a black mane 4 inches long. I am pretty certain this is the same sort of animal which I have seen at the Duke of Devonshire's, and unquestionably a very distinct species from the European stag. I ascended a low hill about 2500 feet above its platform, the lower part covered with trees of enormous size of the same species as on the Columbia. On the summit only low shrubs and small oaks and a species of *Castanea.* This handsome tree I saw at the foot of the hill, but very low, not more than 4 feet high, and being imperfect, from the leaves which are lanceolate, deep glossy-green above and ferruginous underneath, I took it to be a species of *Shepherdia*, but was agreeably disappointed to find my little shrub of the valley change on the mountains to a tree 60 to 100 feet high, 3 to 5 feet in diameter (*Castanea chrysophylla*).[1] Its rich varied foliage, quivering in the wind, clothed to the very roots with wide-spreading branches, and standing alone on the dry knolls or on the crevices of rocks, gives a tint to the general appearance of American vegetation of more than ordinary beauty. As I never saw the American *Castanea*, and having no book, I cannot say whether it is new or not, but am inclined to think it rare ; at least I had a laborious search to find it in fruit, which I did on the very highest peak and only one tree ; fruit little more than half ripe, husk thickly beset with sharp prickles ; found one small lateral shoot with male flowers. I am confident of finding it in perfection on more elevated grounds, and will undoubtedly make every effort to secure such a desirable object. Under its shade is a large evergreen shrub, 4 to 10 feet high—leaves ovate, five-nerved, minutely serrate, upper side glutinous and sweet-scented—which I take to be a fine species of *Clethra*, but not in perfection. Perhaps I may meet with it soon —in the low shady places among moss near rivulets or springs. Abundance of a curious species of *Helleborus*, with ternate leaves, is to be seen. Laid in specimens of all, sticky as they are. Towards dusk I crossed the hill and descended on the other side, and walked along a small serpentine stream for six miles, where I found the brigade encamped in a small cove on its margin. As the weather seemed very unsettled I had my tent pitched and my sundry articles placed secure, and soon afterwards lay down to rest, being much fatigued. By the route the brigade marched six miles.

Tuesday, *10th.*—Morning cool and pleasant, calm and cloudy. Having to cross the same hill I was on yesterday a few miles to the south, and apprehending rainy weather, and the horses poorly off for fodder and ourselves on bad hunting ground, we started at eight o'clock and reached the other side at two P.M., where we encamped in a small woody valley. Besides the *Pinus* usually seen, I observed one not unlike the black spruce,

[1] *Castanopsis chrysophylla*, Sargent, Silva N. Am. ix. p. 3.

very large but without cones. *P. taxifolia*,[1] *P. balsamea*,[2] abound on the lower parts ; *P. resinosa* on the more elevated, all of extremely large dimensions, some 200 to 250 feet high, 30 to 55 feet in circumference. Measured to-day what I have noted as *Arbutus laurifolia*,[3] 3 feet in diameter, but seldom carrying a trunk of these dimensions more than 15 or 20 feet high. I find such trees are only to be seen on dry elevated spots where the soil is a deep rich light loam on a gravelly bottom, and are rarely to be seen in fruit. In open rocky situations where the trees are low and scrubby, abundance of fruit is to be seen, but only on parts where bears are unable to reach the fruit, now half ripe. Observed a few low bushes of *Arbutus Menziesii*, and the fruit of it also destroyed by the same animal. The low herbage on the top of the hill is *Spiraea capitata*[4] ; *Gaultheria Shallon* ; *Berberis nervosa* ; one species of *Salix* with ovate, tomentose leaves ; *Corylus rostrata* ; *Vaccinium* with red berries, the same one so abundant on the banks of the Columbia ; *Epilobium angustifolium* ; and one species of *Carex* near springs or on moist ground. One plant I have been in search of every day during my walks (*Ilex Dahoon*), and am still unable to find it. I hope ere long to see it, if on the range of my present journey. The plant I found yesterday, which I took to be a species of *Clethra*, I now see in perfection on the summit of the hill. I find it will prove a third species different from that so common in the open woods on the higher parts of the Columbia and its branches, and equally different from the deciduous one on the coast. Being so late in the season I could not collect so much seed as I should have wished of this fine plant. Perhaps in the course of my walks I may increase my store ; seeds of *Arbutus laurifolia*[5] I must collect on my return. Found two species of *Caprifolium*,[6] one resembling *C. ciliosum*[7] of Pursh but distinct, with abundance of fine ripe seeds ; a second with small ovate, hirsute leaves, which, if I recollect right, I found in an imperfect state last year on the rocky banks of the Columbia near Oak Point ; of this interesting little plant I could only find six perfect berries. A very slender delicate plant, not more than 3 or 4 feet long, twining itself round low brushwood. The former a strong one and will at least prove a good variety, if not a species, to this handsome tribe of plants. Laid in specimens of all of them. Heavy rain in the afternoon and throughout the night. Marched eleven miles. As I walked all the way and my course was much more circuitous than the brigade, I felt no little embarrassment in making my way through the thicket ; and in addition to what I have mentioned *Pteris aquilina*,[8] 8 to 10 feet high and strongly bound together by *Rubus suberectus*,[9] and several species of decayed *Vicia*, rendered it so fatiguing that every three hundred or four hundred yards called for a rest.

[1] *Pseudotsuga Douglasii*, Mast. in Journ. R. Hort. Soc. xiv. p. 245.
[2] *Abies balsamea*, Mast. *loc. cit.* p. 189.
[3] A. Gray, Syn. Fl. N. Am. ii. I. p. 27, says may be *Prunus caroliniana*, but is indeterminable.
[4] *Neillia opulifolia* var. *mollis*, S. Wats. Bibl. Ind. N. Am. Bot. p. 290.
[5] May be *Prunus caroliniana*, but is indeterminable, A. Gray, Syn. Fl. N. Am. ii. I. p. 27.
[6] *Lonicera*, Benth. and Hook. f. Gen. Pl. ii. p. 5.
[7] *Lonicera ciliosa*, A. Gray, Syn. Fl. N. Am. i. II. p. 16.
[8] *Pteridium aquilinum*, Christensen, Ind. Fil. p. 591.
[9] *Rubus villosus* var. *frondosus*, S. Wats. Bibl. Ind. N. Am. Bot. p. 319.

Wednesday, 11*th.*—Cloudy, with heavy rain, wind, and thunder until noon, when it cleared up for a short time. As the horses stood in need of rest, having gone through much labour without much food, Mr. McLeod thought it better to rest a day. I hung up my blanket and little clothing to dry, which were drenched yesterday, and last night changed the papers of specimens and seeds and exposed them to the sun. I had at three o'clock proceeded to a small wood about a mile from my camp, when I was obliged to return, the rain setting in. I found all my little paper kindly secured in my tent by Mr. McLeod. Rain throughout the night.

Thursday, 12*th.*—Weather still unsteady, heavy showers during the forenoon. Two of the horses having strayed to our last encampment, we had to stay here all day. I took a walk north of the camp. Nothing new came under my notice except a small partridge (very small), 10 inches long, including the tail, which of itself is 2¾ ; back of a light dusky-brown ; breast lead colour ; from the breast to the beak, of a dark red or rather brick colour ; point of the feathers, white ; thighs and inside of the wings, light brown ; beak, remarkably small ; crown of the head, lead colour, with three black feathers in the crown 2 inches long ; from the under part of the beak to the breast, purple red, with a narrow white streak the same length extending from the upper part of the eye and round the back ; legs, naked ; 8 to 12 ounces. This most curious little bird I learn inhabits the upland dry low woods, and in the winter and spring are seen in flocks of forty to a hundred together. Of the nest, eggs, and the number of young I am for the present unable to say. Only five were in the flock and probably one brood which I saw, and although I pursued diligently I was unable to get a second shot. I regret exceedingly that this was so much shattered, the upper beak and right leg being taken by the shot, being too large. Otherwise I should have preserved it. As I have still 4 ounces of small shot, I shall make a point of securing it. The people and horses returned shortly before evening with the horses safe, so should the day be fine we will march early in the morning.

Friday, 13*th.*—Very heavy dew during the night. Morning dull, close, and foggy ; noon clear and fine. Thermometer 66°. Fine bright sky in the evening. Mr. McLeod started this morning, long ere daylight, in quest of a herd of elk he had seen the evening before. After securing my little gleanings I left at seven o'clock, and at nine came to a small rivulet that takes its rise in the mountains to the east and discharges itself into the Umpqua River, south of this about thirty-five or forty miles. Here I found Baptist Mackay, who had gone before yesterday for the purpose of hunting, where Mr. McLeod joined me shortly afterwards. They brought two small deer, a male and female (entire), so I had an opportunity of measuring them. Moving as I am, not more than one day in the same camp, except detained by bad weather, it is beyond my power to preserve them, a thing I much desire. The following are the dimensions of the female : Head—from the nose to the crown, 11 inches ; between the eyes, 3½ ; round the jaws, between the eyes and ears, 17 ; ear, 7 ; length of the neck, 15. Girth at the smallest part, 13 ; length of the body, 31 ; girth round the chest, 33 ; round the belly, 27. Foreleg—depth of the shoulder-blade, 12 ; shoulder-joint to the knee, 11 ; from the knee to the hoof, 11 ; girth of the shoulder-joint, 8 ; at the knee, 5. Hind leg—rump, 18, including the thigh ; from the knee to the hoof, 13½ ; knee, round, 8 ; tail in length, 12. Body, head, neck, and legs of a uniform light grey ; belly, inside of the thighs, and underside of tail, pure white ; lips and tips of the ears, of a darker tinge. In the summer season, after the period of having their young in April and May, they change colour to a light reddish-brown, not unlike the European stag. Three years old, weighing from 90 to 100 lb. without the gut. The buck being gutted, I could not with correctness take it. The proportions seem to agree and differ no way in colour ; they are generally 30 or 40 lb. heavier. The horns are much curved, four, five, six, and the largest, seven pronged ; the present is of the middle size and is four. Extreme length, 16 inches ; root prong, 2¾ long ; middle, 7 ; point, 3¼. The distance between the point of each will show the curve. Between the roots, 1¾ ; between the points of the root prong, 5 ; between the middle, 17 ; point prong, 16 ; extreme point, 9½. Being the only thing in my power to keep, I tied it to my knapsack, and perhaps I may yet obtain good specimens of the animal entire. After resting a short time, Mackay made us some fine steaks, and roasted a shoulder of the doe for breakfast, with an infusion of *Mentha borealis* [1] sweetened with a small portion of sugar. The meal laid on the clean mossy foliage of *Gaultheria Shallon* in lieu of a plate and our tea in a large wooden dish hewn out of the solid, and supping it with spoons made from the horns of the mountain sheep or Mouton Gris of the voyageurs. A stranger can hardly imagine the hospitality and kindness shown among these people. If they have a hut, or failing that, if the day is wet, one of brush-wood is made for you, and whatever they have in the way of food you are unceremoniously and seemingly with much good-will invited to partake. After smoking with a few straggling Indians belonging to the Umpqua tribe, we resumed our route on the banks of the small stream ; track mountainous and rugged, thickly covered with wood in many places ; and in some parts, where *Acer circinnatum* forms the under-wood, a small hatchet or large knife like a hedge-bill is indispensably necessary. Camped at three o'clock five miles further down the stream, having gained this day eleven miles. Mackay's hunter returned this evening from the Umpqua River, but in consequence of some misunderstanding he only brought a few seeds baked on the embers of the pine. As there are in the party two individuals of that nation who both talk the Chenook tongue fluently, in which I make myself well understood, from the questions I have put to them and the answers given, I am almost certain of finding it in abundance. Should I fail, I shall make my way through that very partially known country called ' Clamite,' to the north-east of where I now am, where I will find it without any doubt according to Mr. McDonald, who passed there in September 1825, and this person being present when it came first under my notice, I requested he would look for it in the Cascade range of mountains or through that mountainous country between Mount Hood and another high snowy mountain to the south of it, which I have

---

[1] *Mentha canadensis* var. *glabrata*, A. Gray, Syn. Fl. N. Am. ii. 1. p. 352.

denominated Mount Vancouver. He found it there, being then going on a year's journey ; not knowing where to, and no doubt entertaining views of no very pleasant nature, he of course could not bring me any. Failing my present trip, I shall return in that direction and will probably come on the Columbia at the Great Falls.

Saturday, 14*th.*—Judging from the fine sky of last night and the calm of the night, I did not think of pitching my tent, which I rarely do unless it rains, or has the appearance of rain ; and for my negligence to my own comfort I was this morning, shortly after four o'clock, obliged to rise as it was raining heavily. Continued to rain throughout the day. Several small deer were killed in the morning, and a number wounded which could not be found. The remaining part of our journey being the worst—road being hilly and thickly wooded—and can only be performed in good weather, I arranged my gun and read some old newspapers. Measured a left horn of an elk found by one of the men : 4 feet 1¼ inch long, the two lowest prongs wanting, weighing although dry about 12 lb. This is one of the largest.

Sunday, 15*th.*—Morning dull and cloudy, noon fine, wind variable. Our path being very hilly, rough, and through thick woods, the horses could not proceed. Several small deer were killed in the morning and evening. Towards midday I made a short trip to the neighbouring hills and found one fine strong evergreen shrub, 4 to 15 feet ; leaves obovate, smooth, and ciliated. I regret that not a seed or entire capsule could be found. The only vestige was on the ground and appears to be three-celled and three-seeded ; in rocky and sandy situations, on the low hills ; abundant. Saw several large trees of *Castanea*, but am sorry to say am still unable to find it in perfect fruit. Evening fine.

Monday, 16*th.*—Morning foggy, dull, and raw ; thermometer 42° minimum. At nine o'clock began our day's march, in nearly a westerly course. Passed two miles of open hilly country, intersected by several small streams, where we entered the thick woods. Passed three ridges of mountains, the highest about 2700 feet. Mr. McLeod and I took the lead and were followed by Baptist Mackay and two hunters, hewing the branches down that obstructed the horses from passing. The whole distance not so much as a hundred yards of ground on the same level, and the numerous fallen trees, some of which measured 240 feet long and 8 feet in diameter—I am aware that it could hardly be credited to what a prodigious size they attain. The rain of the two days before rendered the footing for the poor horses very bad ; several fell and rolled on the hills and were arrested by trees, stumps, and brushwood. As I apprehended some accident, I thought it prudent to carry my gleanings on my back, which were tied up in a bear's skin. Crossed the small stream on the banks of which was our last encampment. In the deep valleys on the margin of rivulets a very large fine tree is to be seen, to me perfectly unknown : I think belongs to *Myrtaceae* ; fruit large, globular, covered with a fine thin green skin, enveloping a small nut ; kernel insipid ; flower buds in abundance on the points of the twigs ; but pressed as I am at present, can only carefully preserve them and that will be done by one amply qualified for the task; leaves lanceolate, entire, smooth; wood, fruit, bark, and leaves very aromatic, precisely like *Myrtus Pimenta*, and when rubbed in the hands produces sneezing, like pepper; 40 to 70 feet high, 18 inches to 3 feet in diameter; bark smooth and whitish; young shoots bright green; appears to be a favourite food of squirrels, as I found sundry shells opened. This very splendid evergreen tree will prove, I hope, a most valuable addition to the garden and perhaps may be found useful in medicine or as a perfume. Before I could obtain the seed I had to cut down a large tree, being unable to climb it, the bark being so smooth, which was done at the expense of my hands well blistered. Saw some very large trees of *Castanea*, but none with perfect fruit; also, in addition to what I saw before, abundance of *Ribes sanguineum* and *Rubus spectabilis*, both very luxuriant in the deep shady places near springs and streams. Found one shed horn of Black-tailed Deer: the temptation was too great, so I tied it on the bear-skin bundle. Appears to be a larger animal than the Long-tailed Deer. Every two or three hundred yards called for a rest, and two or three times in the course of the day for a pipe of tobacco, and in this way did I drag over the most laborious and tedious marches I have experienced for many days. On reaching the summit of the last hill the desired sight of the Umpqua River presented itself to our view, flowing through a variable and highly decorated country—mountains, woods, and plains. Half an hour's walk carried us to the banks, where we encamped on the east side in the angle where the little river we marched along to-day falls into it. Arrived at five o'clock, having marched seventeen miles. Having scarcely any fresh food, Mr. McLeod and Mackay went out in hopes of killing deer, while I employed myself chopping wood, kindling the fire, and forming the encampment; and after, in the twilight, bathed in the river: course north-west; bed sandstone; ninety yards broad; not deep, but full of holes and deep chinks worn out by the water. Two hundred yards below is a small rapid in several channels and small grassy islands; will never admit of any barge larger than a ship's jolly-boat, by the numerous rocks that rise above water and the rapids. The distance from the ocean cannot exceed thirty or thirty-five miles, as I observe *Menziesia ferruginea* and *Pinus canadensis*,[1] which keep along the skirts of the ocean. Only a few of the horses now came (eight o'clock) and although the moon is shining, they will not come till to-morrow. Mine is among the number, so that I have nothing to lie down on although very tired. Mr. McLeod returned unsuccessful, having wounded a large deer which perhaps will be found in the morning. Evening fine.

Tuesday, 17*th*—Last night I sat up by the fire until ten o'clock, when we learned that some of the people camped in the valley over the first mountain and some only half-way there, the horses being worn out. Mr. McLeod very kindly gave me his own blanket and buffalo-robe, and reserved for himself two great-coats, as the horse with my articles was one of the furthest behind. Morning dull and heavy, noon fine and clear, wind easterly. Having nothing for breakfast, Mr. McLeod and Mackay and myself went out on the chase. The deer wounded last night by Mr. McLeod was found,

[1] *Tsuga canadensis*, Veitch, Man. Conif., ed. 2, p. 464.

the ball having passed through both shoulders and a second was still necessary before she could be taken. Mackay made a fine shot at the distance of two hundred yards, his ball passing through the chest, upon which the deer took to the water and was swimming to the opposite side when he passed a second in at one ear and out of the other. Arrived at the camp at twelve o'clock, in fine spirits. The deer floated down the stream and was dragged to shore by an Indian boy. The following is its dimensions : From the nose to the point of the tail, 6 feet ; tail, 1 foot ; length of the head, 13 inches ; breadth of the head between the eyes, 5 ; length of the ears, 8 ; girth round the chest, 3 feet 5 inches ; foreleg from the hoof to the shoulder head, 3 feet 5 inches ; hoof to the knee, 1 foot 2 inches ; from the knee to the shoulder, 11 inches ; hind leg, extreme length, 3 feet 5 inches ; hoof to the knee, 15 inches ; knee to the hip-joint, 15 inches ; neck, 15 inches long, girth 2 feet. This is considered one of the largest size, weighing about 190 lb. of meat, very fat. After a comfortable meal between one and two o'clock, I turned my specimens and exposed some of the last gathered seeds and cones of *Pinus taxifolia* [1] and then made a turn down the river, but found nothing different from yesterday. The horses with my articles arrived at four o'clock in a sad condition. The tin-box containing my note-book and small papers broken and the sides pressed close together ; a small canister of preserving powder in a worse state ; and the only shirt except the one now on my back worn by rubbing between them like a piece of surgeon's lint. In the evening, arranged my papers and found nothing materially injured. I am glad I took the precaution of carrying the specimens of seeds and plants on my back, otherwise they would have been much destroyed. The country towards the upper part of the river appears to be more raised and mountainous, and perhaps will afford my wished-for pine being nearer the spot described to me in August 1825 by an Indian while on the Multnomah, in whose smoking-pouch I found some of its large seeds. Should the morning be fine and any provision killed to take with me, I intend to start for a few days. Baptist Mackay has given me one of his Indian hunters, a young man about eighteen years old, as a guide ; of what nation he belongs to he does not know, but tells me he was brought from the south by a war party when a child and kept as a slave until Mackay took him : he is very fond of this sort of life and has no wish of returning to his Indian relations. He speaks a few words of the Umpqua tongue and understands the Chenook, so I will have no difficulty in conversing with this, my only companion. Keep the horns of this large deer, which I will measure at a more convenient season. Evening fine.

Wednesday, 18*th*.—Heavy dew during the night, morning dull and heavy. Before I could get all ready for my march it was eleven o'clock : took my course due south through a broken varied country and crossed the river five miles from our encampment, where there were two lodges and about twenty-five souls, the greater part women and wives of *Centrenose* (an Indian word), the chief of the tribe inhabiting the upper part of the river, and who is at present forty-five miles higher up the river. They

[1] *Pseudotsuga Douglasii*, Mast. in Journ. R. Hort. Soc. xiv. p. 245.

very courteously brought one very large canoe, in which I embarked and swam the horses at the stern, holding the bridles in my hand. I made a stay of a few minutes, and, as I found my young guide to be less conversant in their tongue than I expected, my visit was to me the less interesting. Had some nuts of *Corylus*, roots of *Phalangium Quamash*,[1] and a preparation of meal made from the seeds of a *Syngenesia* already in my possession, with the nuts of my smelling-tree, which are roasted in the embers previous to use. The dress of the men is skins of the small deer undressed, formed into shirts and trousers, and those of the richer sort striped and ornamented with shells, principally marine, which proves our distance from the ocean to be short. The women, a petticoat of the tissue of *Thuya occidentalis*, made like that worn by Chenook females, and a sort of gown of dressed leather, in form differing from the men's only by the sleeves being more open. I had gathered for me a quantity of nuts of my smelling-tree, for which I presented them with a few beads, brass rings, and a pipe of tobacco. The children on seeing me ran with indescribable fear, and on the first interview only one man and one woman could be seen. The others I conceive came on being made acquainted with my friendly disposition. Measured several trees $2\frac{1}{2}$ feet diameter and 60 to 70 feet high : a decoction of the leaves and tender shoots is used by them and is by no means an unpalatable beverage. At 2 P.M. continued my route over a low hill and on the other side of which I was given to understand I should again fall on the river, and by that means save a long circuitous bend of the river, which I found perfectly correct. Just as I was on the bank of the river, a herd of small deer, seventeen in number, rose ; one of the females I shot through the vertebræ on the fore-part of the shoulder and it dropped instantly on the spot. Since I left Fort Vancouver I have seen them frequently run several hundred yards before falling after a ball passing through the heart. As I wanted to ascend as near the high mountains as possible, lying in a south-easterly direction, I sought up and down the river for a fording-place but could find none, and shortly came to the resolution of making a raft, which I did, and after an hour's hard labour, in the course of which my hands were in a sad condition with blisters (and after all I found it by far too small), and finished the labour of this day by kindling my fire and roasting a few ribs of my venison for supper.

Thursday, 19th.—Although the thermometer stood not lower than 41°, yet it was so chilly and raw, with a very heavy dew, that I was under the necessity of rising three times to make up the fire, having only one blanket over me and a small piece of buffalo-skin under, which during the day serves in lieu of a horse-rug. My hands being so bad that I could not use the hatchet, and being only nine miles from Mr. McLeod, I addressed a note to him informing him of my case and sent it by my Indian guide. In the meantime, I took my gun and went out on the chase. Got only one mile from my camp when I wounded a very large buck through the shoulder, and as he was limping away from me I was in hopes of overtaking him, when unfortunately I fell into a deep gully among a quantity of dead wood, in which position I know not. I was on my belly and my face covered with

[1] *Camassia esculenta*, Baker, in Journ. Linn. Soc. xiii. p. 257.

mud when I recovered. I find now, 5 P.M., a severe pain in the chest. Six Indians of the Calapooie tribe assisted me to my camp, and as it would be very imprudent to undertake any journey as I am, I resolved to return to the camp and asked them to saddle my horse and place the things on it, which they readily did. It gave me more pleasure than I can well describe to think I had wherewith for them to eat, and after expressing my gratitude in the best way I could, one came to lead the horse while I crept along by the help of a stick and my gun. On arriving at the Indian lodges I passed yesterday, I found John Kennedy, who had instantly been despatched by Mr. McLeod to make me a raft, and who on learning my case turned and gave me his horse to ride. I had a little tea made me and bled myself in the left foot, and since I feel somewhat relieved. I find eight small deer and two very large bucks have been killed to-day. Evening cool.

Friday, 20th.—Morning dull and foggy, and chilly during the night. I had a restless night and slept but little. I find the pain entirely gone and only a stiffness in my shoulders and back as if under a heavy load. Early in the morning bathed in the river and find myself much better. Waited until midday before some of the horses could be collected, and for Centrenose, the chief, who promised to visit us. Two of his sons came shortly after ten and said he would be from the upper country this evening. Started at twelve on the right bank of river, generally a west course towards the ocean. River circuitous, woody banks, and very rocky, principally sandstone ; country very hilly. Raised one very large male Small Deer which escaped from us although eight shots were fired at him, and some with effect. At three passed over a high, thick-wooded hill, very steep and difficult both to ascend and descend, at the foot of which we came to a small stream and then a low point of thick woods, full of fallen timber and large shattered rocks with numerous mountain-rills. As no one could ride through such a rough country, I was obliged to walk and but little able to endure it. Remained in the rear, and by so doing had the way well pathed for me. I find that what I took to be two different species of *Caprifolium* [1] is only the one found last year on the Columbia at Oak Point ; more hirsute and less luxuriant in open dry situations than in the shady woods, where the leaves are glabrous. If I had not seen the difference on the same plant undoubtedly I should have considered them distinct. Camped in a low semicircular plain, surrounded on all sides by hills, a little before sundown ; find myself much broken by this day's march ; travelled ten miles. Evening cool and dewy.

Saturday, 21st.—As has been the case for the last ten days, very heavy dew during the night with a rawness at sunrise, succeeded by clear sunshine during the day and a westerly breeze after noon. I find myself on the whole much improved this morning, but still stiff as if I had been undergoing great labour. Having to go only about five miles to where the horses must be left, I am in hopes of having a rest for a day or two. Started at ten o'clock, keeping along the right bank of the river nearly a due west course ; passed two bad gullies thickly wooded, and before the horses could pass a road had to be cut, which occupied considerable time.

---

[1] *Lonicera*, Benth. and Hook. f. Gen. Pl. ii. p. 5.

After passing two fine small rich plains, camped shortly before dusk at the west end of the third, at the foot of some high mountains covered with pine. I perceive a hardness in the sky and a softness in the atmosphere for the last three days, with a gradual rise of the thermometer at night and cooler during the day, which indicates our drawing near the ocean. A message was sent to the Indian village a mile further down the river on the same side, to inform them of our arrival. The chief, his son, and about twenty followers accompanied the messengers back in several large fine canoes similar to those used by the Columbian natives, and brought with them a large number of very fine salmon-trout, 2 feet 5 inches long, 10 to 25 round, and some 3½ feet long, of fine quality ; is the same fish as is caught in the small branches of the Columbia in spring and autumn. I learn they are caught by the spear, and, as I understand, the Indians are totally unacquainted with fishing with a net. Evening cloudy.

Sunday, 22nd.—Morning cloudy but pleasant, with a light breath of wind from the west. We had been up not more than an hour when we were again visited by our new Indian friends, who brought for us a fine quantity of salmon-trout, a part of which we had cooked for breakfast and found it excellent. About noon a very large Black-tailed Deer came close to the camp, and was quietly feeding among the horses when Mr. McLeod and Baptist Mackay laid him down with their rifles at the distance of two hundred yards. It is a much larger animal than the Long-tailed, at least a fifth larger, and assuredly a very distinct species. In form the same, except the horns, which are rarely ever more than two-pronged and not curved, standing more erect on the head. The neck is both longer and thicker, and the tail short, 5 to 7 inches. The general colour on the back is a light grey, bluish white on the belly, inside of the forelegs, and thighs ; has a black ring round the nose at the nostrils and the crown between the eyes ; roots of the horns and ears of the same colour ; has a short black mane ; neck and ears same colour as the belly ; tail of course black. Different from the Long-tailed Deer in its range of country and is rarely to be seen on the plains ; is found abundantly on the high ground of and near the Rocky Mountains and on those along the coast, and indeed I might say on almost every hill throughout the country, but seems to prefer a southern climate as it has scarcely ever been seen beyond 48° of N. latitude. By no means such a plentiful animal as the former. Last spring Baptist Mackay brought me from this country a snare made from a grass, as he said, which from its texture I thought would prove a species of *Helonias*. I now find it to be a small species of *Iris*, found abundantly on the low moist rich grounds. I regret that, being so late in the season, it is doubtful if I can glean seeds of it. The snare is used in taking the elk and Long- and Black-tailed Deer, and in point of strength will hold the strongest bullock and is not thicker than the little finger. I observe that the women are mostly all tattooed, principally the whole of the lower jaw from the ear, some in lines from the ear to the mouth, some across, some spotted, and some completely blue ; it is done by a sharp piece of bone and cinder from the fire. It is needless to say it is considered a great mark of beauty ; I have little doubt that such a lady in London would make a fine figure, particularly when a little red and green earth is added to the upper part of the face. Mr. McLeod much engaged all day making arrangements for his journey to the country to the south of this river, where two small rivers and one large one are said to exist. While he is in that quarter I propose, should I keep well, to resume my route towards the headwaters of this river, where I have no doubt of finding out some varieties. Centrenose, the principal chief of the upper country, came to our camp in the afternoon, and in the morning Mr. McLeod is (through his interpreters) to arrange with him to accompany me or send some of his sons. Find myself improving fast. Evening fine, but cloudy.

Monday, 23*rd.*—Morning cloudy and calm. Last night after supper Mr. McLeod kindly spoke to the chief for his son to accompany me to the upper country while he went with himself along the coast, to which he agreed. The road being very hilly, woody, and difficult to pass over, I did not think it necessary to accept of any more horses than what would carry my blanket and paper, which were two—one for my guide and one for the articles. Started at ten o'clock and passed along the same side of the river and crossed at the chief's lodge where I was some days ago. They readily carried me across in the canoe and behaved very civilly. Made a short stay and crossed over the same point of land to the place where I attempted making a raft and without success, where I killed a small doe which gave me a little hope at the beginning of my march. My guide, by kindling a small fire, brought two men and a canoe from their lodge two miles above, round a thick woody point, who instantly took me across and guided the horses to a shallow part of the river where they forded it and received no injury. Proceeded on the opposite side and camped a short distance from the lodge. I could not utter a single syllable, but by signs they kindled my fire, brought me water, nuts, roots of *Phalangium Quamash,*[1] and the sort of meal made of the *Syngenesious* plant spoken of before and some salmon-trout. Finding them not only hospitable but kind in the extreme, I gave them all the flesh of the deer except one shoulder, some presents of beads, rings, and tobacco. Rain in the afternoon but fine and clear in the evening. On the rocky and gravelly shores of the river, I observed a shrubby species of *Lupinus* : leaflets five, seven, nine, lanceolate, silky on both sides ; 2 to 4 feet high ; I could find none in a perfect state.

Tuesday, 24*th.*—Morning cloudy, raw, and dull. My new friends had during the night gone to a small rapid a mile below for the purpose of spearing trout for me and awoke me this morning long ere day to eat. Left my camp at daylight, and passed a low level rich plain four miles long, along the banks of the river, where I entered a thick wood five miles broad and came again on a bend of the river, where I stayed a short time to refresh my horses, being noon ; and although having only made nine miles they were much fatigued by the last five being through deep gullies, rocky and obstructed by fallen timber. About two o'clock resumed my course due east over a bare hill 3000 feet above the level of the river, and on gaining the other side crossed a small stream twenty-five or thirty yards broad, shallow but rapid, where I entered a second point of wood three miles

[1] *Camassia esculenta*, Baker, in Journ. Linn. Soc. xiii. p. 257.

broad, hilly, and an almost impenetrable thicket. On leaving it about five o'clock I was urged to creep along for a feeding-place for my horses, which I found a mile and a half further on, and before I had made my encampment the rain was falling in torrents. Cooked the last of my deer flesh and boiled a few ounces of rice for supper, and, lest I should not see any Indians, I can only afford one meal a day. On rocks and trees in the first point of wood, observed a species of *Vitis*, the first I have seen west of the Rocky Mountains; leaves partially five-lobed, smooth, slightly serrated; wood slender with white bark; destitute of fruit. Marched seventeen miles.

Wednesday, 25th.—Last night was one of the most dreadful I ever witnessed. The rain, driven by the violence of the wind, rendered it impossible for me to keep any fire, and to add misery to my affliction my tent was blown down at midnight, when I lay among *Pteris aquilina* [1] rolled in my wet blanket and tent till morning. Sleep of course was not to be had, every ten or fifteen minutes immense trees falling producing a crash as if the earth was cleaving asunder, which with the thunder peal on peal before the echo of the former died away, and the lightning in zigzag and forked flashes, had on my mind a sensation more than I can ever give vent to; and more so, when I think of the place and my circumstances. My poor horses were unable to endure the violence of the storm without craving of me protection, which they did by hanging their heads over me and neighing. Towards day it moderated and before sunrise clear, but very cold. I could not stir before making a fire and drying part of my clothing, everything being completely drenched, and indulging myself with a fume of tobacco being the only thing I could afford. Started at ten o'clock, still shivering with cold, although I rubbed myself with my handkerchief before the fire until I was no longer able to endure the pain. Shortly after I was seized with a severe headache and pain in the stomach, with giddiness and dimness of sight; having no medicine except a few grains of calomel, all others being done, I could not think of taking that and therefore threw myself into a violent perspiration and in the evening felt a little relieved. Went through an open hilly country some thirteen miles, where I crossed the river to the south side near three lodges of Indians, who gave me some salmon such as is caught in the Columbia and at this season scarcely eatable, but I was thankful to obtain it. Made a short stay and took my course southerly towards a ridge of mountains, where I hope to find my pine. The night being dry I camped early in the afternoon, in order to dry the remaining part of my clothing. Travelled eighteen miles.

Thursday, 26th.—Weather dull and cloudy. When my people in England are made acquainted with my travels, they may perhaps think I have told them nothing but my miseries. That may be very correct, but I now know that such objects as I am in quest of are not obtained without a share of labour, anxiety of mind, and sometimes risk of personal safety. I left my camp this morning at daylight on an excursion, leaving my guide to take care of the camp and horses until my return in the evening, when I found everything as I wished; in the interval he had dried

---

[1] *Pteridium aquilinum*, Christensen, Ind. Fil. p. 591.

my wet paper as I desired him. About an hour's walk from my camp I was met by an Indian, who on discovering me strung his bow and placed on his left arm a sleeve of racoon-skin and stood ready on the defence. As I was well convinced this was prompted through fear, he never before having seen such a being, I laid my gun at my feet on the ground and waved my hand for him to come to me, which he did with great caution. I made him place his bow and quiver beside my gun, and then struck a light and gave him to smoke and a few beads. With my pencil I made a rough sketch of the cone and pine I wanted and showed him it, when he instantly pointed to the hills about fifteen or twenty miles to the south. As I wanted to go in that direction, he seemingly with much good-will went with me. At midday I reached my long-wished *Pinus* (called by the Umpqua tribe *Nàtele*), and lost no time in examining and endeavouring to collect specimens and seeds. New or strange things seldom fail to make great impressions, and often at first we are liable to over-rate them ; and lest I should never see my friends to tell them verbally of this most beautiful and immensely large tree, I now state the dimensions of the largest one I could find that was blown down by the wind : Three feet from the ground, 57 feet 9 inches in circumference ; 134 feet from the ground, 17 feet 5 inches ; extreme length, 215 feet. The trees are remarkably straight ; bark uncommonly smooth for such large timber, of a whitish or light brown colour ; and yields a great quantity of gum of a bright amber colour. The large trees are destitute of branches, generally for two-thirds the length of the tree ; branches pendulous, and the cones hanging from their points like small sugar-loaves in a grocer's shop, it being only on the very largest trees that cones are seen, and the putting myself in possession of three cones (all I could) nearly brought my life to an end. Being unable to climb or hew down any, I took my gun and was busy clipping them from the branches with ball when eight Indians came at the report of my gun. They were all painted with red earth, armed with bows, arrows, spears of bone, and flint knives, and seemed to me anything but friendly. I endeavoured to explain to them what I wanted and they seemed satisfied and sat down to smoke, but had no sooner done so than I perceived one string his bow and another sharpen his flint knife with a pair of wooden pincers and hang it on the wrist of the right hand, which gave me ample testimony of their inclination. To save myself I could not do by flight, and without any hesitation I went backwards six paces and cocked my gun, and then pulled from my belt one of my pistols, which I held in my left hand. I was determined to fight for life. As I as much as possible endeavoured to preserve my coolness and perhaps did so, I stood eight or ten minutes looking at them and they at me without a word passing, till one at last, who seemed to be the leader, made a sign for tobacco, which I said they should get on condition of going and fetching me some cones. They went, and as soon as out of sight I picked up my three cones and a few twigs, and made a quick retreat to my camp, which I gained at dusk. The Indian who undertook to be my last guide I sent off, lest he should betray me. Wood of the pine fine, and very heavy ; leaves short, in five, with a very short sheath bright green ; cones, one 14½ inches long, one 14, and one 13½, and all containing fine seed. A little before this the cones are gathered by the Indians, roasted on the embers, quartered, and the seeds shaken out, which are then dried before the fire and pounded into a sort of flour, and sometimes eaten round (*sic*). How irksome a night is to such a one as me under my circumstances ! Cannot speak a word to my guide, not a book to read, constantly in expectation of an attack, and the position I am now in is lying on the grass with my gun beside me, writing by the light of my Columbian candle—namely, a piece of wood containing rosin.

Friday, 27th.—My last guide went out at midnight in quest of trout with a flare and brought one small one in the morning, which I roasted for breakfast. He came two hours before day in great terror and hurry, and uttered a shriek. I sprang to my feet, thinking the Indians I saw yesterday had found me out, but by gesture I learned he had been attacked by a large grizzly bear. I signed to him to wait for day, and perhaps I would go and kill it. A little before day Bruin had the boldness to pay me a visit, accompanied by two whelps, one of last year's and one of this. As I could not consistently with my safety receive them so early in the morning, I waited daylight and accordingly did so. I had all my articles in the saddle-bags and the horse a mile from the camp, when I mounted my own, which stands fire admirably and rode back and found the three feeding on acorns under the shade of a large oak. I allowed the horse to walk slowly up to within twenty yards, when they all stood up and growled at me. I levelled my gun at the heart of the mother, but as she was protecting one of the young, keeping them right before her and one standing before her belly, my ball entered the palate of the young one and came out at the back part of the head. It dropped instantly, and as the mother stood up a second time I lodged a ball in her chest, which on receiving she abandoned the remaining live young and fled to an adjoining hummock of wood. The wound was mortal, as they never leave their young until ready to sink. With the carcase of the young one I paid my last guide, who seemed to lay great store by it. I abandoned the chase and thought it prudent from what happened yesterday to bend my steps back again without delay. So I returned and crossed the river two miles further down, and camped for the night in a low point of wood near a small stream. Heavy rain throughout the day.

Saturday and Sunday, 28th and 29th.—Both days very rainy, and having very little clothing and impossible to keep myself dry night or day obliged me to make all the exertion in my power to reach the camp near the sea, and being under the necessity of leading my horse the whole distance, he being greatly fatigued and the road daily getting worse by the continual rain. Camped on Saturday evening at my second crossing-place, but could get no food from the Indians, the bad weather preventing them from fishing, the river being much swollen. Boiled the last of my rice for supper, without salt or anything else, and had but a scanty meal. At daybreak on Sunday I resumed my march and went on prosperously until I came to the large woody hill half-way between our first march on the north bank of the river, when one of the poor horses fell and descended the whole height over the dead wood and large stones, and would have been inevitably dashed to pieces in the river had he not been arrested by being wedged fast between two large trees that were lying across the hill. I immediately tied his legs and head close to the ground to keep him from wrestling, and with my hatchet I cut the lower tree and relieved him, having received but little injury. I felt over this occasion much, for I got him from Mr. McLoughlin and it was his favourite horse. Reached the camp in the dusk, where I found Mr. Michel Laframboise, the Chenook interpreter, and an Indian boy, who told me the Indians had been troublesome since the brigade of hunters left him on Monday. He kindly assisted me in pitching my tent, gave me a little weak spirit and water, and then made a basin of tea, which I found very refreshing. Very heavy rain during the night.

30th.—Last night about ten o'clock several Indians were seen round our camp all armed, and of course instead of sleep we had to make a large fire, leave the camp a little distance, and hide in the grass to watch. An hour and a half before day a party of fifteen passed us, crashing among the grass towards our fire ; we immediately fired blank shot and scared them. Returned to the camp and made some tea and ate a little dry salmon for breakfast, and as I had not a single bit of dry clothing and it still raining, I sat in my tent with a small fire before the door the whole day.

Tuesday, 31st.—Heavy showers, with south-west wind off the ocean. Cold and raw. Brought wood in the morning for fuel and some branches of pine and Pteris aquilina [1] for bedding. At noon an Indian who had undertaken to guide two of the hunters to a small lake twenty or thirty miles to the south-east of this, returned to our camp and brought on his back one of their coats and had in his possession some of the hunting-implements, and looked altogether very suspicious : for the present, as we do not understand their language, we pay no attention ; perhaps he has stolen and not murdered them. Kept up our watch as usual; find myself greatly fatigued and very weak. Were not troubled during the night.

Wednesday, November 1st.—Heavy rain until two o'clock. In the afternoon Baptist Mackay returned from the coast, who tells me he hardly ever experienced such bad weather; he had not a dry day. We felt a little relieved to think our small party getting strong, particularly such a one as Mackay, as he will soon procure us fresh food. Evening cloudy.

Thursday, 2nd.—Baptist Mackay went out in the morning and very fortunately killed a fine large doe Long-tailed Deer, which he brought home on his horse at noon. I was glad to stand cook, and ere 4 P.M. I had a large kettle of fine rice soup made, and, just as we were sitting down to eat, thirteen of the hunters came in sight in five canoes and of course were invited to partake. I find this evening pass away agreeably to the eleven preceeding, and although the society at many times uncouth, yet to have a visage of one's own colour is pleasing ; each gave an account of the chase in turn. I find myself stand high among them as a marksman and passable as a hunter.

Friday, 3rd.—Early in the morning made a trip about twelve miles below this place in hopes of meeting my companion, Mr. McLeod, whom

[1] Pteridium aquilinum, Christensen, Ind. Fil. p. 591.

I daily expected. From this place as far as I went the banks are steep, rocky, and thickly wooded with pine, the same as is found on the Columbia near the ocean. The lowest part of the river I saw might be from seven to eight hundred yards wide ; tide flows up the river twenty-seven miles, and fifteen from the sea rises 4 feet. Returned at six o'clock in the evening. Whole day very heavy rain. Collected a fine shrub in ripe fruit : berries in a thick close long raceme, round and somewhat pointed at the top, two-seeded in a soft blood-red pulpy juice ; skin hard, brittle and pubescent ; leaves opposite entire, pubescent underneath ; petiole short, ovate, smooth ; a handsome evergreen shrub, 4 to 10 feet high ; wood hard and brittle, with a whitish bark ; abundant on rocky places. I learn it is also plentiful to the north, in the country inhabited by the Killimuck Indians and also to the south, but is not eatable. Found in abundance *Vaccinium ovatum*, loaded with fruit : berry small, jet black when perfectly ripe, and yields a great quantity of thin watery bloody juice, but exceedingly pleasant acid ; fruit in clusters at the point of the branches ; gathered a large paper of seeds and took a few twigs as specimens, as also the preceding. Still unable to find *Castanea* in a perfect state, as well as *Ilex Dahoon*. At a small village nineteen miles from the sea, on the right bank of the river, consisting of seven or eight lodges, I got a few berries of a species of *Vaccinium*, very large, fully as large as marrow-fat peas, light blue colour (nearly azure), has on pressure scarcely any acid, gives a large quantity of thick jelly of the same colour as the skin. Whether the shrub be evergreen, large or small, I am unable to learn, and the only thing respecting it is that it grows on the mountains.

Saturday, 4th.—I had not been at our camp more than an hour last night when I had the satisfaction of being joined by Mr. McLeod from his travels to the southward. He informs me that this river (the Umpqua) at its confluence with the ocean is about three-quarters of a mile broad and has a shallow sand-bar and much broken water at the flow of the tide ; will not admit ever of any shipping. He journeyed along the sea-beach for twenty-three miles, when he came to a second river, similar in size to this one and also affording the same sort of salmon and salmon-trout. At its mouth are numerous bays : some of them run considerable distances through the country, which is by no means so mountainous as that north-wards, and in one of the said bays he pursued his route to the south in a canoe for twenty miles, where he came to a third river, a little smaller than the others, but by the Indian account takes its waters a long distance in the interior. Abounds with the same fishes. Here for the present his expedition stops until he has his party all forward. By his account from the Indians a large stream of water falls into the sea, perhaps about sixty miles still further to the south, where the natives are said to be very numerous : one of his linguists, who has seen the Columbia and the new river, says it is much larger than it. Mr. McLeod tells me the country on the coast assumes a very different appearance from that on the Columbia. In the inter-mediate distance between here and there, and to the north half-way between the last two rivers, which will be found about 41° North latitude, the genus *Pinus* is no longer to be seen ; on the banks of rivers and on the mountains the fine-smelling tree is more plentiful than on this river, that it takes the place of the pines : he measured some 12 feet round, and 70 to 100 feet in length, and the lightest breath of wind to stir the leaves sends a fragrance through the whole grove. All the natives like those here had never before seen such people as we are, and viewed him narrowly and with much curiosity ; but hospitable and kind in the extreme. Kindled his fire, assisted in making his encampment, glad and pleased beyond measure on receiving a bead, ring, button, in fact the smallest trifle of European manufacture for their services. Have the same clothing and houses as those in this neighbourhood. Mr. McLeod tells me that two of his men are going to Fort Vancouver with a despatch on Monday morning, and as the season is far spent and the rainy weather set in, and at the same time doubtful if he will have any more communication before I should start on March 1st for the other side of the continent, I have made up my mind to return, and shall retain a grateful recollection of the kindness and assistance I have uniformly had from this gentleman. (Recollect on your arrival in London to get him a good rifle gun as a present.)

Sunday, 5th.—Morning dull and cloudy, heavy rain at noon. Tied up my little parcels in bear-skin to preserve them from the rain.

Monday, 6th.—Heavy rain until noon, with a high westerly wind detained all day in consequence of it.

Tuesday, 7th.—The rain last night fell in torrents but moderated at daybreak. As good weather could not be looked for at this late season of the year, I resolved on beginning my march. Started at ten o'clock A.M. with John Kennedy, an Irishman, and Fannaux, a Canadian, and nine horses. Mr. McLeod expressed his regret to see me leave him with such a small stock of food and that not of the best quality : a few dried salmon-trout, which were purchased of the Indians, and a small quantity of Indian corn and rice mixed together, which was brought from Fort Vancouver. In all, a week's food for two persons ; but at this season I hope to find abundance of wildfowl, failing meeting with small deer, so that there is little to be feared as to starving. As the late rains had rendered the high hill impassable for bonded horses, we were under the necessity of carrying our baggage up the river in three small canoes. Camped twelve miles up the river near two Indian lodges and had from the Indians some salmon-trout. Towards dusk it became fine and fair, with clear moon-light, which gave us an opportunity of drying our clothing.

Wednesday, 8th.—Cold and chilly during the night, thick fog in the morning. Went out on the chase before daylight, returned at ten o'clock unsuccessful. Detained waiting for the canoes till twelve o'clock, when we had the horses caught, loaded, and proceeded on our journey. Were much annoyed by the lacing of the saddle-bags stretching and becoming slack ; every three or four hundred yards had to be tightened. Passed our first encampment on the Umpqua at two o'clock, when I took my deer horns and specimens of wood which I had left three weeks ago. Camped in a small low circular plain at the commencement of the wooded hill ten miles from the Umpqua, our horses much fatigued and scarcely any grass for them to eat. As we were all unsuccessful to-day, we cannot afford but one meal a day. About midnight one of Baptist Mackay's dogs came to the camp; I found him in the morning at his accustomed place, asleep at my feet. Having no opportunity of sending him back, I allowed him to proceed until the return of the men.

Thursday, 9th.—River la Bische, which we found on our journey outward rarely above 4 or 5 feet deep, was now quite unfordable and the hill so slippery that we had to make a new path, which was very difficult from the immense quantity of low brushwood that we found over all the woody parts of the country. Several of the horses were so reduced and weak that we apprehended we should be under the necessity of leaving them. Both mine, although to appearance the most powerful, gave up : the one going light, the other with only 45 lb., my blanket, and collection. Six hours hard walking took us over the hill to the low plains, which we crept slowly along and camped in the evening at our encampment which we left on October 13th. Day fine, dry, and clear, succeeded at night by a heavy dew. The men proposed to have a sort of soup for supper made of pounded *Camass*, or the roots of *Phalangium Quamash*[1] of Pursh, to which I agreed, they observing it was very fine. I had not more than two spoonfuls when, with its sweet sickening taste together with the exertion I had made during the day, I became very sick and did not sleep during the whole night. Saw several deer, but could not get a shot at them.

10th.—Long ere day I was up by the fire and anxiously wishing for the morning, and certainly wished for a little tea, the greatest and best of comforts after hard labour. My horses being extremely weak, I got one of Kennedy's to ride, and Fannaux placed my articles on one of his and allowed mine to go light for a day or two to recover. Very heavy dew during the night ; the morning became overcast shortly after daylight, when the rain began to fall in torrents for the whole day ; like yesterday we experienced hard labour in crossing the second hill, although by no means so high nor so difficult. It being on the same track that I had already passed over, nothing new came under my notice. Still unsuccessful in the chase ; killed only one goose. Being anxious to reach our camp of October 7th, the nearest good place for allowing the horses to feed, but night stealing on us too soon, and endeavouring to complete our march after dark, we went off the path and had to camp near a small brook, the rain still falling in torrents. With great difficulty we pitched my tent, but could not make any fire by reason of the wind and rain ; we soon crept below our blankets for the night.

Saturday, 11th.—Last night, after lying down to sleep, we began to dispute about the road, I affirming we were two or three miles off our way, they that we were quite close to our former encampment ; all tenacious of our opinions. The fact plainly this : all hungry and no means of cooking a little of our stock ; travelled thirty-three miles, drenched and bleached with rain and sleet, chilled with a piercing north wind ; and then to finish the day experienced the cooling, comfortless consolation of lying down

---

[1] *Camassia esculenta*, Baker, in Journ. Linn. Soc. xiii. p. 257.

wet without supper or fire. On such occasions I am very liable to become fretful. Before sleeping we had agreed to go to a small lake seven miles further on, next day, where we hoped to find wildfowl and give the horses some rest. At daybreak I started on foot for the lake, leaving the men to bring up the horses; but being, as I have already observed, off our way the preceding evening, I had only walked about three miles when I perceived myself again off the road. The day being cloudy and rainy, and having no compass, I thought it prudent to return to the camp, which I did and found they had started, but by which course I could not say. I looked about and readily found our camp of October 7th and then proceeded by the old route. About midday I was met by Kennedy, who had gone to the lake by a new way and not finding me there became alarmed about my safety, and had come in search of me, leaving Fannaux to take care of the new camp. On reaching the plain three miles from the camp at 4 P.M., I proposed to go in search of wildfowl if he would go and assist Fannaux with the encampment; we did not part without my getting strict caution about going astray a second time. By six o'clock I had three geese and one duck, and on my way home, when I observed a large flock a little to the left of my path, I laid down my hunt, gun-slip, and hat to approach them, and after securing one returned in search of my articles, but was unsuccessful in finding them, although I devoted two hours to it. Reluctantly I gave it up and proceeded to the camp, and as the night was exceedingly dark I would have had some difficulty in finding it had they not made signals with their guns to guide me. Close to the camp fired among a cloud of ducks that were flying over my head and killed one ! I was hailed to the camp with ' Be seated at the fire, Sir,' and then laughed at for losing myself in the morning, my game and other property in the evening. There is a curious feeling among voyageurs. One who complains of hunger or indeed of hardship of any description, things that in any other country would be termed extreme misery, is hooted and browbeaten by the whole party as a *pork-eater* or a young voyageur, as they term it; and although in many instances I have observed they will endure much privation through laziness, and not unfrequently as a bravado, to have it said of them they did so-and-so, I found in this instance my men very willing to cook the fowls and still less averse to eating them. Heavy rain.

Sunday, 12th.—Very heavy rain, with a high westerly wind, during the whole night. In the grey of the morning I returned back in quest of the articles I had mislaid last night and readily found all safe, and an additional goose a short distance from the other three, one which I had killed without knowing it. To-day could afford myself and people breakfast (!) and then started, the weather having become more moderate. At two o'clock passed Longtabuff River, which falls into the Multnomah; and instead of following the old track, which we conceived more circuitous, we took a more northerly direction along the west bank of the river, intending to cross the Multnomah 150 miles above where we left our old camp on that river. Nine geese were killed, seven by me and two by Kennedy, which with what were killed the day before made us tolerably independent. Camped on the edge of a small lake, where there was abundance of wildfowl. Country open, rich, level, and beautiful. Marched thirty-four miles. Frequent hail and showers during the day.

Monday, 13*th.*—During last night four geese were killed by random shots, they sitting in thick shoals on the lake. Our firing and the smoke from our fire attracted several Indians to our camp belonging to the Calapooia tribe, who had very little food and had come to beg a little. I was glad in being able to relieve them, and as none of us knew the way one of them undertook to guide us to the crossing-place and to procure for us a canoe at the same time. As both my horses were unable to proceed further, I resolved to leave them with the Indians until the men should return. At the crossing-place the river is about 150 yards wide, deep, and very rapid, and occupied two hours crossing with the horses and property. Continued our northerly course over an extensive plain intersected by narrow belts of wood and groups of low oaks : soil deep rich alluvial deposit, on a bottom of clay and gravel. At mid-afternoon we came to a small stream flowing in a westerly course to the Multnomah, banks thickly covered with alder, poplar, ash, and willow, very steep, with deep water Spent the remainder of the day looking for a crossing-place, but found none ; camped on the south bank. Drizzly rain in the morning, cloudy and dull throughout.

Tuesday, 14*th.*—Early this morning two Indians fortunately came to our camp and informed us that we could cross the river on a fallen tree and the horses could swim at an old traverse, a little below. This we found correct and they for a small compensation assisted us. Continued our journey through the same sort of country as yesterday, and as usual put up at dusk near a small rivulet on a rocky bottom of fine white sandstone, near which is to be seen the channel of a river of large dimensions. Rainy all day.

Wednesday, 15*th.*—Light rain. In the morning I left the camp at daylight in search of game, leaving the men with the horses and being scarce of them and at the same time weak, I chose to walk. On arriving at Sandiam River, which falls in the Multnomah, a stream of considerable magnitude, we found the village deserted and no canoes. A raft could soon have been made, but from the rapidity of the current we could not guide it across. Therefore we looked up and down for the most suitable place to swim. The men chose to swim on their horses, I alone. Fannaux in the midst of the stream, in spurring on his horse, imprudently gave the bridle a sudden jerk, when rider and horse went hurling down before the current ; fortunately he extricated himself from the stirrups, and of course had to adopt my plan of swimming alone. I had articles of my clothing and my bedding drenched, but what gave me most pain was the whole of my collection being in the same state. Proceeded on and found an Indian village only two miles further on, with plenty of canoes. Camped about three o'clock, being fain to give my collection and clothing time to dry, which employed me all the evening. Killed no game ; gained about eighteen miles. To-night, from constant exposure to the wet and cold, my ankles are swollen, painful, and very stiff.

Thursday, 16*th*.—This morning we started in hopes of reaching our old camp on the Multnomah and would have done so had we not mistaken the path. Being unacquainted with the country, and lest we should go wrong, we chose to keep along the bank of the river, which took up more time than if we had taken the proper path, nearly a straight line. At two o'clock were met by *Tochty* (or Pretty) one of the Calapooia chiefs, who directed us on the right way and said that we should find canoes on the Multnomah, a few miles above his house. Our horses failing us, we had to camp ten miles from the place where we were to leave them. In the morning keen frost.

Friday, 17*th*.—This morning was much colder than yesterday. Started shortly before the men in quest of deer, having nothing to eat. Saw several but could not get near enough by the crisping noise among the frozen grass. Went down on the high banks of the river to two Calapooia lodges, where I was kindly treated by the inmates. The only article in the way of animal food was a small piece of the rump of Long-tailed Deer, which the good woman on seeing I stood in need of food had without loss of time cooked for me. The greater part of it was only the bare vertebræ, which she pounded with two stones and placed it in a basket-work kettle among water and steamed it by throwing red hot stones in it and covering it over with a close mat until done. On this, with a few hard nuts and roots of *Phalangium Quamash*,[1] I made a good breakfast. After paying my expenses with a few balls and shots of powder, and a few beads, I resumed my walk towards the end of my journey, five miles distant. At one o'clock I reached our old camp on the Multnomah, where I found the men, who had come on at a quick pace with the horses. Intending now to complete the remainder of our journey by water, our next step was to look for a canoe. One of the men went up a few miles to an Indian village, the other stayed to take care of the camp while I went down the east bank of the river in search of Etienne Lucien, one of the Canadian hunters, who, as I learned, had camped about seven miles below. Returned two hours after dark, unable to find him. Killed a fine large buck, the half of which I brought to the camp and hung the other on a small tree out of reach of the wolves, lest we should stand in need ; if not, to give it to some poor Indian. Fannaux joined us at nine o'clock, unable to find a canoe. Overcast at noon, light rain remainder of the day.

Saturday, 18*th*.—Last night an old woman and her son came to the camp who had been at the Falls twenty-five miles below buying salmon. They informed me that Lucien's family was below at their camp. With much persuasion I obtained the loan of her small canoe and despatched one of the men to procure a larger one, sufficient to carry us all. Shortly after mid-day he returned with one in which we all embarked and proceeded on our route to their camp, the only place we could put up at. On our arrival in the evening I found Mr. James Birnie and B. La Zand the Columbian guide, and a party of six men, who had arrived there in the interval on their way to the Umpqua River, where I had just left. He kindly provided a comfortable supper consisting of venison steaks, a few potatos, and a basin

[1] *Camassia esculenta*, Baker, in Journ. Linn. Soc. xiii. p. 257.

of tea, which he had brought from Fort Vancouver. Having had no tea for some time before, it prevented me from sleeping.

Sunday, 19th.—At four in the morning I started with the guide and seven men in a boat, and by a hard day's work arrived at Fort Vancouver at eight in the evening, having been absent two months.

Monday, 20th.—Received by the express across the continent letters from Jos. Sabine, Esq., dated February 20, Mr. S. Murray of Glasgow, and my brother. In the list of my acquaintances are a Mr. Simpson, a lieutenant in the Royal Navy, who left London in February, and a Mr. McMillan, one of the resident partners of the Fur Company, who gave me a friendly welcome. Received a very punctual note from Mr. J. G. McTavish, Chief Factor, York Factory, Hudson's Bay, acknowledging the safe arrival of my tin box of seeds sent last spring by Mr. McLeod, and at the same time informing me that it should be duly forwarded to England the same season by the return of their ship. Learned with regret that my box containing the remainder of the collection made during the summer and left at Kettle Falls in August had been omitted to be sent down ; hope that it is safe. Remainder of the day employed drying seeds, papers, and clothing.

21st to 30th.—From fatigue and constant exposure to the rain and cold, my ankles, which were slightly inflamed, swelled accompanied by an acute pain, and obliged me to remain within doors for nine days. Arranged what plants I had collected, dried and put up seeds, made a small packing-box, and repaired some of my shoes. Weather mild but very changeable, generally rainy ; failing that, cloudy heavy weather, westerly winds prevailing.

Friday, December 1st to the 8th.—Being a little recovered, I employed myself forming a collection of the various timbers that compose the forests, at the same time gleaning a few Cryptogamic plants.

December 9th to 25th.—My time lying heavy on my hands, I resolved on visiting the ocean in quest of Fuci, shells, or anything that might present itself to my view. Hired a canoe of some Chenooks who were here on a trading excursion. Mr. McLoughlin sent one of his men with me, who with two Indians formed my party. Two days took me to Fort George, the old establishment, where I slept and waited until the wind abated, before I could cross the river to the north side. Scarcely had I been ashore when the wind began to blow a strong gale from the south-west and the rain falling in torrents ; one of the most dreadful nights I ever witnessed. About midnight I was awakened by the breaking of the surge on the shore and the crashing of the drift-wood pile above pile ; and the sea rising so suddenly and so unusually high, in an instant dashed my canoe to pieces and obliged me to strike my tent at midnight and retire back into the wood In the morning when the storm abated I went to the house of Cassicass, son of the chief Com Comly, and borrowed a canoe and proceeded along Baker's Bay ; crossed the portage over Cape Disappointment to the bays near Cape Shoalwater of Vancouver, which I gained in two days' march, both rainy. Another short day's march took me to the house of my old Indian friend, Cockqua, who greeted me with that hospitality for which he is justly noted by people of the Establishment and his countrymen. He regretted that dry salmon and berries of *Gaultheria Shallon* was all the variety he could offer me. The boisterous weather had obliged the wild-fowl to seek more sheltered situations than his neighbourhood afforded, and it was too rough to venture on fishing. They were subsisting only on these articles. The salmon is very bad, lean in the extreme, killed in the small creeks in September, October, and November, in the spawning season : when dried resembles rotten dry pine-bark. Having nothing but this to subsist on, I was seized with a most violent diarrhœa, which reduced me in four days unable to walk. The weather giving no proofs of improve-ment—and from my increasing weakness I became alarmed lest it should prove dysentery—I abandoned the idea of prolonging my stay. There-fore in the morning I set off for the Fort, having obtained one duck to make a little broth. Three days took me to the village of Oak Point, where the Indians had that day caught ten sturgeon. Learning I stood in need of some, they instantly told me I could have none unless I should give either my hatchet or coat : being neither willing nor could I encourage such on principle, I declined bargaining. I offered tobacco, powder, and ball, and my knife for as much as would be one meal for myself and people, which they refused, but as I was a good chief and liked them they would let me have it if I would give the handkerchief from my neck for one small bit, and seven buttons off my coat for a second bit of the same size, which I did ! I have heard of people put to many shifts to live, but never in my life was I in such a hard case in bargain-making. He had my blessing and promise of a sound flogging should I ever meet him in a convenient place. I slept half a mile above the village and the following morning started early with the tide and a light air of wind. Put ashore for two hours a short distance below the Multnomah and waited the tide, when I again embarked ; went on all night, and arrived at Fort Vancouver on Christmas Day at midday, having gleaned, like my trip in the same quarter last year, less than any journey I have had in the country. Collected one specimen of *Pinus* (2nd, see Pursh), a low, scrubby, small tree, on marshy ground, rare ; one variety *Ledum latifolium* ; *Kalmia* sp. ; *Vaccinium* sp. ; *Oxycoccus*, on low marshy ground ; with a few mosses.

December 26th to 31st.—Weather dry and pleasant. On Saturday a keen northerly wind, with appearance of frost. Occasionally out with my gun hunting, gathering woods, and mosses. Got the blacksmith to make me a mineral hammer, as my other was left in the interior. Soon recovered from my sickness by a change of food.

1827

Monday, January 1st.—Morning dull but dry. The New Year was ushered in by a discharge of the great guns at daybreak. Day spent much to my satisfaction : after breakfast took a ride on horseback and carried my gun ; returned at dusk to dinner. The evening, like many I have passed in N.W. America, lay heavy on my hands.

Tuesday 2nd to Wednesday 31st.—On the 5th, heavy rain and sleet, with a south-east wind, succeeded at night by keen frost, 10° above zero.

On the 7th, 8th, and 9th snow, with little intermission : a regular fall of 18 inches to 2 feet over the whole country. The forest presents a most dismal appearance, the immense pines loaded with snow and their wide, spreading branches breaking under their load. This to me irksome, being prevented from going out, the snow too soft for snowshoes. On the 22nd slight thaws during the day, showers of hail and rain, sometimes freezing at night. To pass away the time I copied some notes of the Chenook tribe of Indians.

Thursday, February 1st to Wednesday, 28th.—The changeable weather of the last month continued until the 10th of this, when we had a second fall of snow, 15 inches deep, which lay until the 25th, and after that, frequent rains and gusts of wind. Killed a very large vulture, sex unknown. Obtained the following information concerning this curious bird from Etienne Lucien, one of the hunters who has had ample opportunity of observing them. They build their nests in the thickest part of the forest, invariably choosing the most secret and impenetrable situations and build on the pine-tree a nest of dead sticks and grass ; have only two young at a time ; egg very large (fully larger than a goose-egg), nearly a perfect circle and of a uniform jet black. The period of incubation is not exactly known ; most likely the same as the eagle. They have young in pairs. During the summer are seen in great numbers on the woody part of the Columbia, from the ocean to the mountains of Lewis and Clarke's River, four hundred miles in the interior. In winter they are less abundant : I think they migrate to the south, as great numbers were seen by myself on the Umpqua river, and south of it by Mr. McLeod, whom I accompanied. Feeds on all putrid animal matter and are so ravenous that they will eat until they are unable to fly. Are very shy : can rarely get near enough to kill them with buck-shot ; readily taken with a steel trap. Their flight is swift but steady, to appearance seldom moving the wings ; keep floating along with the points of the wings curved upwards. Of a blackish-brown with a little white under the wing ; head of a deep orange colour ; beak of a sulphur-yellow ; neck, a yellowish-brown varying in tinge like the common turkey-cock. I have never heard them call except when fighting about food, when they jump trailing their wings on the ground, crying ' Crup Cra-a,' something like a common crow. The remainder of the month heavy rains.

Thursday, March 1st to Friday, 9th.—On Friday the 2nd made a journey to the sea in company with Mr. E. Ermatinger and returned on Friday week. My object was to procure the little animal which forms their robe, but am sorry to say was disappointed in consequence of one of the principal men of the village, a cousin of my Indian friend Cockqua, dying the night before I arrived, when on such occasions it is extremely difficult to get them to do anything. Promises were made me that they should be brought to the establishment in time for the ship. Heavy rains, with light westerly winds.

Saturday 10th to the 19th.—Learning that the 20th of this month was the day fixed for starting on my journey across the continent, commenced packing my collections to go by sea in the first vessel for England, and

R

have made a tin box for as large a portion of the seeds as I could think of carrying on my back across the mountains ; in this one will go my journals. Three boxes packed by the 17th, and should any occur in the interim, Mr. McLoughlin will cause an additional one to be made. Left room in one for my plants that were omitted to be sent down in the autumn. Had my very small clothing made ready, paid my debts, and received a copy of the amount. Weather fine until the 17th. My old friend Mr. McLeod and party returned from the south as they were obliged to relinquish their journey owing to the heavy rains. He informs me that several streams fall into the ocean beyond his survey in the beginning of winter when I was with him, but the famous river so much spoken of by the natives for its size is by no means so large as represented. About 300 yards wide, bold rocky banks, deep and very clean, navigable for small vessels thirty miles from the sea ; has a sand-bar at its mouth. Found the natives hostile, one of his party being killed by them.

Tuesday, 20th.—Showery all day. Preparations being made for the annual express across the continent ; by five o'clock in the afternoon I left Fort Vancouver in company with Mr. Edward Ermatinger for Hudson's Bay, Messrs. McLoughlin, McLeod, Annance, and Pambrun for the interior. We were accompanied to the riverside by the few remaining individuals who constituted my little society during the winter, where we wished each other a long farewell—I glad that the time was come when my steps should once more be bent towards England. I cannot forbear expressing my sincere thanks for the assistance, hospitality, and strict attention to my comfort which I uniformly enjoyed during my stay with them—in a particular manner to Mr. McLoughlin (Chief Factor). Camped at sundown four miles above the establishment.

Wednesday, 21st.—Morning cold and raw. Started at 4 A.M. and breakfasted at nine on Point Vancouver ; continued our route at ten. Camped on the Cascade portage at dusk. Purchased a fine salmon-trout, weighing about 15 lb. Drizzly rain in the evening ; nothing particular occurred during the day.

Thursday, 22nd.—Heavy rain at daybreak and showery until noon. Finished the portage and took breakfast by ten. Camped half-way between the Grand Rapids and Dalles, on a sandy bluff. High wind during the night.

Friday, 23rd.—Morning cloudy and cool; noon fine, clear, and pleasant. Breakfasted at Thomson's portage, where I found several species of *Umbelliferae* in blossom ; *Lilium pudicum* [1] ; *Cruciferae*, annual, seed-vessel nearly orbicular, three to four seeded, maculate on one side ; leaves radical, hastate, smooth ; flowers faint white and red ; on moist ground, near springs ; rare. This is the only plant I had not seen before. Made the portage over the Lower Dalles by three o'clock, and the Upper or Little Dalles by five, and lest we should be annoyed by the Indians on the Falls, four miles higher up, we camped on the gravelly beach of the river. Wood, being scarce, wherewith to boil our kettle, was purchased. Watched all night.

[1] *Fritillaria pudica*, Baker, in Journ. Linn. Soc. xiv. p. 267.

Saturday, 24th.—Started at five and crossed the Falls portage at nine, where we breakfasted. In the interval the boat was gummed and otherwise repaired, being slightly injured. One of the boats being returned to Fort Vancouver, and disappointed in not finding horses, a party intending to go by land, Mr. McLoughlin and Mr. McLeod started on foot and were shortly followed by Mr. Pambrun and myself. Early in the day the boats being favoured with a following breeze enabled them to use a sail, which gave great assistance, the stream being rapid and very strong. Embarked at one o'clock, as the road was bad and unable to keep pace with the boat. Came up with those who started in the morning, but did not embark, horses being promised by Indians residing twelve miles above the Falls. Camped on the north side of the river, seven miles above Day's River, at dusk. Were joined by our friends, who walked all day. The servant, Overy, who had waited behind for the purpose of bringing up the horses, came to us two hours after dark, having in his hand eight or ten broken arrows which he wrested from an Indian who threatened to put one through him if he did not allow himself to be pillaged. He might have laid him dead on the spot, but prudently chose to allow him to walk away, being rewarded previously with a heavy flogging and deprived of his bow and arrows. Five horses at the camp and the owner agreed to go to Wallawallah. Very high wind from the south-west during the night.

Sunday, 25th.—The Indian who was engaged last night chose this morning to change his mind. He got a comfortable supper and a whiff of the pipe, which perhaps was all he wanted. Three of us went off on foot and three in the boat until nine o'clock, when I took my turn until dark. High wind in the morning, calm at noon. Saw many butterflies and swallows. Camped on the north bank, nine miles below the big island.

Monday, 26th.—At daylight went off on foot over a point of land and met the boat at the lower end of the big island at nine o'clock, when we took breakfast ; and having to cross over to the south channel, the north being too shallow, I embarked for the remainder of the day. Put ashore at two at the upper end of the island, where we discovered that Mr. McLoughlin's gun had been left at our breakfast-place, and being loth to lose it, having some celebrity attached to it (Sir Alexander McKenzie used it on both his former journeys), Overy, another Canadian, and an Indian were despatched for it ; in the meantime we halted for them. Saw nothing new. Observed that the two species of *Cactus* found abundantly over the plains are used as food by baking them in the same manner as *Phalangium*.[1] Purchased some horseflesh of the Indian, on which we supped. Very high wind during the whole night. Took my turn of watching and cooking by the kettle with Mr. McLeod.

Tuesday, 27th.—At daylight Dupond the Canadian returned, and told us that Indians had been at our breakfast-place and carried off the gun, and Overy had gone in quest of them and informed us he would soon follow. With difficulty four horses were hired of a Kyuuse Indian, when about eleven o'clock three started overland, leaving a horse for Overy to follow. I walked along the banks of the river in quest of plants, but found nothing

[1] *Camassia esculenta*, Baker, in Journ. Linn. Soc. xiii. p. 257.

different from what I had seen. Killed a male grouse of the plain, a last year's bird, and although neither so large nor well plumed as many I have seen, I could not help skinning it, lest I should not get a better one. The gizzard is large but smooth in comparison with others of the tribe. The windpipe is exceedingly large, and strong (fully stronger than a goose). Their voice I cannot learn. Saw two others, but could not secure either. Camped on an island seven miles below the establishment of Wallawallah.

Wednesday, 28*th*.—Started at five and reached the establishment at eleven, where we found our friends, who came overland and arrived there last night. Stayed until three o'clock, during which time I changed what few plants I had gleaned and put my grouse in order. Camped three miles below Lewis and Clarke's River. Evening fine.

Thursday, 29*th*.—Morning dull and heavy. About an hour after we had embarked three large flocks of grouse rose from the gravelly shore of the river, but out of reach of my gun, that part of the river being shallow, which prevented the boat from getting near them. Observed them in groups of eight or ten, dancing, most likely holding their weddings. Vegetation much later than last year. *Lilium pudicum*[1] the only plant in blossom on the plains. Blue Mountains, where I was last year, enwrapped in snow to their base. High south-westerly wind until noon. Camped above the commencement of the clayey hills, seven miles.

Friday, 30*th*.—Heavy rain during last night, which continued until mid-afternoon. Walked all day ; nothing worthy of notice occurred. Saw three grouse which escaped, being unable to keep our guns dry. Camped in the bend of the river, fifteen miles below the Priest Rapids. Much fatigued and my feet painful from the gravel and shattered rocks, having nothing but shoes of deerskin dressed—that is, the hair off and smoked with rotten wood. Had a fine camp : plenty of firewood, which enabled us to dry our clothing.

Saturday, 31*st*.—Morning fine and pleasant. Crossed the river to the low gravelly grounds below the rapid, which is thickly covered with *Tigarea*[2] and *Artemisia*, where clouds of grouse were flying round us. At this season they appear to be more shy than in the autumn. Observed them in groups as before, dancing. The males spread the tail like a fan and puff up their breast or pouches to as large a size as the whole body, and like the pigeon singing their song, which I listened to with much pleasure. The voice is ' hurr-r-r-r " hoo—hurr-r-r-r " hoo,' a very hollow, deep, melancholy sound. The female I have heard call only when rising from the ground, which is ' Cack—cack—cack,' like the common pheasant. The flesh is fine, but not so white as many others. Four cocks were killed, two by Mr. McLeod with his rifle and two by myself with heavy shot : all too much injured for preserving. Met Mr. A. McDonald from Thomson's river, who returned with us to go to Okanagan. The river being lower than ever it has been observed by Baptist Latand, the guide, considerable difficulty was experienced dragging the boat over

---

[1] *Fritillaria pudica*, Baker, in Journ. Linn. Soc. xiv. p. 267.
[2] *Purshia*, Benth. and Hook. f. Gen. Pl. i. p. 617.

the shallow rocks. While they were in the rapid I took a turn round the ground and found a species of *Allium*, 2 to 6 inches high, pink and white flowered, in gravelly and rocky places. Killed a hen pheasant, which I shall skin to-morrow, and some curlews. Camped in the middle of the rapid at dusk. The wind was so high that little fire could be kept in during the night.

Sunday, April 1st.—Skinned my pheasant in the morning and break-fasted at the upper end of the rapid, when, as usual, I resumed my walking until dusk. Passed high steep rocks of fine white marble and granite ; took a fragment of each, and as no plants came under my notice picked up a few minerals. Camped on the south side of the river, seventeen miles above the rapid.

Monday 2nd to Thursday 5th.—The river flowing through a more mountainous country, and further to the north, scarcely a vestige of vegetation can be seen, only the gravelly bank and north side of the river, all the ground covered with snow. Walked along the banks of the river picking up any mineral that seemed curious : found some very fine pebbles. Arrived at Okanagan on Thursday, a little before dusk.

Friday, 6th.—Fine, clear, and pleasant. At two o'clock I alone em-barked in the boat to go round the big bend, a day and a half's journey, being much fatigued and my feet very painful, blistered, and blood-run, having walked eleven days. My fellow-travellers remained to come over the point on horseback. Parted with Mr. A. McDonald, from whom I have had much information, assistance, and hospitality. Camped fourteen miles above the establishment. I intended to have left the pair of grouse here, but not being perfectly dry I was afraid they would fall a prey to insects.

Saturday, 7th.—Continued my route and passed the Dalles at midday ; a dangerous part of the river during the time of high water. Trees 3 feet in diameter lay on the rocks, 43 feet above the present level, at the narrow est place 57 yards ; placed there by the water. The current passes with amazing velocity. Killed a few of the common stock duck. Camped at the big stone, or more properly a natural column, about 30 feet high, 900 in circumference, composed of trap, clink, lime, and gravel. Many of these columns are distributed over the country, some octagon, hexagon, and circular. Some are solely trap rock with a portion of iron. In that part between the present and former bed of the Columbia they are numerous. Mr. Ermatinger being the only one in the overland party who knew the place of rendezvous, mistook it and went twelve miles further up, where they remained an hour after dark for me ; fearing an accident had befallen me in the Dalles, they rode down to my camp and roused me from a sound sleep at half-past nine. Supper was made then, and we laid down at eleven. Had good jokes at losing the way.

Sunday 8th to Thursday 12th.—Passing over the same country that I did last year, and being at the same season, nothing additional occurred. Laid in a few duplicates of plants collected last year ; generally walking but taking the boat to pass the perpendicular rocks. The high grounds around the junction of the Spokane River and Kettle Falls covered with snow.

The last three days' journey afforded good sport with the gun among the small pheasant, curlew, and black or mountain grouse, basking on the shores of the river. In company with Messrs. McLoughlin and McLeod, I arrived at Fort Colville on the Kettle Falls in the morning, having travelled on foot seventeen miles. We were most cordially welcomed by my old and kind friends, Messrs. Dease and Work. Hung out my papers, examined the seeds; found my box with all my articles quite safe which was omitted to be sent down last autumn.

Friday 13th to Tuesday 17th.—Weather changeable : hail, snow, and rain, wind northerly. The first night of my arrival, I had the great misfortune to get my pair of grouse devoured, the skins torn to pieces by the famished Indian dogs of the place. Although they were closely tied in a small oilcloth and hung from the tent-poles, the dogs gnawed and ate the casing, which were leather thongs. Grieved at this beyond measure. Carried the cock bird 457, and the hen 304 miles on my back, and then unfortunately lost them. Wrote a note to Mr. Archibald McDonald at Okanagan to endeavour to procure for me a pair against the sailing of the first vessel for England. Mr. Work showed me a pair of *Mouton Blanche* of the voyageurs, male and female, skins in a good state of preservation. Is the same animal which I saw in Peall's Museum at Philadelphia, brought by Lewis and Clarke. The male is large, 200 to 250 lb. weight. The female considerably smaller, a purer white colour, at the same time finer, the beard less, and the horns shorter. Also a pair of Black-tailed Deer, male and female, likewise in a good state. The former killed on the high mountains twenty miles higher up the river, during the late heavy snows, by the Indians on snowshoes, with their bows. The latter abounds in all the mountainous country in this neighbourhood and is killed abundantly in the same manner. The same gentleman had a solitary skin of the small wolf of the plains, a singular variety and curious from its being the deity or god of the Flathead tribe of Indians. Perhaps I might have got the whole off Mr. Work, but knowing them to have been procured at the particular request of Mr. Ganny[1] in London, I of course could not ask for them. Mr. Sabine, I hope, will get them through that channel. Took a single specimen of each plant not already sent to England, and packed in one of my old journals to save room. Packed the remainder to be sent to England and left the few minerals collected on my journey upward in charge of Mr. Work. Gathered a few bulbs of *Claytonia lanceolata,*[2] *Lilium pudicum,*[3] and roots of *Erythronium grandiflorum.* Although in a bad season for removal, I cannot forbear making a trial. Made a memorandum for Mr. McLoughlin regarding the final packing of two boxes at Fort Vancouver to be placed on the ship's invoice as ' dry plants, seeds, preserved animals, and articles relating to natural history,' for the Horticultural Society of London. Made a note to be read to my Chenook friend Cockqua, regarding skins of *Arctomys,* which I was unable to get when there. In order that Mr. Sabine may know of the ship's arrival in England, that the collection

---

[1] Name almost illegible.—ED.

[2] *Claytonia caroliniana* var. *sessilifolia,* S. Wats. Bibl. Ind. N. Am. Bot. p. 117.

[3] *Fritillaria pudica,* Baker, in Journ. Linn. Soc. xiv. p. 267.

may come to hand without delay, wrote the following note to him to go by sea :—

Fort Colville, Columbia River
April 16th, 1827.

DEAR SIR,—I beg to inform you that the Hudson's Bay Company's ship has arrived, on board of which are four boxes for the Society, containing the total collection made by me in N.W. America.—I am dear Sir, your most obedient servant, D. DOUGLAS.

To JOS. SABINE, Esq., &c. &c.

Made a note to the commander of the vessel as to treatment. By mid-afternoon of Tuesday, preparations being made for our departure, I in company with Mr. McLoughlin and Mr. McLeod took an airing on horseback and returned at dusk to dinner. About nine o'clock at night I was convoyed to my camp, about a mile above the establishment, where we pitched in order that no time would be lost in starting in the morning by them, who spent a few minutes with us and then returned. Having now just bid farewell to my Columbian friends, I cannot in justice to my own feelings refrain from acknowledging the kindness shown to me during my stay among them, a grateful remembrance of which I shall ever cherish. My society now is confined to Mr. Edward Ermatinger, a most agreeable young man who goes to Hudson's Bay with us and seven men— four Canadians and three Iroquois Indians. Our next stage is Jasper House, in the Rocky Mountains, distant about 370 miles. Laid down to sleep at 2 A.M.

Wednesday, 18th.—Overslept ourselves this morning and were not up until daylight, when we hurriedly pushed off lest we should be seen by our old friends, who left us last night. A shower of snow fell during the night, which continued throughout the morning. Thermometer 28°, wind northerly and very piercing. General course of the river northerly, bounded on both sides by high mountains : many places rugged, granite, iron, and trap rock. *Pinus taxifolia,*[1] *P. resinosa* on the hills to their summits, and *P. Larix* [2] in the valleys, the size of neither so large as lower down. The river at many places is narrow, 70 to 100 yards broad, and the numerous rocks and high gravelly banks forming points render it rapid and very laborious to ascend. Passed the Dalles at 8 P.M. and was shortly over- taken by an Indian with a letter that was forgotten. The only plants in flower are *Lilium pudicum* [3] and a species of *Pulmonaria* already in the collection. Camped at dusk eight miles above the Dalles on the left- hand side of the river. Travelled twenty-nine miles. Mr. Ermatinger, during the time of boiling the kettle, favoured me with some airs on the flute, which he plays with great skill. Noon cloudy.

Thursday, 19th.—Just as I laid down my pen at midnight a heavy shower of snow commenced and continued for two hours. Three inches deep in the morning ; found it very local, for five miles higher up there was none. Like many others, this day's journey admits of little variety.

---

[1] *Pseudotsuga Douglasii,* Mast. in Journ. R. Hort. Soc. xiv. p. 245.

[2] *Larix occidentalis,* Mast., *loc. cit.,* p. 218.

[3] *Fritillaria pudica,* Baker, in Journ. Linn. Soc. xiv. p. 267.

Country more mountainous and rugged, the timber smaller. Ten miles from our camp, about eight o'clock A.M. passed Flathead River, a stream not more than 30 yards broad at its entrance, but throws a large body of water into the Columbia. The entrance is cascades, 9 or 10 feet high, over which the water is dashed, which has a fine effect, issuing as it were from a subterranean passage, both sides being high hills with large pines overhanging the stream. The headwater of this stream was passed by Lewis and Clarke in their tour across the continent. I am informed by Mr. P. Sogden, who possesses more knowledge of the country south of the Columbia than any other person, that its source is a small lake in the Rocky Mountains, which discharges water to both oceans : from the east end is the headwater of one of the branches of the Missouri, and one, as I have observed, is a feeder of the Columbia. Took breakfast two miles and a half above it on the opposite side at nine, where we stayed our usual time, half an hour. From the high grounds on the bank of the river, as far as the eye can behold, nothing is to be seen but huge mountains, ridge towering above ridge in awful grandeur, their summits enwrapped in eternal snow, destitute of timber, and no doubt affording but a scanty verdure of any sort. Lower down the scene is different : rugged perpendicular cliffs of granite and scattered fragments which from time to time have been hurled from their beds in masses too large and weighty for anything to withstand. At the foot the timber is larger. In addition to those seen yesterday, one *Pinus Larix* [1] of small growth and a species of scrub pine which I saw last year on the high grounds between Kettle Falls and Spokane with small round cones, leaves short and in pairs. *P. taxifolia* [2] and *P. resinosa* smaller, on the latter more of the eatable moss. *Pinus Strobus* abounds and although very lofty I have not seen one exceed 2 feet in diameter. Now that I see this I think the large species found on the Umpqua River and south of it has considerable affinity to it, but at the same time still specifically distinct. On the moist grounds in the valleys and shore of the river, birch is seen of larger dimensions than any that has yet come under my notice. The *Populus* is *P. tremula*, *Corylus* is seen as underwood to *Betula*. Of herbaceous plants *Ranunculus* sp., *Claytonia lanceolata*,[3] and *Erythronium grandiflorum* in flower ; a few of the latter I laid in. General course of the river northerly ; scarcely a mile without a rapid. Camped on the right at the foot of a high mountain remarkable for its being circular. Morning cool ; noon clear, fine, and pleasant. Sky beautiful at sunset, the snowy summits of the hills tinted with gold ; the parts secluded from his rays are clothed with cloudy branches of the pine wearing a darker hue, while the river at the base is stealing silently along in silvery brightness or dashes through the dark recesses of a rocky Dalle. How glad should I feel if I could do justice to my pencil (when you get home, begin to learn). Last night I forgot to say, a small stream four miles below our last camp falls into the river—called White Sheep River, from the antelopes found on its banks—a few miles back from the Columbia. Also I must observe

---

[1] *Larix occidentalis*, Mast., in Journ. R. Hort. Soc. xiv. p. 218.
[2] *Pseudotsuga Douglasii*, Mast., *loc. cit.*, p. 245.
[3] *Claytonia caroliniana* var. *sessilifolia*, S. Wats. Bibl. Ind. N. Am. Bot. p. 117.

that Mr. Work presented me with a nightcap made of the hair and wool of that animal, netted by an Indian girl, and a pair of inferior snowshoes called bear's paws. Travelled thirty miles.

Friday, 20th.—Slight frost in the morning. The tent being wet and partly covered with snow from the preceding night, a small fire had to be kindled in it before it could be folded. Passed, about a mile above our camp, McGillivray's or Cootanie River, also a stream of some magnitude, rapid, and very clear water. This is said to be a good route across the mountains, but from the hostile disposition manifested by the natives inhabiting the higher parts of the Saskatchewan, the Athabasca portage is preferred, being free from such visitors. Five miles above it the Columbia gradually widens to a lake, one to two and a half miles broad, some places very deep, having bold perpendicular rocks ; at other places small bays with gravelly or sandy beach with low points of wood. The scenery to-day is fine, but not so broken, the hills fully as high and more thickly wooded ; high snowy peaks are seen in all directions raising their heads to the clouds. Took breakfast at 8 A.M., gained then nine miles. Course of the river then north-west and by north-west, and north-east. About 10 P.M. a light breeze sprang up which enabled us to use a sail, which slackened during the middle of the day, but freshened up again in the afternoon. Noon clear and fine. Intended to have arranged a few words of the Chenook language, but was molested out of my life by the men singing their boat-songs. That small species of *Juniperus* found last year at Kettle Falls is the only plant that came under my notice which I did not see yesterday. Camped at dusk on a low sandy point on the left side, four miles from the upper end of the lake. Our distance this day is about forty-seven miles.

Saturday, 21st.—Shortly after dusk last night an Indian and his two children came to our camp and sold a small piece of venison and a few small trout, 10 to 14 inches long, of good quality, and some small suckers, so common in the lower parts of the river. I learn that sturgeon is in the lake, but is not fished by the Indians. Morning clear and fine, wind easterly, which greatly impeded our progress. Started at daylight and continued our route along the north shore. At seven passed a camp of Indians, consisting of three families, from whom three pair of snowshoes, such as I obtained at Kettle Falls, were purchased. Reindeer (Cariboux of the voyageurs) it would appear are found in abundance in the mountains : not fewer than a hundred skins were in this lodge. They are killed readily during the deep snow with the bow. From Mr. A. R. McLeod, who spent several years on the McKenzie River, I learn there are two varieties of this animal, a larger and a smaller. The latter abounds in the high latitudes of the north. The former, or larger, is the one found here and differs from that east of the mountains in no respect except the hair a little darker colour and somewhat curled on the belly and inside of the thighs. The large hoof which this species has (not observed in any other of the genus) is a proof of the wise economy of Nature, given it to facilitate its tedious wanderings in the deep snows. At 10 A.M. put ashore at the upper end of the lake to breakfast, where we stayed three-quarters of an hour. Instead of four miles, as I observed last night, I found it to be eleven. Here were four Indians gathering from the pines a species of lichen, of which they make a sort of bread-cake in times of scarcity. In their camp were horns of Black-tailed deer and one pair of Red, or stag, the first I have seen since I left the coast. The last eleven miles of the lake due east. Drift-wood is seen on the rocks, 10 feet above the present level. On leaving the lake the river returns to its natural breadth about 150 yards, and continues with a swift current for a mile, when it again gradually widens out into a second lake, neither so broad nor deep as the one already mentioned. For fifteen miles the shores are low sand and gravel, with low points or necks of land chiefly wooded with *Pinus Larix* [1] and *P. canadensis* [2]; under their shade a species of *Lunaria* is just peeping through the ground. *Pinus resinosa* is no longer seen on the banks. This part the course is west of north. At the termination of the fifteen miles the country becomes still more broken; bolder shores. The timber is of more diminutive size and the lake widens out into bays on both sides, about three to four and a half miles broad. The canoes of the natives here are different in form from any I have seen before; the under part is made of the fine bark of *Pinus canadensis*,[2] and about 1 foot from the gunwale of birch-bark, sewed with the roots of *Thuya*, and the seams neatly gummed with resin from the pine. They are 10 to 14 feet long, terminating at both ends sharply and are bent inwards so much at the mouth that a man of middle size has some difficulty in placing himself in them. One that will carry six persons and their provisions may be carried on the shoulder with little trouble. Weather at noon pleasant; thermometer 55° in the shade; chilly towards evening. Camped on the right near a high rock of pure white marble. A mile on each side of the lake stumps and entire dead trees stand erect out of the water; by some change in Nature the river has widened. The same thing occurs ten miles above the Grand Rapids, 148 miles from the sea. Our distance to-day is thirty-one miles.

Sunday, 22nd.—In the grey of the morn, at four o'clock, we were on the water and pleasantly pursuing our journey, it being calm and the lake fine and smooth. Slight frost during the night; noon fine and warm but cloudy, which continued throughout the day. Crossed over to the right side and passed two points a little beyond the latter in a gravelly bay where we stopped to breakfast at nine, having already gained fifteen miles. This part of the lake has bold rocky shores. Course north and by east. Four miles further on the right there is a remarkable rock 240 feet high, perpendicular, of blue limestone on a substratum of granite. From this point one of the most sublime views presents itself: nine miles of water about five miles in breadth, having on the left a projecting point resembling an island and a deep bay on the right, with lofty snowy peaks in all directions. Contrasted with their dark shady bases densely covered with pine, the deep rich hue of *Pinus canadensis* [2] with its feathery cloudy branches quivering in the breeze, and the light tints but more majestic height of *Pinus Strobus* exalting their lofty tops beyond any other tree of

[1] *Larix occidentalis*, Mast. in Journ. R. Hort. Soc. xiv. p. 218.
[2] *Tsuga canadensis*, Mast., *loc. cit.*, p. 255.

the forest, imparts an indescribable beauty to the scene. A river seems to flow into the bay on the east. At the end of the lake, at the foot of a high and steep hill, were three Indian lodges. Camped on the land. Purchased of them a little dried reindeer-meat and a little black bear, of which we have just made a comfortable supper. They seem to live comfortably, many skins of Black-tailed, Rein, and Red deer being in their possession. I purchased a little wool of *Mouton Blanche* as a specimen of the quality of the wool ; gave seven balls and the same number of charges of powder for it. (Get a pair of stockings made of it.) Continued our route, leaving the lake at 4 P.M., the river being due north. Very shallow, 2½ to 3 feet deep, 200 yards broad, with a fine gravelly bottom ; the banks low and covered to the water's edge with wood : poplar and birch of large dimensions on the brink, with brushwood of *Cornus* and *Symphoricarpos*. This part of the river has low banks and in many places long sandbanks with large quantities of dead timber buried and bound together in the sand. Observed flocks of a small bird fluttering in the pines resembling the English wren but somewhat smaller : has a sweet chirping voice and hangs by the claws, head down, from the cones of the pines. Camped at dusk on a high point of wood (the channel of the river being here covered with snow) on the right hand. A Sunday in any part of Great Britain is spent differently from what I have had in my power to do. Day after day without any observance (except date) passes, but not one passes without thoughts of home. Plants not observed before are *Pinus nigra*,[1] *Linnaea borealis*, and *Pentstemon* (shrubby species), which I found last year on the hills at Kettle Falls ; *Asplenium trichomanoides*[2] and a species of *Polypodium*. *Pinus Strobus* and *Thuya occidentalis* with *Betula* increase in size the farther we go, while all others decrease. Our distance this day is twenty-nine miles of the lake and seven of the river, equal to thirty-six miles.

Monday, 23rd.—As usual, started in the grey of the morning, about four o'clock. Banks of the river low and shallow with very gravelly banks ; on the low points *Thuya* predominates, some enormously tall. I measured one 157 feet and another 204 feet. Breakfasted on the right-hand side at nine ; gained nine miles. Purchased some fish of a woman, consisting of three kinds—grey and red suckers, and white mullet, the latter of fine quality. Continued our journey at ten. The river and country the same until five o'clock P.M., being then about fifteen miles further, where the river takes a sudden bend to the north-east and to all appearance loses itself in the mountains. At this place and for two miles higher a scene of the most terrific grandeur presents itself ; the river is confined to the breadth of 35 yards—rapids, whirlpools, and still basins, the water of a deep dark hue, except when agitated. On both sides high hills with rugged rocks covered with dead trees, the roots of which being laid bare by the torrents are blown down by the wind, bringing with them blocks of granite attached to their roots in large masses, spreading devastation before them. Passing this place just as the sun was tipping the mountains and his feeble

---

[1] *Picea nigra*, Mast., *loc. cit.*, p. 222.
[2] ? *Asplenium platyneuron*, Christensen, Ind. Fil. pp. 126, 136.

rays now and then seen through the shady forests, imparts a melancholy
sensation of no ordinary description, filling the mind with awe on beholding
this picture of gloomy wildness. At the head of this (I would almost
say subterranean) passage there is a very dangerous rapid, where the
water falls 9 feet over large stones, to pass which took all our united
strength : two in the boat guiding her with poles and seven on the line.
Carried all my articles, lest evil should overtake them. Here it becomes a
little broader, the shores also rocky, and owing to the deep snow is dangerous
and fatiguing to walk. In the narrows the rocks are micaceous granite,
blue limestone, and white marble with red veins. Camped two miles higher
up and being still in the midst of rapids were obliged to camp on the right
side, among large stones and snow. Morning mild ; noon calm and
warm, 65° in the shade ; evening fine but chilly. This night from exertion
I can hardly write. No new plants to-day. Progress, twenty-eight
miles.

Tuesday, 24*th*.—Scarcely anything worthy of notice occurred this day,
the country wearing the same mountainous appearance as yesterday.
Started at daylight, the river flowing from west of north, shoally and very
rapid. Gained eight miles before breakfast. Continued through the same
sort of country throughout the day. Camped on the right-hand side, on
the edge of the thick wood, where we had some trouble in finding a dry
spot from the melting of the snow. Morning mild and cloudy ; light
drizzly rain at noon which still continues (now half-past seven). Progress,
twenty-three miles. The only plants of to-day are the species of
Shepherdia found last year at Kettle Falls of diminutive growth, and the
same *Aralia* or *Panax* found on the Cascade Mountains. Saw two beavers
in the water, neither of which could be killed. Likewise a few geese and
some ducks of two species, both plentiful on the coast.

Wednesday, 25*th*.—The rain last night ceased about twelve o'clock.
Morning dull but pleasant ; the mornings and evenings appear long.
The high mountains on the banks of the river screen the sun's cheering
influence from us until eight A.M. which is withdrawn shortly after four P.M.
Stayed for breakfast at nine at the foot of the Dalles des Morts ; gained
eight and a half miles ; general course, west of north. Here the river is
confined to the space of 35 to 50 yards broad, the current exceedingly
rapid and obstructed by large stones ; one place in the middle of the
narrows it is dashed over the shattered rocks, which it leaves in foam for
the distance of 40 yards below. Reached the head of the Dalles, about a
mile and a half long, shortly after noon, but not without much labour
and anxiety. Carried all my valuables (seeds and notes) on my back along
the rocks ; my other articles brought by one of the men. This place,
which is looked on as one of the most dangerous parts of the whole river,
derives its name from a melancholy circumstance which occurred a few
years ago to a party of men who were ascending as we now are. They had
the misfortune to have their craft dashed to pieces among the rocks when
by a supernatural exertion all escaped to shore, where they endured a short
respite from death, more unsupportable than immediate death itself :
without food, without arms, scarcely any clothing, being stripped, and three

hundred miles from any assistance. In that state, between the hope of life
and dread of death, which must be trying to all mankind, two agreed to
make an effort to save themselves by endeavouring, before their strength
would fail them, to reach Spokane, the nearest establishment, which they
did in such a state as readily bespoke the misery they had endured. One
of the veterans I have seen, an Iroquois, by name Francis, one of the best
boatmen. The remaining six being divided in opinion could come to no
resolution as to what step should be taken, and no doubt, as is the case in
such trying circumstances, became insensible to their safety. All died
except one, who, it is supposed on good foundation, supported his dreadful
existence on a forbidden fare, having previously imbrued his hands in the
blood of his companions or companion in suffering. This, be it as it may,
could not be brought home to him in point of law, and the wretch was sent
out of their service to Canada. Camped on the right-hand side of the river
on a sandbank, having gained ten and a half miles above the Dalles. Progress nineteen miles. Nothing in the way of plants this day. The rocks,
micaceous granite. Warm during the day, evening cool; high snowy
mountains in the distance at forty miles north. Will prove, I hope, the
dividing ridge of the continent.

Thursday, 26th.—Keen frost last night; obliged to rise twice to make
fire. Morning clear, noon fine, warm 70°, afternoon chilly. Breakfast
on the left; gained seven miles. Continued our journey through the
same sort of country as the preceding days. No plants came under my
notice: wood smaller growth; *Pinus Strobus* becomes rare. Snow in
many places six feet deep. Banks of the river steep and rugged down to
the channel, course north and very rapid. Could find no suitable place to
take up lodgings. Obliged to lie down on the shore (the only place clear
of snow) among the stones, but, being fatigued, glad of any place.
Twenty-three miles progress; walked in the morning, the snow having
a crust. High mountains seen in every direction. A ridge seen lying
north-east: I take to be the Rocky Mountains. Passed those seen last
night, which I took for them. Experienced a most violent headache the
greater part of the day, occasioned by the cold during the night. Walked
until in a state of perspiration, which gave me relief. Minimum heat 27°.

Friday, 27th.—Keen frost last night, morning clear, wind easterly.
Started at daylight, having enjoyed but a comfortless night's rest among
the stones on the shore of the river. Course north ; country the same as
yesterday. Breakfasted on the left side at nine; gained eight miles.
After resuming our journey for 300 yards, at a short turn of the river one
of the most magnificent prospects in Nature opened to our view. The
daily wished for dividing-ridge of the continent, bearing north-east, distant
six miles. The sight of the mountains is most impressive. Their height
from the level of the river from 6000 to 6500 feet, two-thirds covered with
wood, gradually diminishing to mere shrubs towards the confines of eternal
snow. One rugged beyond all description, rising into sharp rugged
peaks ; many beyond the power of man to ascend, being perpendicular
black rocks distinctly seen, having no snow on their surfaces. On the
right, rising from the bed of Canoe River, the northern branch of the

Columbia, they seem to be most rugged ; on the left, rising from the bosom of the Columbia, stands a peak much higher than the former, with a smooth surface.   Although I have been travelling for the last fifteen days surrounded by high snowy mountains, and the eye has become familiar to them and apt to lose that exalted idea of their magnitude, yet on beholding those mentioned impresses on the mind a feeling beyond what I can express. I would say a feeling of horror.   Arrived at the boat encampment 12 A.M. a low point in the angle between the two branches, the Columbia flowing from the east and Canoe from the north ; the former sixty yards wide, the latter forty, but very rapid.   At low water, as it is at present, the former has three channels, the latter two, which are not seen at high water, the space at that season being a perfect circle about six hundred yards diameter. Around the camp on the point the woods are *Pinus taxifolia*,[1] *P. canadensis*,[2] *Thuya occidentalis, Populus*, sp., all of large growth.   The underwood, *Cornus, Corylus, Juniperus*, and two species of *Salix* not yet in flower. *Linnaea borealis* found on the coast, and on the highest mountains a species of *Lilium* just peeping through the ground.   Examined the seeds in my tin-box and found them in good order ; repacked them without delay and at the same time tied up all my wardrobe, toilet, &c., which is as follows : four shirts (two linen and two flannel), three handkerchiefs, two pair stockings, a drab cloth jacket, vest and trousers of the same, one pair tartan trousers, vest and coat ; bedding, one blanket ; seven pairs of deer-skin shoes, or as they are called, moccasins ; one razor, soap-box, brush, strop, and one towel, with half a cake of Windsor soap.   In addition to these I was presented with a pair of leggings by Mr. Ermatinger, made out of the sleeves of an old blanket-coat or capot of the voyageurs. This, trifling as it may appear, I esteem in my present circumstances as very valuable.   When the half of these my sole property is on my back, the remainder is tied in a handkerchief of the common sort.   Now that I conceive my wanderings on the Columbia and through the various parts west of the Rocky Mountains to be over, I shall just state as near as possible their extent :—

In 1825.

	Miles
From the ocean to Fort Vancouver, on my arrival in April . .	90
In May, to and from the ocean to Fort Vancouver . . .	180
In June, to and from the Great Falls	210
In July, to and from the ocean and along the coast . . .	216
In August, journey on the Multnomah River	133
In September, to the Grand Rapids	96
On the mountains of the Grand Rapids	47
In October and November, to the sea	90
In the same, trip to Cheecheeler River or Whitbey Harbour . .	53
Ascending said river	65
Portage from it to the Cowalidsk River	35

---

[1] *Pseudotsuga Douglasii*, Mast. in Journ. R. Hort. Soc. xiv. p. 245.
[2] *Tsuga canadensis*, Mast., *loc. cit.*, p. 255.

Descending the latter   40
An allowance of my daily wanderings from Fort Vancouver, my
    headquarters   850
                                                    —————
                                                 2105

In 1826.

In March and April, from Fort Vancouver to the Kettle Falls . 620
In May, journey to Spokane 150
In June, from Kettle Falls to the junction of Lewis and Clarke's River 414
In June, journey to the Blue Mountains 190
In July, a second to the same 137
In July, ascending Lewis and Clarke's River to the north and south
    branch 140
A third journey to the Blue Mountains from that place . . 103
From Lewis and Clarke's River to Spokane 165
From Spokane to Kettle Falls 75
In August, from Kettle Falls to Okanagan by land . . . 130
From Okanagan to Fort Vancouver 490
In September, October, and November, from the Columbia to the
    Umpqua River and the country contiguous thereto . . 593
To the ocean and the bays north of the Columbia in December . 125
Daily allowance from my places of rendezvous 600
                                                 —————
                                               3932

In 1827.

In March and April, the whole chain of the Columbia from the ocean
    to the Rocky Mountains 995
                                               —————
                                Total . . 7032

My notes will show by what means it was gained.

Saturday, 28th.—Last night cold : minimum heat 18°, maximum, 51° ;
obliged to rise during the night to make fire. Delayed commencing my
journey, Mr. E. being employed laying the boat and other articles *en cache*,
until 8 A.M., when we breakfasted and took leave of the main body of the
Columbia in a due east course. Passed a low point of wood of a mile and
entered a swamp about three miles long, frequently sinking to the knees in
water, which was doubly fatiguing from the thin ice on its surface, too
weak to bear us up. Crossed a deep muddy creek and entered a second
point of wood of an uneven hilly surface. At eleven obliged to have
recourse to my bears' paws or snowshoes, the snow 4 to 7 feet deep, being
then soft by the sun's influence. Much annoyed throughout the day
by their lacing or knotting becoming slack by the wet, and being little
skilled in the use of them, now and then I was falling head over heels,
sinking one leg, stumbling with the other ; they sometimes turning
backside foremost when they became entangled in the thick brushwood.
Passed several rivulets which were only seen at the rapid places, all the

other places being covered with snow. Nothing different as to country. Observed the following plants : *Salix*, male and female, in flower, 6 to 10 feet high, on the margins of the creeks ; flowering being covered with snow, took specimens of it. *Juniperus Sabina*, var. *procumbens*, on dry spots in conjunction with *Arbutus Uva-ursi*,[1] *Betula nana*, and *Ledum*, in the high tufts of grass in the marshes. *Pyrola secunda*, *P. media*, and *P. umbellata*[2] with *Lycopodium alpinum* on the dead stumps, and on the ground, growing luxuriantly on decayed leaves. *Berberis Aquifolium* is seen of diminutive size in the shady wood ; *B. nervosa* seems a stranger. Observed dead stems of what I took to be *Helonias* at Kettle Falls and the *Ribes* with slender spiny shoots found at the same place inhabiting edges of springs and streams, also small. The *Vaccinium* with large amber-coloured berries, frequenting the Blue and Cascade Mountains, strong and vigorous. A low spiny-stemmed rose is seen, but rare. The timber is *Pinus Banksiana*, *P. Strobus*, *P. rubra*,[3] *P. taxifolia*,[4] *P. canadensis*,[5] and *Thuya*, neither remarkable for size. Camped on the west side of the middle branch of the Columbia at two o'clock; progress nine miles. Of animals saw a small Bunting, the whole body of a uniform light brown, except the wings, which were a dirty-white ; beak short, thick, white. Also the Blue-crested Jay, so common on the coast. Saw two Squirrels about the size of the English one, of a light chocolate colour, feeding on the seeds of the pines. A large Wolverene visited our camp in the evening, but escaped before a shot could be put his way. To-day is a scene of some curiosity even to myself, and I can hardly imagine what a stranger would think to see nine men, each with his load on his back (food and clothing), his snow-shoes in his hand, starting on a journey over such an inhospitable country : one falling, a second helping him up, a third lagging and far behind, a fourth resting smoking his pipe, and so on. Mr. Ermatinger handsomely offered that all my articles should be carried and I to go light. This I could not accept seeing him with his load, and although I was perfectly satisfied as to their safety, yet I could not but carry what has already cost me some labour and anxiety. I therefore took all the seeds in the tin-box and journals, secured in an oilcloth, weighing 43 lb., my wardrobe and blanket carried by one of the men. Somewhat tired, my shoulders painful from the straps. Evening fine.

Sunday, 29*th*.—Morning clear, minimum heat 23°, maximum 43°. Started at four, being refreshed by a sound sleep, in an easterly course for six miles, in the course of which made seven traverses across the river from one point to another on the channel, which is from three to five miles broad, and at high water during the summer forms an inland sea covering the whole valley from the foot of the mountains. Turned to the north-east four and a half miles over the same sort of ground, making seven more fords ; water in several 2½ to 3½ feet deep, current swift and strong, but not rapid. Did not require snowshoes, the snow being hard with a strong

[1] *Arctostaphylos Uva-ursi*, A. Gray, Syn. Fl. N. Am. ii. i. p. 27.
[2] *Chimaphila umbellata*, A. Gray, *loc. cit.*, p. 45.
[3] ? *Picea rubra*, Veitch, Man. Conif., ed. 2, p. 450.
[4] *Pseudotsuga Douglasii*, Mast. in Journ. R. Hort. Soc. xiv. p. 245.
[5] *Tsuga canadensis*, Mast., *loc. cit.*, p. 255.

crust of frost. Found the cold piercing, alternately plunging to the middle
in water 35° Fahrenheit and skipping with my load to recover my heat
among the hoar frost. At 9 A.M. entered a point of wood where the snow
was 4 to 7 feet deep, with a weak crust not strong enough to support us.
Obliged to put on my bears' paws ; path rough, and in addition to the
slender crust, which gives the traveller more labour, were dead trees and
brushwood lying in all directions, among which I was frequently caught.
Towards noon, the snow having become soft and we weary with fatigue,
camped on the brink of a river, where no time was lost in making a little
breakfast, every person's appetite being well sharpened by our walk.
Travelled in the wood four and a half miles, course north. Progress to-day
fifteen miles. Saw large number of geese on the banks of the river in
the valley, but none were killed. Killed one female wood-partridge, I
think not different from that on the coast, only a deeper brown and red
above the upper latchet of the eye ; does not seem shy. I wanted to take
it alive ; stood till within two yards and then fluttered among the dead
leaves when I placed a little lead in her body. The *Aralia* or *Panax*
found on the Cascade Mountains here grows 8 to 12 feet high and proportionally strong : gathered a few seeds of it, perhaps will grow. On the
dry places of the valley passed through *Potentilla fruticosa*, and *Dryas*
species are found in abundance. *Betula*, sp., a tree 40 to 60 feet high,
18 inches to 2½ feet diameter, in the moist parts of the woods ; of this the
canoes are made. In the twilight last night the Wolverene paid us a
second visit, when I gave a few shots which he thought he could carry,
which he did in consequence of it being dark, he secreting himself in some
hole under the root of a tree. Evening fine. Made a pair of socks out
of the legs of a pair of old stockings ; the feet being worn, took the skirts
of my coat to wrap round my toes instead of socks. Strict economy here
is requisite : my feet, ankles and toes very painful from the lacing of my
snowshoes ; otherwise, well and comfortable, lying in a deep hole or pit
among the snow on a couch of pine branches with a good fire at my feet.
If good weather visits us, we are thankful ; if bad, we make the best of
a bad situation by creeping each under his blanket, and, when wet, dry it
at the fire.

Monday, 30*th*.—Maximum heat 43° at my camp 700 feet on the
mountain, at 4 P.M. in the valley 22°. In the grey of the morning we
resumed our route on snowshoes in the wood about three-quarters of a mile.
Entered a second valley, course north-east. Rested after having travelled
two and a quarter miles, in the course of which we made seven fordings
over the same river that we crossed yesterday. Continued in the same
course for the distance of four miles more until reaching the east end,
making four fordings more. Here the stream divides into two branches,
that on the left flowing from the north, that on the right due east. Took
our course in the angle between the two, north-east, entering a thick wood
of the same kinds of timber already noticed. *Pinus balsamea* [1] more
abundant and of greater size. After passing a half mile in the wood
reached the foot of what is called the *Big Hill*, also thickly wooded. Steep

---

[1] *Abies balsamea*, Mast. in Journ. R. Hort. Soc. xiv. p. 189.

and very fatiguing to ascend, the snow 4 to 6 feet deep in the higher spots.
The ravines or gullies unmeasurable, and towards noon becoming soft,
sinking, ascending two steps and sometimes sliding back three, the snow-
shoes twisting and throwing the weary traveller down (and I speak as I feel)
so feeble that lie I must among the snow, like a broken-down waggon-
horse entangled in his harnessing, weltering to rescue myself. Obliged
to camp at noon, two miles up the hill, all being weary. No water ; melted
snow, which makes good tea ; find no fault with the food, glad of anything.
The remainder of the day is spent as follows : On arriving at a camp, one
gathers a few dry twigs and makes fire, two or three procuring fuel for the
night, and as many more gathering green soft branches of *Pinus balsamea* [1]
or *canadensis* [2] to sleep on, termed ' flooring the house,' each hanging up
his wet clothing to the fire, repairing snowshoes, and arranging his load
for the ensuing day, that no time may be lost ; in the morning, rise, shake
the blanket, tie it on the top, and then try who is to be at the next stage
first. Dreamed last night of being in Regent Street, London ! Yet far
distant. Progress nine miles.

Tuesday, May 1st.—This morning our fire that was kindled on the snow
had sunk into a hole 6 feet deep, making a natural kitchen. Minimum
heat 2°, maximum 44°, on the highest part of the big hill. Started at daybreak, finding the snow deeper and the trees gradually diminish towards the
summit ; laborious to ascend. Went frequently off the path in consequence of not seeing the marks on the trees, being covered with the snow.
Reached the top at ten, three miles, where we made a short stay to rest.
Course north-east. Descended in the same direction and came on the
river which we left two days before. Passed in the valley two small level
spots clear of wood and one low point of wood of small trees, *Pinus nigra* [3]
and *P. Banksiana*, where we camped at midday, being unable to proceed
further from the deep soft snow. Progress seven miles. Mr. E. killed on
the height of land a most beautiful male partridge, a curious species ;
small ; neck and breast jet black ; back of a lighter hue ; belly and
under the tail grey, mottled with pure white ; beak black ; above the
eye bright scarlet, which it raises on each side of the head, screening the
few feathers on the crown ; resembles a small well-crested domesticated
fowl ; leaves of *Pinus nigra* [3] in the crop. This is the sort of bird mentioned
to me by Mr. McLeod as inhabiting the higher parts of Peace and Smoky
Rivers. This, however, is not so large as described. Perhaps there may
be two varieties. Said also to be found in Western Caledonia. This
being the first I have seen, could not resist the temptation of preserving it,
although mutilated in the legs and in any circumstances little chance of
being able to carry it, let alone being in a good state. The flesh of the
partridge remarkably tender when new killed, like game that has been killed
several days ; instead of being white, of a darkish cast. After breakfast at
one o'clock, being as I conceive on the highest part of the route, I became
desirous of ascending one of the peaks, and accordingly I set out alone on

---

[1] *Abies balsamea*, Mast. in Journ. R. Hort. Soc. xiv. p. 189.
[2] *Tsuga canadensis*, Mast., *loc. cit.*, p. 255.
[3] *Picea nigra*, Mast., *loc. cit.*, p. 222.

snowshoes to that on the left hand or west side, being to all appearance the highest. The labour of ascending the lower part, which is covered with pines, is great beyond description, sinking on many occasions to the middle. Half-way up vegetation ceases entirely, not so much as a vestige of moss or lichen on the stones. Here I found it less laborious as I walked on the hard crust. One-third from the summit it becomes a mountain of pure ice, sealed far over by Nature's hand as a momentous work of Nature's God. The height from its base may be about 5500 feet : timber, 2750 feet ; a few mosses and lichen, 500 more ; 1000 feet of perpetual snow ; the remainder, towards the top 1250, as I have said, glacier with a thin covering of snow on it. The ascent took me five hours ; descending only one and a quarter. Places where the descent was gradual, I tied my shoes together, making them carry me in turn as a sledge. Sometimes I came down at one spell 500 to 700 feet in the space of one minute and a half. I remained twenty· minutes, my thermometer standing at 18° ; night closing fast in on me, and no means of fire, I was reluctantly forced to descend. The sensation I felt is beyond what I can give utterance to. Nothing, as far as the eye could perceive, but mountains such as I was on, and many higher, some rugged beyond any description, striking the mind with horror blended with a sense of the wondrous works of the Almighty. The aerial tints of the snow, the heavenly azure of the solid glaciers, the rainbow-like hues of their thin broken fragments, the huge mossy icicles hanging from the perpendicular rocks with the snow sliding from the steep southern rocks with amazing velocity, producing a crash and grumbling like the shock of an earthquake, the echo of which resounding in the valley for several minutes. On the rocks of the wood were *Menziesia caerulea* [1] ; *Andromeda hypnoides* [2] ; *Lycopodium alpinum* ; *L.* sp. unknown to me ; dead stems of *Gentiana nivalis* ; *Epilobium* sp., small ; *Salix herbacea* ; *Empetrum nigrum*, fruit in a good state of preservation underneath the snow ; *Juncus triglumis* ; *J. biglumis*, with a few *Musci*, *Jungermanniae* and lichens.

Wednesday, 2nd.—My ankles and knees pained me so much from exertion that my sleep was short and interrupted. Rose at 3 A.M. and had fire kindled ; thermometer 20°. Started at a quarter-past four through a gradually rising point of wood which terminated three hundred yards below the highest part of the pass in the valley. An hour's walking took us to one of the head springs of the Columbia, a small lake or basin twenty yards in diameter, circular, which divides its waters, half flowing to the Pacific and half to the hyperborean sea—namely, the headwaters of the Athabasca River. A small lake, about 47° of N. latitude, divides its waters between the Columbia and one of the branches of the Missamac, which is singular. This being a half-way house, or stage, I willingly quickened my pace, now descending on the east side. This little river in the course of a few miles assumes a considerable size and is very varied. There are two passes, one four miles from its source and one seven, when it finds its way over cascades, confined falls, and cauldrons of fine white and blue lime-

---

[1] *Bryanthus taxifolius*, A. Gray, Syn. Fl. N. Am. ii. I. p. 37.
[2] *Cassiope hypnoides*, A. Gray, loc. cit., p. 36.

stone and columns of basalt, like the feeders of the Columbia at the deep passes of the mountain ; where the torrents descend with furious rapidity it spreads out into a broad channel bounded by the mountains. The descent from the east is much greater than from the west, the mountain more abrupt and equally rugged. Found the snow eight miles below the ridge gradually diminish. The heat increases and the quantity of snow on the east not equal to the west. Passed on the right a very high (perhaps 4000 feet) perpendicular rock with a flat top, and three miles lower down on the same side two higher ones, rising to peaks about a mile apart at the base with a high background which appears two-thirds glacier, and in the valley or bosom of the three, columns and pillars of ice running out in all the ramifications of the Corinthian order. From the mouth of the valley of this awesome spectacle a passage is seen more like the crater of a volcano than anything else : stones of several tons weight are carried across the valley by the force of the current during the summer months. The change of climate is great and the herbage equally so. No more huge specimens of *Thuya* nor *Pinus taxifolia* [1] or *P. Strobus, Acer* nor *Berberis*, so abundant only a few miles on the other side. *Pinus nigra* [2] and *P. Banksiana* here take the lead, a few bare species of *Salix* and under the shade *Ledum palustre* with carpets of *Sphagnum*. In the dry hilly parts *Ledum buxifolium*,[3] *Arbutus Uva-ursi*.[4] The only bird seen on the extreme height of land is a small light dun Jay, who, with all the impudence peculiar to most of his kindred, fluttered round our camp last night picking any food thrown to him. Of the structure of the mountains I cannot speak ; it is worthy of notice, however, that all I have yet seen here and west of the Rocky Mountains have a dip of from 30° to 45° south-west. I do not recollect a single exception. Blue and micaceous granites, limestone trap, and basalt are the most common. Sand or freestone I have not yet seen. Halted at 10 A.M. to breakfast, and the snow being here now, fifteen miles from the ridge, intend to go on in the evening. At 2 P.M. started ; very warm, 57°. Passed through the valley for three miles and then entered a rocky point of wood ; the river confined into a very narrow space and rapid. On the dry gravelly shores *Dryas octopetala* and another species with narrow entire leaves : perhaps this may be *D. integrifolia*. Low wood, very wet and difficult to pass over, *Aralia* sp., a low shrub, 2 feet high. Went off my way, being before the others three miles, but as fortune would have it, just as the sun was creeping behind the hill, I observed smoke about a mile east of me. Without any loss of time I soon made up to it and found Jacques Cardinal with eight horses, who had come to meet us. An hour after one of the men came up, and shortly afterwards we heard several shots fired, which I knew to be signals for me, which obliged me to send the man on horseback to say I had arrived at the Moose encampment. Old Cardinal roasted, on a stack before the fire, a shoulder of *Mouton gris*, which I found very fine. He had a pint copper kettle patched in an ingenious manner, in which he

---

[1] *Pseudotsuga Douglasii*, Mast. in Journ. R. Hort. Soc. xiv. p. 245.
[2] *Picea nigra*, Mast., *loc. cit.*, p. 222.
[3] *Leiophyllum buxifolium*, A. Gray, Syn. Fl. N. Am. p. 43.
[4] *Arctostaphylos Uva-ursi*, A. Gray, *loc. cit.*, p. 27.

was boiling a little for himself; this with a knife was all the cooking-utensils. He observed he had no spirit to give me, but turning round and pointing to the river he said 'This is my barrel and it is always running.' So having nothing to drink out of, I had to take my shoulder of mountain sheep and move to the brook, helping myself as I found it necessary. Informed that Dr. Richardson had in February arrived at Cumberland House ; that Captain Franklin had met a ship in the North Sea ; and that Mr. Drummond, who spent last summer in this neighbourhood, had in November gone to Fort Edmonton on the Saskatchewan River. Finding none of my travellers come up, Cardinal gave me his blanket, reserving for himself the skin of a Reindeer. Mountains on all sides still as high and uneven, but with less snow ; no glaciers and more wood. Crossed the river fifteen times in three places ; two half full of water, very rapid and full of large stones.[1] This day marched twenty-five miles.

Thursday, *3rd.*—Shortly after daylight Cardinal went with his horses and brought up my companions at seven o'clock, when we took breakfast and had our little articles tied on two of the horses, and proceeded on the bed of the river where it was still covered with ice and snow, over points of wood, low marshes, and low hills. The path was extremely difficult from the dead wood and the ground still soft from the melting of the snow. General course north-east, at our camp to the main branch of the Athabasca River, a rapid stream seventy yards broad, where it is joined by the one on the banks of which we had descended. Crossed it at the junction, the course being then north ; descended the east bank. Intended to put up at the usual camp, but finding the horses and land better than expectation, they proceeded to the end of the portage on horses, I following with the gun in search of birds. Arrived at a small hut called the Rocky Mountain House at half-past six o'clock much fatigued. Progress this day thirty-four miles. Killed one partridge, the same as that found on the height of land, but I regret it was too much destroyed to preserve. Cardinal tells me my bird is small, that they are generally a fifth larger ; perhaps this is a young bird. Fired at a cock bird, of a light grey with a black ruffle and top like the common wood-pheasant, but I think different ; unfortunately he escaped, although I brought him to the ground. Saw two hens afterwards, which I took to be the same species. Of plants, two specimens of a hand-some *Anemone* : flowers large, faint blue and white; destitute of leaves ; seems to be the first flowering plant ; in dry rocky places. Minimum heat 25°, maximum 51°. A little above the camp is a small lake a mile and a half broad, with a beautiful plain on each side. Soil light and gravelly ; the valley seven miles, oval shape, commanding a fine view of the mountains. Evening fine.

Friday, *4th.*—This morning I was glad and somewhat relieved to know that the mountain portage was completed and our journey for three days would be water communication—namely, to [Fort] Assiniboine. Embarked at daybreak in two birch canoes, and being light went down the stream rapidly. The river banks are low, many places narrow, and widens out to long narrow shallow lakes full of sand shoals. Mountains gradually

[1] This sentence reads thus quite clearly in MS.—ED.

become lower, more even, and more thickly wooded.   Took breakfast on
a small low sandy island in the upper lake, where we were joined by a
hunter having in his possession a very large female sheep, cut up in quarters,
only killed about an hour before on the subalpine regions of the mountains,
where I am informed they are in abundance.   Hair short, coarse, and very
thick, of a uniform light brown on the neck, head, back, and sides ; belly
a dirty white.   I would judge at 170 lb. weight.   Continued our route and
passed on the right a high rugged range of mountains, and five miles lower
on the left some of lower elevation, seemingly the termination of the
dividing ridge.   Arrived at Jasper House, three small hovels on the left
side of the river, at two o'clock, where we put up to refresh ourselves for the
remainder of the day.   Minimum heat 29°, maximum 61°.   Fine and
warm.   The country to the south undulating and woody ; on the north
low and hilly, with even surface, also woody, with a most commanding and
beautiful view of the Rocky Mountains on the west and east.   The difference of climate is great and the total change of verdure impresses on the
mind of the traveller an idea of being, as it were, in a different hemisphere
more than in a different part of the same continent, and only a hundred
miles apart.   Obtained from J. Cardinal a pair of ram's horns, which he
considers the largest, and the skull, but I regret the lower jaw is wanting.
Had some of the much talked-of white fish for supper, which I found good,
although simply boiled in water, eaten without sauce or seasoning, hunger
excepted, not so much as salt, afterwards drinking the liquor in which it was
boiled ; no bread.   To-night comfortable.   Nothing in the way of plants
this day.   Observed one species of *Ribes* on the dry banks of the river
(prickly), not yet in blossom.   *Salix* two species, one large with red bark,
6 to 10 feet high, on the sandy banks of the river, and another low one in
the same place.

   Saturday, 5th.—Last night an old violin was found at our new lodgings,
and Mr. E.'s servant being something of a performer nothing less than
dancing in the evening would suit them, which they kept up for a few
hours.   This may serve to show how little they look on hardship when past ;
only a few days ago, and they were as much depressed as they are now
elated.   Morning fine ; minimum heat 29°, maximum 62° ; wind easterly.
At daybreak embarked in our canoe and after passing the sandy shores and
shoals of the lake went rapidly before the stream.   This day admits of little
variety.   The river is 100 to 140 yards broad, shallow, full of rapids, and
although the canoe drew only about nine inches of water, yet ere 2 P.M. we
were under the necessity of putting to shore for the purpose of repair.
The banks in the upper parts on quitting the mountains are high, gravelly,
or sandy clay on a stratum of sandstone lower down, as far as we have
come.   The banks are low and gravelly, covered with only two species of
Pinus—*P. nigra* [1] and *P. Banksiana*—on the shore.   *Betula* and *Alnus* of
diminutive growth.   Last view of the Rocky Mountains closed at 11 A.M.,
distant forty miles.   The water of this river is muddy and small in comparison with the clear majestic Columbia, and it deserves to be noticed that
no stream that flows into it has the clear water of itself.   This river

   [1] *Picea nigra*, Mast. in Journ. R. Hort. Soc. xiv. p. 222.

abounds with geese and ducks : several were killed, and in the evening the Northern Chia, with his wild but mellow voice aroused our camp. The egg of this bird is about the size of a goose egg, greenish-blue. I am in hopes of having some sent from the Columbia. General course of the river south. Camped on the left bank at sundown, having gained ninety-three miles.

Sunday, 6th.—Windy during the night, minimum 34°, maximum 56°. Wind continuing throughout. Started at sunrise and resumed our voyage with progress until breakfast, having gained about twenty-seven miles. Found the current less and more ice on the banks of the river. Proceeded only three miles further when we overtook Mr. George McDougall and four men on their way from Western Caledonia. Had suffered great hardship passing the mountains from hunger ; had been nine days coming from Jasper House, which we left yesterday following the ice as it cleared. Obliged to put up with him until four o'clock, when the ice made a rapid move and we embarked and made six miles more and again obliged to put up for the night. Country the same as yesterday, nothing of interest except to myself. Burnt my blanket and great toe at the fire last night.

Monday, 7th.—At daylight started and proceeded a few miles, when we were again detained by the ice. Made a scanty breakfast of some geese killed the preceding day. After remaining two hours and seeing no likelihood of the ice giving way, a short portage was made over the ice into the main channel, which was open and had a very strong current. Proceeded on our voyage until noon, when we were obliged to put to shore, being overtaken by a large flotilla of ice the whole breadth of the river, which continued until two o'clock. Embarked a second time in hopes from the great quantity of ice that passed the river would be cleared below and that we should meet with no other obstruction. At mid-afternoon we passed a large moose-deer standing on the banks of the river ; having only one deranged gun in our canoe a sight of him was all that was had—did not seem shy and approached within sixty yards. At sundown arrived at [Fort] Assiniboine, where we were received by Mr. Harriott. The whole distance of this river, from Jasper House to this place, 184 miles, admits of no variety, seeing one mile gives an idea of the whole : the banks are low marshy or clay mixed with gravel. The wood of diminutive growth— *Pinus alba,*[1] *Alnus, Betula.* The underwood *Corylus* and *Mespilus, Neottia, Linnaea,* and *Cypripedium* and a thick carpet of *Sphagnum* and *Hypericum.* Horizontal beds of coal abound in the banks of the river, under rotten slate. Maximum heat 59°, minimum 37°, wind easterly.

Tuesday, 8th.—Provisions being scarce and from the hostile disposition of the Indian tribes in the south it was deemed unsafe to go in so small a party, we intended to wait for the people from Lesser Slave Lake who are hourly expected. As Mr. McDougall was going down the river with the intention of procuring food for the men from Columbia and Western Caledonia, I accepted an invitation merely to see the country, being yet too early for affording me any plants, and by doing this I may put myself in possession of some birds. Started at nine o'clock, having a scanty

[1] *Picea alba,* Mast., *loc. cit.,* 221.

but the best breakfast the place afforded, in one birch canoe and eleven men. Country low and marshy. *Pinus* rare, *Betula* and *Populus* abound. Shores of the river covered with ice piled on each other in large masses. Did not kill a single animal of any description. Camped on the left or north side of the river at dusk. Gained forty-seven miles. Rain in the evening. No supper. Cloudy and sultry during the day, maximum heat 63°, minimum 40°.

Wednesday, 9th.—Morning clear and fine; minimum heat 40°, maximum 61°. Proceeded on our course, still unsuccessful in procuring food by our own exertions. At eight o'clock came to the camp of a Nipissing hunter who had the day before killed a small black bear, which he gave us with some half-dried beaver-meat, on which we made a hearty meal and then resumed our route down the river as far as the junction of Slave Rise River, which flows out of Lesser Slave Lake into the Athabasca River. We had not been here more than two hours when we were joined by the party from Lesser Slave Lake, headed by John Stuart, Esq., Chief Factor of the district, who received me in the most friendly manner. During the time the canoes were repairing, a dinner of reindeer steaks was prepared when we embarked and ascended the Athabasca again. Country the same as yesterday, low and marshy.

Thursday, 10th.—Minimum heat 36°, maximum 40°. At daybreak a heavy fall of snow, which continued throughout the day until dusk, in consequence of which we were under the necessity of remaining in our camp, which was very bad, being on the low ground among *Equisetum hyemale*. Mr. Stuart, I find to have a more intimate knowledge of the country than any person I have yet seen, and a good idea of plants and other departments in natural history. He was the first individual who crossed the Rocky Mountains and established Western Caledonia in 1805, and the same year reached the Pacific at Fraser's River near Puget Sound, and has since been over a vast extent of country in these parts, first explored by Sir A. McKenzie. He has been also on the Columbia. He informs me, from a letter received from Dr. Richardson, dated Fort Resolution, on Great Bear Lake, of the return of the whole expedition without having reached Icy Cape. That a connected survey of 13° of longitude to the west of Mckenzie's River has been made, but from the hostile disposition of the Esquimaux they found it impenetrable and that on the navigation opening they should all be at Cumberland. From the information he gives me of the opportunities that canoe travelling affords of collecting subjects of natural history, I have abandoned my idea of going either to Montreal or New York, and agreeably to the plans pointed out by Mr. Sabine last year as the better way, I shall sail from Hudson's Bay. By doing this I can remain six weeks at some place in the interior, and still be in time for the ship. By going through Canada nothing could be done and my trip would no doubt be expensive. Perhaps I may be enabled to go to Swan and Red Rivers overland from Carlton House, a journey of twenty days. This I am told will depend on the route taken by the Stone Indians, who are hostile, and if in that direction would be deemed unsafe. Learned that Mr. McDonald, the person who had in charge my box of seeds addressed to be left at Fort Edmonton on the Saskatchewan River, had endured much misery descending the Athabasca, the ice being taken before he had made good half his journey. In company with him Mr. Drummond. Hope my box is safe (do not relish botanist coming in contact with another's gleanings).

Friday, 11*th*.—Minimum heat 25°, maximum 45°. Continued our route at daylight, cheerless and comfortless ; a uniform fall of snow of 12 inches. Camped at the junction of Pembina River, a muddy stream of considerable size, a hundred yards broad. Here the party divided, some ascending it and some going to Assiniboine.

Saturday, 12*th*.—Minimum 29°, maximum 42°. Continued our route. Nothing occurred.

Sunday, 13*th*.—Minimum 32°, maximum 55°. Close and cloudy. By making an early start ten miles was gained to breakfast ; shortly afterwards we left the canoe and cut over a low point of wood and arrived at Assiniboine at two o'clock. Mr. Stuart killed a male partridge, the same species that I saw in the bosom of the mountains, called by him White Flesher—different from the common ruffled grouse—too much destroyed for preserving. Make some small slug and procure a pair of this fine bird.

14*th*.—Morning dull and cloudy ; minimum heat 36° ; maximum 61°. Loud peals of thunder at noon with showers of hail. At two o'clock had all my things tied up and crossed the river. Mr. S., I must mention, gave me two horses, one to carry my collection and one to ride, but being averse to that he went light. Took our course in a south direction, through a low marshy country, low points of poplar and birch, and on the higher places low pines. Path very bad, sinking to the knees in mud and kept by the still frozen soil at the bottom from going further. Passed through a point of wood and camped five miles from the river ; killed a female partridge of the species killed on the Rocky Mountains ; differs not in size. The colour on the back a little lighter, on the breast not so jetty black. Could not preserve it. Regret it.

Tuesday, 15*th*.—This uninteresting wretched country affording me no plants, at daylight I took a gun and went in quest of partridges ; killed a pair of White Fleshers and a hen of a different species. The male a beautiful bird. Only one of the three, the hen of the White Flesher, was worth skinning. Had eggs—small, pure white, about the size of the common pigeon. Pine leaves, and leaves of birch in the crop. The flesh is remarkably white. The breast-bone uncommonly high and very short, differs only from the male in the ruffle being less conspicuous as to colour and size. The hen of the other species was about the same size, of a dark red. Only one shot struck it, but, as misfortune would have it, through the head. Camped at 'Two Rivers' much fatigued. Observed two species of *Ribes* (one in Order or Section *Ribes* and one in *Grossulariae*) ; both common west of the mountains on the margin of mountain rivulets. Minimum heat 29°, maximum 53°. Showery, with hailstones of large size.

Wednesday, 16*th*.—Continued our route through the same sort of country as yesterday, still bad road and nothing in the way of plants.

Breakfasted in a low wet plain near a narrow lake, where we made a stay of a few hours to refresh the horses. Passed through a low point near a creek bank, in many places by old beaver dams. Entered a second plain and camped in the wood at the south end. Rainy with thunder. Minimum 38°, maximum 48°.

Thursday, 17*th.*—Morning raw and unpleasant. Minimum heat 33°, maximum 52°. Heavy showers during the forenoon. Two hours' walk through an excessively bad road took us to Paddle River, a rapid muddy creek, thirty yards wide and at present swollen over its banks from the melting of the snow. A raft was constructed and two men swam across and pulled it over by a rope ; by this tedious operation we got all over in the space of three hours. This place affording no fodder for the horses, went on until we came to a low plain at midday, when we, as soon as refreshed in the afternoon, continued through a thick woody country intersected by narrow lakes until we reached Pembina River at dusk, and regretted at not finding the canoes which we parted from six days ago.

Friday, 18*th.*—Morning fine and pleasant. Three large rafts being made, sufficient to carry all the baggage and persons in one trip, the river being too broad and rapid to return with one, we all crossed except Mr. Stuart, who intends to await his canoes. Went before the brigade in quest of birds, but still unfortunate. Path a little better than the preceding days ; ground high and not so thickly wooded. Shortly after two o'clock arrived at Eagle Lake, six or eight miles long, three to four broad, in which are caught grey sucker, white fish, and pike or jack fish, where we must stay to procure wherewith to complete the remainder of our journey to the Saskatchewan. Thirty-one were caught before dusk on the first visit of the net—of pike and sucker but no white fish. Some were also had from a small barrier in the stream flowing out of the lake.

Saturday, 19*th.*—The fishermen during last night caught only sixty fish—grey sucker or carp of the voyageur and jack fish. The former, 2 to 3 lb., the latter 8 to 12 ; both good eating. Minimum heat 39°, maximum 64° ; pleasant. Found a species of *Fumaria,* annual or biennial, in blossom ; flowers yellow ; on the edge of the lake ; saw only one plant ; laid in specimens. In my walk round part of the lake in wood killed a pair of White Fleshers, male and female. The former being a fine bird, and having time on my hands, I preserved it. Found a nest with six eggs, of the species which I killed a few days since ; the hen escaped me. Eggs small, bright dun colour. Minimum heat 31°, maximum 63°.

Sunday, 20*th.*—High wind during the night ; minimum heat 36°, maximum 61°. Cloudy towards even. Received a friendly note from Mr. Stuart, whom I have just left, that should I arrive at Fort Edmonton and find that a few days can be spent to advantage, to wait and descend the Saskatchewan in his boat. A sufficiency of fish being caught for a part of the journey, raised camp at ten, keeping a south course. Road very bad, much worse than any yet gone over ; passing the numerous swamps often sinking nearly to the middle in mud and water. Some of the horses obliged us to camp earlier than expectation, being broken down with fatigue and having to pass a thick wood five miles long where no fodder could be had. Camped on the outskirts on north side. The distance to the establishment being only about forty miles, I intend, should the day be fine, to start on foot in order that a few days may be had to collect before the brigade comes up ; at the same time I am most anxious to learn the fate of my packet of seeds. Perhaps I may go in one day. This day's march is nine miles.

Monday, 21st.—At daylight, four A.M., started on foot accompanied by an old Nipissing Indian who had spent many years west of the Rocky Mountains. Although to appearance upwards of seventy years of age, I found him a most excellent walker. Passed a deep muddy swamp a mile broad, and entered a thick point of pine of five more, when I was informed the laborious part of the journey was over. At seven met five men and twenty-five horses going to meet the brigade. They offered me a horse, but I chose to walk, thinking that the horses might be all required to bring up the baggage. Continued my route along Lake Bowland to the south end, where two men had been sent on to fish. Having been unsuccessful and no breakfast, my stay was short. Crossed a deep narrow creek and walked along a low moist meadow through which the just-mentioned creek descended for four miles when the country became very different : a fine undulating ground with clumps of poplar and willow on low parts, *Mespilus canadensis* [1] on the dry spots intermixed with Rose and *Rubus*, both shy in growth, the country being from time to time burned by the Indians. Passed the small deep rivulets by means of throwing down two trees. All the hollow parts of the plains overflowed with water—to all appearance shallow lakes. Appears to have at one time abounded in Red and Long-tailed deer, many horns being strewed over the ground, the horns and skulls of buffalo lying in all directions. At three o'clock came to Sturgeon River, a small deep muddy stream but at this season large, the banks overflowed. My hatchet being small, two hours were spent making a raft. I would not have lost three minutes in crossing, but my poor old guide was afraid the chilliness of the water would injure him, having perspired much, and on his account I assisted him in raft-making. Being then only nine miles from Fort Edmonton on the Saskatchewan, my spirits revived and I hastily tripped over the ground and passed many muddy creeks and shallow sheets of water, wading to the middle. Night creeping in on me, my view of the country gradually disappeared. At eight, on reaching a rising eminence unexpectedly, I heard the evening howl of the sledge-dogs, which to me was sweet music, and perceived fires in some lodges which I knew to be near the establishment. Being all over with mud, I returned half a mile to a small lake, stripped and plunged myself in and then comforted myself with a clean shirt which I carried on my back in a bundle. I was most kindly received by Mr. John Rowand, and had supper prepared for me of fine moose-deer steaks, which were most acceptable after a walk of forty-three miles through a most wretched country without having anything to eat. I found Mr. F. McDonald here, who took charge of my box last year. I now learn it had sustained injury, it having been broken.

[1] *Amelanchier canadensis*, S. Wats. Bibl. Ind. N. Am. Pl. p. 272.

Will see it in the morning. I must mention the particular attention of
Mr. Rowand, and the manner of conferring it.   Thinking that they would
be better to have the paper changed, he had them in his presence examined
by Mr. Drummond, who was there at that time, but who is now at Carlton
House.   This was kind.   Morning frosty ; minimum 30°, maximum 49° ;
showery, with high wind.

Tuesday, 22nd.—Fatigued so much with yesterday's walking that no
sleep could be had ; rose at daybreak and had my box opened ; found the
seeds in much better order than could be expected from the trouble the
person had before he reached this place.   Only eighteen papers had
suffered, amongst which I am exceedingly sorry to say is *Paeonia*.   This
one of the finest plants in the collection.   It often happens that the best
goes first.

Rubus sp.; Lewis and Clarke's River.

Rubus sp.

Uvularia, perennial.

Anemone sp. ; perennial.

Syngenesia sp. ; perennial.

Oenothera sp.; annual; small; flowers white.

Vaccinium sp.; on the summit of the high mountains.

Malva sp.; perennial; tall.

Hexandria, Monogynia, 339 of 1825.

Polygynia, Dipentagynia.

Phlox, different from *Phlox speciosa.*

(152) *Lupinus*, flowers golden-yellow.

Ribes sp.; on the mountains of Lewis and Clarke's River.

(165) *Pentstemon* sp. : *P.* on the mountains of Lewis and Clarke's River.

(99) *P.* sp.; perennial.

(71) *Pentstemon* sp.; shrubby.

Seeds or plants should be enclosed in soldered tin-boxes to prevent
wet or moisture and placed in strong wooden boxes.   Fortunately my
shirts were in the box, so they absorbed the moisture.   However, from my
very small stock being entirely rotten I can at the moment ill spare them.

Wednesday 23rd to Saturday 26th.—The lateness of the season affording
nothing in plants, *Viola canadensis, V. pedata*, and a few *Gramineae* were the
only plants found here in a perfect state—no mosses.   Killed one male
small pheasant of the plains, which was preserved.   This is the same so
common on the dry sandy wastes of the Columbia.   The country here
is undulating, low stunted pines, on the banks of the river, *Populus, Betula,*
and *Salix*, and a scanty herbage of herbaceous plants.   A fine young
Calumet Eagle, two years old, sex unknown, I had off Mr. Rowand ; brought
from the Cootanie lands situated in the bosom of the Rocky Mountains
near the headwaters of the Saskatchewan River.   His plumage is much
destroyed by the boys, who had deprived him of those in the tail that were
just coming to their true colour.   Many strange stories are told of this bird
as to strength and ferocity, such as carrying off young deer entire, killing
full-grown Long-tailed deer, and so on.   Certain it is, he is both powerful
and ferocious.   I have seen all other birds leave their prey on his approach,

manifesting the utmost terror. By most of the tribes the tail feathers are highly prized for adorning their war-caps and other garments. The pipe-stem is also decorated with them, hence comes the name. Abundant at all seasons in the Rocky Mountains, and in winter a few are seen on the mountainous country south of the Columbia on the coast. Are caught as follows: A deep pit is dug in the ground, covered over with small sticks, straw, grass, and a thin covering of earth, in which the hunter takes his seat; a large piece of flesh is placed above, having a string tied to it, the other end held in the hand of the person below. The bird on eyeing the prey instantly descends, and while his talons are fastened in the flesh the hunter pulls bird and flesh into the pit. Scarcely an instance is known of failing in the hunt. Its ferocity is equal to the grisly bear's; will die before he lose his prey. The hunter covers his hands and arms with sleeves of strong deerskin leather for the purpose of preventing him from being injured by his claws. They build in the most inaccessible clefts of the rocks; have two young at a time, being found in June and July. This one had been taken only a few days after hatching and is now docile. The boys who have been in the habit of teasing him for some time past having ruffled his temper, I took and caged him with some difficulty. Had a fresh box made for seeds and another for my journals, portfolio, and sundry articles. Could find no lock to put on it. The river here is broad, four hundred yards, high, clayey, and muddy banks, water muddy. Coal is found in abundance.

Saturday 26th to Thursday 31st.—Last night, before we should part with our new friends, Mr. Ermatinger was called on to indulge us with a tune on the violin, to which he readily complied. No time was lost in forming a dance; and as I was given to understand it was principally on my account, I could not do less than endeavour to please by jumping, for dance I could not. The evening passed away pleasantly enough; breakfasted at five o'clock and embarked in Messrs. Stuart and Rowand's boat with all my baggage and went rapidly before the stream. Day warm and pleasant. Put ashore in the dusk to cook supper, and as the Stone Indians had for the last twelve months manifested hostile intentions it was deemed unsafe to sleep at a camp where fire was. We therefore embarked, had the boats tied two and two together, and drifted all night. Finding this mode of travelling very irksome, never on shore except a short time when cooking breakfast, always dusk before a second meal, I began to think this sort of travelling ill adapted for botanising. Breakfasted at Dogrose Creek, where I found *Ribes hudsonianum* (Richardson in Frankl. 2nd Journ. App. p. 6). The country here changed much for the better; small hills and clumps of poplar and small rocks. Just in the dusk of the evening had a fine chase after two Red deer swimming in the water, and on following in the boat both were killed; smaller than those west of the mountains. Saw a huge grisly bear (unsuccessful in killing him) and a number of small plain wolves. Passed Fort Vermilion, an abandoned establishment, and Bear and Red Deer Hills, where the country becomes pastoral and highly adorned by Nature. Soil dry and light, but not unfertile.

On Wednesday at sunrise five large buffalo bulls were seen standing on a sandbank of the river. Mr. Harriott, who is a skilful hunter, debarked and killed two, and wounded two more ; all would have fallen had not some of the others imprudently given them the wind, that is on the wind side. Fifty miles further down a herd was seen, and plans laid for hunting in the morning. Some deer were killed this evening and some of the Prong-horned antelope of the plains. Skinned one but unable to preserve the hair on. This little animal is remarkably curious in his disposition ; on seeing you he will at first give three or four jumps from you, return slowly up to within a hundred or a hundred and fifty yards, stand, give a snort, and again jump backwards. A red handkerchief or white shirt—in fact, any vivid colour—will attract them out, and hunters crawl to them on all fours, raising the back like a quadruped walking, and readily kill them.

June 1st.—A party of hunters went out at daylight after the herd of animals seen last night. Most willingly I followed them, not for the purpose of hunting but gathering plants. Found *Phlox Hoodii* in flower, or rather I might say declining. Laid in specimens—some *Diadelphia* and *Gramineae.* Returned well pleased ; supped earlier than usual, and again embarked. Mr. Harriott and Ermatinger and three hunters went off to the opposite side to a herd and killed two very large and fine animals. Seeing their boat at the side of the river and no one in it, gave us to know they had all gone for the meat and we put to shore. A party from our boats was sent off to help them. Accompanied by Mr. F. McDonald, they readily were guided to their companions by calls, and found Mr. H. and E. pursuing a bull that had been wounded, in which he joined. The animal, which had suffered less injury than was expected, turned and gave chase to Mr. McDonald and overtook him. His case being dreadful coming in contact with such a formidable animal and exasperated, seeing that it was utterly impossible to escape, he had presence of mind to throw himself on his belly flat on the ground, but this did not save him. He received the first stroke on the back of the right thigh, and pitched in the air several yards. The wound sustained was a dreadful laceration literally laying open the whole back part of the thigh to the bone ; received five more blows, at each of which he went senseless. Perceiving the beast preparing to strike him a seventh, he laid hold of his wig (his own words) and hung on ; man and bull sank the same instant. His companions had the melancholy sensation of standing to witness their companion mangled and could give no assistance—all their ball being fired. Being under cloud of night, and from what had taken place, his life could not be expected. One returned and acquainted the camp, when each with his gun went off to the spot. On arriving some of the half-breed hunters were in a body to discharge their guns at him, when I called out to Mr. Harriott not to allow them to fire all together; that one well-directed shot was enough and by firing more Mr. McDonald if alive might fall by one, being close if not under the bull. He agreed to it, but while giving orders to some that he depended on, a shot went off by accident without doing any injury to anyone, and had the unexpected good fortune to raise the bull, first sniffing his victim, turning him gently over, and walking off. I went up to him and found life still apparent, but quite senseless. He had sustained most injury from a blow on the left side, and had it not been for a strong double sealskin shot-pouch, with ball, shot, wadding, &c., which shielded the stroke, unquestionably he must by that alone have been deprived of life, being opposite the heart. The horn went through the pouch, coat, vest, flannel, and cotton shirts, and bruised the skin and broke two ribs. He was bruised all over, but no part materially cut except the thigh—left wrist dislocated. My lancet being always in my pocket like a watch, I had him bled and his wounds bound up, when he was carried to the boat ; gave twenty-five drops of laudanum and procured sleep. In hopes of finding Dr. Richardson no time was lost to convey him to Carlton. The following day several more were killed, but from what I had seen my desire of seeing such dreadful brutes cooled. I continued gleaning plants when the limited times occurred. At 2 P.M. on the 3rd arrived at Carlton House and was received with politeness by Mr. Pruden. Here I found Mr. Thos. Drummond had come down to meet Dr. Richardson in spring. The doctor is now below at Cumberland House. In the evening had an account of his travels and progress and informed me had received a note from Mr. Sabine concerning *Phlox Hoodii.* He appears to have done well. I must state he liberally showed me a few of the plants in his possession—birds, animals, &c., in the most unreserved manner.

Monday, June 4th.—Accompanied by Mr. Drummond, I made a trip contiguous to the establishment and was guided to several habitats by him. I learned with regret that my anticipated journey overland to Swan and Red Rivers could not be accomplished. In the first place, two horses would be requisite to carry my papers, blanket, and food—unsafe to have one in the event of dying ; in the next place, it was uncertain in what direction the Stone Indians were, and in the event of their meeting me mine would beyond any doubt be a done career. One of the Canadian servants was four weeks ago murdered within four miles of the house, his gun and horses taken, and his body left stripped. The villain who committed this horrid deed was, I am informed, kept during the winter in food, being an object of pity and his family starving ; and on his quitting in spring manifested his ingratitude by perpetrating the foulest of crimes. Therefore with regret I had, on the advice of the persons in authority, abandoned it and proceeded to Norway House, where perhaps an opportunity may offer of visiting Red River in preference to going by Canada or the States, both being monstrously costly. I would have preferred to remain at Carlton had not the following considerations presented themselves : first, had I remained it might be looked on as an encroachment by him, and as there was no opportunity of a passage by the Company's boats I would have to solicit a passage of Captain Back on his way from Great Bear Lake. This I should have done in the most unreserved manner, being well convinced, if in his power or in the power of any officer of the expedition, it would be granted ; but from the embarrassed state of their boats with their own collection, which I learn with pleasure is grand, it is doubtful if they could accommodate me. For these reasons I now go to Cumberland.

Tuesday 5th to Saturday 9th.—The route from Carlton to Cumberland is so well known from the description of the Arctic voyageurs that anything from my pen is unnecessary. The journey admits of little variety. Eighty miles below Carlton at a high bank on the left-hand side of the river, called ' The Women's Encampment,' the pastoral and rich verdure is no longer seen ; instead low thick marshy woods of *Pinus Banksiana*, *P. rubra* [1] and *alba*,[2] *Betula*, *Populus*, and *Salix*—*Carices* and *Gramineae* in the marshes. Very unsteady rainy weather with high winds. During the short turns on shore picked up a few plants. Arrived at Cumberland at 5 P.M. on Saturday and was kindly welcome by Mr. J. Leith. Here I was greeted by Dr. Richardson, safe from his second hazardous journey from the shores of the Polar Sea. Every man must feel for the hardship and difficulties which he endured and overcame, and the successful termination of the perilous undertakings. What must be most gratifying is extricating themselves from the formidable Esquimaux without coming to violence. The doctor has a splendid herbarium and superior collection in almost every department of natural history. On telling him I wished to remain at Carlton, he observed that Captain Back would have willingly given me a passage. Informed him of my intention of going to Red River and sailing from Hudson's Bay ; approved of it much. Could do nothing going to Canada.

10th.—Morning and part of the forenoon spent looking over Mr. Drummond's collection from the Rocky Mountains. Many fine Alpine plants. New *Dryas*, *Potentilla*, *Juncus*, *Salix*, *Saxifraga*, *Menziesia*, and a superior collection of *Musci*. Considering the opportunities, he had many fine plants all arranged together. The Doctor not yet brought out from Great Bear Lake.[3] The remainder of the day was spent in looking round the establishment. Ground a perfect marsh and thickly wooded of the same species as along the river. As the doctor goes direct to England, kindly offers to carry letters to be left below at Norway House if I am gone to Red River before his arrival at that place.

Monday 11th to Saturday 16th.—The country throughout presents the same uniformity. Thick low wet woods and muddy banks. Gathered a species of *Salix* and *Carex*. No place for botanising. Towards dusk on the second day reached the head of the Grand Rapid and walked down through the wood while the boats descended—one unfortunately struck on some rocks so that they reached the shore with some difficulty. All next day spent drying the cargo and repairing the boat. Had a ramble in the woods and procured a few things which will be noticed below. Killed a fine large male pelican and preserved the skin. The mischievous boys tore the neck and otherwise injured it. Killed a small plover and preserved it also. High wind and sleet during the whole night and following day. Did not rise until midday. Moderated at sundown, when we embarked and entered Lake Winnipeg. Slept none. Charmed by the mournful cries of the Northern Diver. At eight o'clock put to shore and

---

[1] *? Picea rubra*, Veitch, Man. Conif., ed. 2, p. 450.
[2] *Picea alba*, Mast. in Journ. R. Hort. Soc. xiv. p. 221.
[3] So in MS.—ED.

breakfasted; shore cold, pure white limestone. Found on some of the small rocky islands abundance of gulls' eggs. Towards noon the wind rose very high, producing a high swell, and being near the shore in such broken water, we were under the necessity of lying to. Had much difficulty to land, the water breaking over the boats. Gathered a few *Musci*. The following morning at daylight embarked and having a fair wind went speedily along. Passed Mossy Point, a part of the lake with steep muddy banks and rotten moss 3 to 4 feet deep on the top. Gained the old establishment of Norway House at 1 P.M. where we took some breakfast and at two resumed our route to the new one, eighteen miles below on Jackfish River, where we arrived at 8 P.M. Here I found my old friend Mr. John McLeod, who last year carried my letters across from the Columbia, also J. G. McTavish, Esq., from whom I had much kindness and who desired to know in what way he would be of most service to me. Received a letter of Jos. Sabine, Esq., London, 10th March: good news, the vessel from the Columbia arrived safe; collection sustained no injury. A letter from William Booth, Mr. Murray, Dr. Hooker of Glasgow, and my brother, the latter affording me but news of a melancholy cast.

June 17th.—This morning at daylight George Simpson, Esq. (Governor), arrived from Montreal, who I state with pleasure gave sufficient testimony of his friendly attentions and kind offices. Seeing me perhaps rather indifferently clothed, he offered me some linen, &c., which I refused, at the same time extremely indebted to him. Changing the paper of my plants, &c. The following were collected on my descent of the Saskatchewan River from Fort Edmonton to Norway House:

(1) *Linum* sp.; perennial; flowers blue; in dry, light, elevated soils; in solitary plants, 1 foot to 18 inches high.

(2) *Anemone* sp.; perennial; in moist woods among moss; plentiful.

(3) *Uvularia* sp.; perennial; in shady woods; this plant extends over the whole continent; found in all woods; very strong in elevated places.

(4) *Convallaria* sp.; perennial; abundant in all woods and gravelly grounds.

(5) *Erigeron* sp.; perennial; flowers white; on moist grounds; rare.

(6) *Dodecatheon Meadia*; in marshy grounds; abundant.

(7) *Rubus stellatus*; on dry upland open woods on peaty soils.

(8) *Cerastium* sp.; on the dry gravelly plains; abundant.

(9) *Androsace* sp.; annual; abundant in the same places.

(10) *Gramineae* sp.; perennial; abundant in all low places, banks of rivers, &c.

(11) *Carex* sp.; a low plant, 8 to 10 inches high at most; as well as the following species:

(12) *Carex* sp.; perennial; a minute plant.

(13) *Carex* sp.; perennial; 1 foot to 14 inches high; in woods at Cumberland.

(14) *Sisyrinchium anceps* [1]; abundant on the dry elevated ground at Carlton.

---

[1] *Sisyrinchium angustifolium*, Baker, Handb. Irid. p. 124.

T

(15) *Lithospermum canescens*; on the plains in gravelly soils; 1 foot high; flowers bright yellow.

(16) *L. decumbens* [1]; in the same situations; a small plant.

(17) *Pentstemon* sp.; perennial; flowers fine azure-blue; leaves glaucous; this in every respect agrees with that so common on Lewis and Clarke's River; the Columbian only a stronger plant; Carlton.

(18) *Allium* sp.; flowers white; on the plain at the Eagle Hills.

(19) *Hedysarum* (?) sp.; imperfect, not yet in flower; a strong plant; on the dry soils near Carlton House.

(20) *Actaea* (?) sp.; perennial; in the shady woods; abundant; 2 feet high.

(21) *Juncus arcticus*; in the marshes.

(22) *Dioscorea* (?); in the plain near Carlton, among the bushes; rare.

(23) *Caltha palustris*; in marshy ground.

(24) *Plantago*; a variable plant; leaves lanceolate, five to seven ribbed, pubescent, smooth; in all low grounds near creeks, 6 to 12 inches.

(24 [*bis*]) *Cineraria* (?) sp.; perennial; a small plant, in moist situations; imperfect; rare.

(25) *Sonchus* sp.; perennial; a low plant, in moist ground near springs.

(26) *Polygala Senega*; in dry, open, elevated grounds, in single tufts.

(27) *Cruciferae*, annual; on the gravelly banks of rivers; found in the Columbia last year; the Columbian are stronger and more abundant.

(28) *Umbelliferae*, perennial; flowers yellow; leaves glaucous; low plant, in dry light elevated grounds, with the following:

(29) *Umbelliferae*, perennial; leaves pubescent.

(30) *Umbelliferae*, perennial; flowers yellow; leaves cordate; in moist open woods; plentiful.

(31) *Fragaria* sp.; perennial; flowers white; in open gravelly soils; abundant.

(32) May differ from the former; in woods; plentiful.

(33) *Pulmonaria* sp.; perennial; strong plant, 1 foot to 18 inches high; in shady woods.

(34) *Viola pedata* (?); in dry high grounds; abundant.

(35) *Viola canadensis* (?); a variable species, found in all situations, in the low brushwood in fertile soils; 8 inches to 1 foot high.

(36) *Viola* sp.; flowers blue; common on the plains in the upland grounds of the Columbia River.

(37) *Ribes* sp.; flowers small, brown colour; abundant in woods.

(38) *Ribes*; flowers faint yellow, in large racemes; abundant on the outskirts of woods; a fine plant, 5 feet high.

(39) *Ribes hudsonianum* (Richardson's *nigrum*); this will, I find, agree with the one found on the Kettle Falls of the Columbia; this much smaller.

---

[1] *Lithospermum hirtum*, A. Gray, Syn. Fl. N. Am. ii. i. p. 205.

(40) *Ribes* ; flowers faint white ; abundant in conjunction with the former.

(41) *Viburnum* sp. ; a low shrub ; in woods ; flowers white.

(42) *Prunus* sp. ; common on the banks of rivers, with the following :

(43) *Prunus* sp. ; a more slender shrub than the former.

(44) *Salix* sp. ; male and female ; a shrub 10 to 14 feet high ; creeks and springs ; abundant.

(45) *Conradia* [1] (Nuttall) ; abundant on the plains ; this I found on the Columbia in abundance.

(46) *Acer* sp. ; a small shrubby tree ; Carlton and Cumberland ; sugar is made from the species in small quantities.

(47) *Mespilus canadensis* [2] : abundant anywhere.

(48) *Betula* sp. ; tall tree, canoe birch ; abundant in the woody parts of the Saskatchewan.

(49) *Mentha viridis* ; in mossy woods.

(50) *Corallorrhiza innata* ; in shady pinewoods ; abundant.

(51) *Carex* sp. ; small, in low open woods.

(52) *Rubus* ; abundant at Cumberland ; sometimes with five leaves.

(53) *Aralia* sp. ; perennial ; a strong plant ; in shady woods ; abundant.

(54) *Cerastium* sp. ; perennial ; on the plains ; abundant in all soils.

(55) *Ribes* sp. ; leaves smooth ; peduncle slender ; flowers brown ; abundant in the woods ; thrives well in moist situations.

(56) *Salix*, male and female ; 10 to 15 feet high ; in all wet grounds at Cumberland.

(57) *Leontodon* sp. ; abundant everywhere on the plains at Carlton.

(58) *Ranunculus* sp. ; perennial ; in marshy grounds around Cumberland ; abundant.

(59) *Trientalis americana* ; in woods ; plentiful.

(60) *Draba* sp. ; annual ; abundant around Carlton.

(61) *Thlaspi* sp. ; annual ; in the same situations.

(62) *Sisymbrium* sp. ; annual ; a tall strong plant ; found sparingly on the upper parts of the river.

(63) *Didynamia*, perennial ; imperfect ; rare ; near Carlton.

(64) *Astragalus* sp. ; small ; flowers fine purple-blue ; 4 to 8 inches high ; very abundant, in thick clumps, on the upper parts of the river.

(65) var. *alba* of the preceding ; rare.

(66) *A. campestris* [3] ; a fine plant, 12 inches high ; in rocky and gravelly places ; this is also found on the higher parts of the Columbia.

(67) *A. succulentus* [4] (Franklin's App.) ; abundant from Bear Hills to Carlton.

(68) *Oxytropis argentata* [5] ; abundant on the plains. I have marked the following :

---

[1] *Macranthera*, Benth. and Hook. f. Gen. Pl. ii. p. 971.
[2] *Amelanchier canadensis*, S. Wats. Bibl. Ind. N. Am. Bot. p. 272.
[3] *Oxytropis campestris*, Ledeb. Fl. Ross. i. p. 591.
[4] *Astragalus caryocarpus*, S. Wats. Bibl. Ind. N. Am. Bot. p. 192.
[5] ? *Oxytropis nana*, S. Wats. *loc. cit.*, p. 246.

(69) —— ; it appears to differ in having a less ragged pubescent stem; flowers darker and calyx larger.

(70) *Astragalus* sp.; perennial; decumbent; leaves pinnate, slightly pubescent; flowers large, nearly pure white; a very fine plant and appears new; abundant above the Red Deer Hills on the Saskatchewan.

(71) *Hedysarum Mackenzii* (Richardson App.); a fine plant; on the Saskatchewan.

(72) *Hedysarum* sp.; perennial; a strong plant, 18 inches to 2½ feet high; rare.

(73) *Astragalus* (?); flowers white; plant villous; Carlton House, abundant; a fine plant.

(74) *Vicia* sp.; perennial; small flowers blue and white; on the dry plains.

(75) *Vicia* sp.; a large plant; in woods; plentiful with the following:

(76) *Vicia* sp.; perennial; flowers purple.

(77) *Thermopsis* sp.; perennial; abundant in all dry soils.

(78) *Phlox Hoodii*; on the plains; found only a few plants in perfect blossom; thrives on rock where there is lime.

(79) *Corydalis* sp.; annual; flowers purple and yellow; plant glaucous; on rocks at Norway House.

(80) *Vaccinium buxifolium* [1]; on rocks.

(81) *Saxifraga tricuspidata*; in rocky situations; abundant.

(82) *Gramineae*, perennial; on rocks.

(83) *Carex* sp.; small; at the Grand Rapid, Lake Winnipeg; plentiful.

(84) *Calypso borealis*; in shady woods.

(85) *Primula* sp.; in open woods, where ground has been burned, thrives well.

(86) *Cerastium* sp.; perennial; small.

(87) *Polygala* sp.; small; both abundant at the Grand Rapid.

(88) *Salix* sp.; male and female; a low plant; on the dry open woods, Grand Rapid.

(89) *Salix* sp.; male and female; a shrub 5 to 10 feet high; Grand Rapid.

(90) *Alnus* sp.; glutinous; on the Grand Rapid.

(91) *Pentandria, Monogynia*; a low shrub; on moist ground; flowers green.

(92) *Salix* sp.; male and female; leaves tomentose; a strong willow; Grand Rapid.

(93) *Salix* sp.; female; Grand Rapid.

(94) *Fumaria* sp.; annual; glaucous; flowers yellow; abundant everywhere.

Monday, July 2nd.—By three o'clock P.M. everything was in readiness for our departure and the canoe in the water, but being lumbered much more than expectation a place for Augustus could not be found; he had therefore to remain at Norway House. Took under my charge a packet of letters for the Red River Settlement and a box containing Church ornaments for the Roman Catholic Bishop. Sent the Calumet Eagle to

[1] *Gaylussacia brachycera*, A. Gray, Syn. Fl. N. Am. ii. i. p. 19.

Hudson's Bay by a Mr. Ross, wild fowl and other meat being scarce, and as he will not eat fish I was unable to keep him at the latter place. Placed the white-headed one under the care of a woman attached to the establishment, until my return. Left my sundry articles gleaned in my descent of the Saskatchewan River; and the roots or bulbs brought from the Columbia being still fresh and nearly dry I halved, placing the one in a well-secreted place in the wood, contained in a folded piece of birch bark, fearing the mice may find them; the other in a paper bag, hung up to the roof of the house with some bird-skins. Started at four; pleasant. Camped at 8 P.M., eighteen miles above the establishment. Rain and loud peals of thunder at dusk and during the night.

Tuesday, 3rd.—Morning cool and calm. Started at a quarter-past three A.M. and went on very prosperously, there being no swell on the lake. Took breakfast on a small rock, where an hour was spent, when we again resumed our route until dusk. Shores of the lake low muddy peat, sandy, gravelly, with numerous shallow pools or small lakes behind; at a few places granite is seen. The wood is small: *Pinus alba*,[1] *P. Banksiana*, and *P. rubra*,[2] *P. balsamea*,[3] *Populus trepida*,[4] *Betula*, and several species of *Salix*. Gathered two species of *Corydalis*, one yellow-flowered, glaucous leaved, found on the Assiniboine portage, and one upright-growing strong species with purple flowers. Laid in specimens of *Potentilla* and *Gramineae*. Camped on a small rocky island.

Wednesday, 4th.—Loud wind last night which increased so much towards midnight that the tent was nearly blown down; and the rain beating in on us, while the Captain supported the poles in the inside, the Doctor and I went in search of large stones to lay on the sides, being as I observed camped on a rock and pegs of no use. Before we had accomplished this we were well drenched, and as the fire was washed out each crept under his blankets until day. In the grey of the morn it moderated and we proceeded at 5 A.M. and went on for the space of four hours, when a strong head wind and a heavy surge obliged the canoe to take shelter a second time in a small rocky bay. These stoppages give a few moments for collecting, shifting, and drying paper. At three o'clock the wind ceased and the lake being calm we resumed, and camped at dark on a small rocky island near Pigeon River. Laid in some *Poa*, *Carex*, and *Potentilla*.

Thursday, 5th.—Had a fine camp last night; preferred sleeping on the rock close by the fire, where there was a fanning breeze, than to be annoyed by mosquitoes. Morning windy, detained until ten o'clock; in the interval laid in *Phlox linearis* [*sic*]. Obliged to put in a second time, into a small muddy creek surrounded on all sides by water. Gathered nothing; dried papers at the fire. This part of the shore is low and marshy.

Friday, 6th.—Light airs of wind and drizzly rain until mid-afternoon. Made a good day's march. Camped on a small island; found *Cerastium* sp., perennial; *Apocynum* sp., perennial; *Silene*, annual. For the first time

---

[1] *Picea alba*, Mast. in Journ. R. Hort. Soc. xiv. p. 221.
[2] *? Picea rubra*, Veitch, Man. Conif., ed. 2, p. 450.
[3] *Abies balsamea*, Mast. in Journ. R. Hort. Soc. xiv. p. 189.
[4] *Populus tremuloides*, Sargent, Silva N. Am. xi. p. 158.

I have observed *Quercus*, stunted trees of *Q. obtusiloba*, also small trees
of *Fraxinus pubescens*. Evening fine, with heavy dew.

Saturday, 7th.—Started at 5 A.M. and went on until nine, when the wind
increased so much that we could no longer proceed, although the lake here
is not more than three and a half miles broad. The waves were heavy and
broke with great violence on the shores, which are white limestone ; no
alteration of the conditions throughout the day. Found in the woods
Pyrola rotundifolia, *P. chlorantha*, *P. uniflora*[1] and *P. secunda*, *Cypripedium
pubescens*, *Vicia* sp., *Orchis* sp., *Solanum* sp. Annoyed by the smoke
while engaged drying paper, the wind blowing with great violence.

Sunday, 8th.—Started at 6 A.M. and passed several high limestone
cliffs. Took breakfast on a low sandy shore where, in small still waters,
was abundance of *Utricularia* in blossom. Remained on shore an hour and
then proceeded for twelve or fourteen miles when a stay was made for
changing linen. Arrived at the establishment on the River Winnipeg
(Fort Alexander, or Basch), the Riviera of the voyageurs. We were
welcomed there by Mr. John McDonald, a brother of the person who
crossed the Rocky Mountains last autumn ; he was also here on his way
to Canada. Became acquainted with the Rev. Mr. Picard, of the Roman
Catholic Mission at Red River, on his way to Canada.

Monday, 9th.—Early in the morning had a large fire made for drying
paper and had all my plants changed before breakfast. Wrote a short
letter to Governor Clinton of New York, saying I should sail from Hudson's
Bay for England. About 10 A.M. Captain Franklin and Dr. Richardson
started in their canoe for Canada, and took with them as passenger Mr.
Picard, he having been disappointed in going by the Company's barges or
canoes. Feel obliged to Captain Franklin (good man) ; will see Mr.
Sabine. Made several walks round the establishment and collected a few
plants. Several heavy showers accompanied by thunder and lightning.
The scenery of this place is fine, rich, and very beautiful ; well-wooded ;
low, level country ; soil fertile, deep alluvial loam, with a heavy sward of
herbage. Requested of Mr. John McDonald the favour of hiring me a
small canoe to carry me to Red River. Neil McDonald, a person who
accompanied Captain Franklin, offered to carry me in his, but found it too
small for our luggage. The Indians being camped a considerable distance
from the place and all at this season being much engaged, I had hired for
me a Canadian, who agreed to carry me for the sum of four dollars and his
food. Saw that his canoe was in repair in the course of the evening and
made preparations for starting in the morning.

Tuesday, 10th.—High winds during the night and morning from the
lake ; delayed until ten o'clock, the swell being too heavy for such a small
canoe. Being provided with provision for myself and man, I took my
leave and descended the river to the lake ; pleasant clear weather. Proceeded along the western shore close to the land, which is low, well-wooded ;
many places overflowed. The lake at this season being high, nothing
worthy of notice occurred ; saw no plants ; observed flocks of passenger
pigeons. Camped at dusk on a gravelly beach near a small creek, and was

[1] *Moneses uniflora*, A. Gray, Syn. Fl. N. Am. ii. I. p. 46.

visited by some Indians, of whom I purchased some birch bark for my specimens, serving instead of pasteboard.

Wednesday, 11th.—At six in the morning I embarked, the wind which blew violently during the night having moderated. Proceeded with the canoe through the low overflowed woody points of the lake, finding my way among some large species of *Arundo, Scirpus, Typha,* and *Carex.* At ten came to a low projecting point, which I was unable to double without the risk of wetting my papers, and rather than do so I put to shore, made myself some breakfast, overhauled the new-laid-in plants, exposed others to the sun, and took a short turn in the woods. Unable to paddle any further myself, yesterday's labour having put both my hands in sheets of blisters. At half-past two the wind moderated and I rounded the point and inclined my course a little more to the west again, among the brushwood and grass, which greatly shields off the wind. Aquatic birds appear to be rare : have now been two days afloat and only seen a few ducks. About 6 P.M. entered Red River, five or seven miles above the entrance of the lake, having as I stated before cut the overflowed points. The stream is considerable, 250 to 300 yards broad, deep and muddy ; banks, low ; deep-black alluvial earth ; thinly wooded. A few trees of *Alnus, Platanus, Quercus alba,* and *Acer* along the banks. Laid in some fine strong *Gramineae, Utricularia, Polygonum,* and *Asclepias.* Evening close and cloudy. Much annoyed during the forepart of the night by mosquitoes.

Thursday, 12th.—Morning cool, with a heavy dew ; started at 3 A.M. up the river. At sunrise passed several thinly planted low houses, with small herds of cattle wandering from the folds ; humble and peasant-like as these may appear to many, to me—who have been no sharer of civilised society for a considerable time past—they impart a pleasant sensation. At seven took breakfast two miles below the rapid, where I left the canoe and my luggage to go by land, taking with me my boards and paper. Strangers in this quarter appear to be few : scarcely a house I passed without an invitation to enter, more particularly from the Scottish settlers, who no doubt judging from my coat (being clothed in the Stewart or royal tartan) imagined me a son from the bleak dreary mountains of Scotland, and I had many questions put to me regarding the country, which now they only see through ideal recollection. Appear to live comfortable and have the means of subsistence by little exertion. Walked along the right or north bank of the river : about two o'clock passed the Church Missionary establishment and heard the bell ring for the boys to assemble to school ; found two at play, one about four years, the other six years of age. I inquired if they were scholars and had answer, ' Yes, sir.' ' What book do you read at school ? ' The elder, who was the spokesman, said ' I read the parables, and he (pointing to the younger) reads " Tom Bowles." ' This all pleasing, I presented them with a few trifling articles, when as soon as they manifested their thanks by a low bow they galloped off to their companions, who flocked round them to hear their story. About a mile further on passed a large windmill from which Fort Garry appeared, situated at the junction of Assiniboine River with the river among some wide-spreading oaks, and on the opposite side the Roman Catholic church and Mission establishment, both forming a fine effect. Called at Fort Garry and presented myself to Donald McKenzie, Esq., the Governor of the colony, who received me with great kindness. While a basin of tea was preparing at my request, a large tureen of fine milk was placed on the table, which I found excellent. I handed him Governor Simpson's note, but found that a note was unnecessary with Mr. McKenzie. His conversation to me is the more acceptable from the intimate knowledge he possesses of the country west of the Rocky Mountains. In 1819 he ascended the Missouri River and crossed the continent to the mouth of the Columbia with an American party ; was the companion of Messrs. Nuttall and Bradbury as far as they accompanied the expedition up the former river. He has travelled largely through the country south of the Columbia, in the interior, behind the Spanish settlements, and like all who share in such undertakings, shared in the fatigues and hardships attendant on these expeditions. But his was more than usual, being the first who ventured on these untrodden wilds. He has since recrossed by the Columbian route. Had a visit paid me by Spokane Garry, an Indian boy, native of the Columbia, who is receiving his education at the Missionary school. He came to inquire of his father and brothers, whom I saw ; he speaks good English ; his mother tongue (Spokane) he has nearly forgotten. Sent the box and letters under my care to the Bishop. Evening fine.

Friday, 13*th*.—Arranged plants ; dried paper, and took a short walk in the evening ; found a few plants not in the collection. Showery, thunder and lightning.

14*th*.—Made a short excursion up the Assiniboine River and returned to breakfast. At eleven o'clock Monseigneur J. N. Provenchier, the Bishop of the Roman Catholic Mission, the Rev. Theophilus Harper, and Mr. Buchier, a young ecclesiastic, called on me and made a long stay. The Bishop speaks good English, but with that broken accent peculiar to foreigners. Mr. Harper speaks the English language with as much fluency as his native tongue, French. They conversed in the most unreserved affable manner and made many inquiries concerning the different countries I had visited. I have some reason to think well of their visit, being the first ever paid to any individual except the officers of the Hudson's Bay Company. I am much delighted with the meek, dignified appearance of the Bishop, a man considerably above six feet and proportionally stout ; appears to be a man of the most profound acquirements, seen only through the thick rut of his great modesty. Resumed my walk in the evening, still adding some things to the collection.

Sunday, 15*th*.—At church. Heard a sermon from the Rev. David T. Jones, minister of the English church. There being no timepiece for the colony and the habitations widespread, the hour of the day is guessed by the sun. The service being begun half an hour before I got forward, in consequence of missing the proper path, the clergyman, seeing me from one of the windows, despatched a boy to fetch me on the proper path. This struck me as the man of the world who, in the parable, was compelled to go to the feast by the person stationed on the wayside. After service

Mr. Jones received me with every demonstration of kindness, and politely invited me to his house and said that I should be no stranger during my stay.   Returned home at midday and put some plants in order.   Very warm and cloudy.

Monday, 16*th.*—Arranged the following plants collected on my way from Norway House :

(107) *Cucumis* sp., annual; flowers white; leaves five-lobed; in shady places, banks of rivers; abundant western shores of Lake Winnipeg and Red River.

(108) —— (?) aquatic; abundant in still waters, in all shallow muddy bays of Lake Winnipeg.

(109) —— (?) aquatic; abundant with the former.

(110) *Populus trepida,*[1] Lake Winnipeg.

Betula papyracea, in some parts this attains a great size, 18 inches to 2½ feet diameter, 40 to 80 high; Lake Winnipeg.

(111) *Alnus* sp.; a common tree, with the two former; this is no doubt *A. glutinosa;* attains a considerable size.

(112) *Carex* sp.; small, in all dry shady woods, among *Hypnum;* Lake Winnipeg.

(113) *Stellaria glauca,* on the limestone and granite rocks of Lake Winnipeg; found in conjunction with *Potentilla tridentata.*

(114) *Ribes* sp.; leaves somewhat five-lobed, serrate, smooth; flowers erect, faint yellow, tinted with purple; berry appears to be hirsute; abundant; a strong bush.

(115) *Gnaphalium* sp.; perhaps dioecious; on the granite rocks.

(116) *Poa* sp.; perennial; a low grass, in dry light or rocky situations.

(117) *Carex* sp.; abundant in dry open woods.

(118) *Poa* sp.; perennial; a fine low-growing plant in rocky places; abundant at Norway House, Lake Winnipeg.

(119) *Caprifolium*[2] sp.; leaves smooth, glaucous under; flowers yellow-rose; a slender plant; outskirts of woods, Norway House.

(120) *Vaccinium* sp.; flowers yellowish-white; in woody places; abundant at Norway House.

(121) *Ledum latifolium;* in marshy grounds; abundant among *Sphagnum.*

(122) *Vaccinium buxifolium;* in dry open and rocky situations; abundant at Norway House.

(123) *Rosa* sp.; flowers fine pink colour, fragrant; a low, slender, twiggy plant; in woody places.

(124) *Umbelliferae,* 18 inches to 2½ feet high; in damp open woods and low meadows; plentiful.

(125) *Shepherdia,* in fruit; abundant on the River Winnipeg, forming the underwood on the banks of creeks; 3 to 5 feet high; in fruit.

(126) *Hippophaë,* in fruit; found in the same places with the former plants; 4 to 6 feet high; more slender than it.

---

[1] *Populus tremuloides,* Sargent, Silva N. Am. ix. p. 158.
[2] *Lonicera,* Benth. and Hook. f. Gen. Pl. ii. p. 5.

(127) *Hedysarum* sp.; perennial; flowers white; 18 inches to 2½ feet high; in open places, River Winnipeg; abundant on the high grounds of the Columbia, also very plentiful on the plains of Red River.

(128) *Astragalus* sp.; perennial; flowers faint yellow; a strong growing species, 2 to 3 feet high; in large tufts on fertile soils, River Winnipeg and plains of Red River.

(129) *Lathyrus* sp.; perennial; leaves glaucous; flowers beautiful dark purple-red; abundant on the banks of rivers, in fertile soils.

(130) *Stachys* sp.; banks of river and creeks; 18 inches to 2½ feet high.

(131) *Arenaria* sp.; perennial; on rocks and dry grounds; plentiful.

(132) *Dracocephalum*, perennial; abundant near all old camps.

(133) *Juncus* sp.; perennial; in moist ground near creeks and streams.

(134) *Chenopodium* sp.; annual; banks of streams; abundant.

(135) *Carex* sp.; a fine plant; in marshy grounds with all the former species; found on the Winnipeg River.

(136) *Carex*, perennial.

(137) *Carex*, perennial.

(138) *Carex* sp.; perennial; a strong plant, 2 to 3 feet high; in all overflowed grounds.

(139) *Gramineae*, annual; on moist soils; a fine grass, 2 to 3 feet high; this I found very sparingly on the north-west coast.

(140) *Utricularia* sp.; in still pools and bays of Winnipeg; abundant.

(141) *Solanum* sp.; annual; flowers faint white and yellow; stem somewhat hispid; leaves pubescent; found only one plant on the south shore of Lake Winnipeg.

(142) *Acer* sp.; a low slender plant 8 to 14 feet high.

(143) *Xylosteum*[1] sp.; flowers yellow; a low shrub, 2 to 3 feet high; in rocky places, shores of Lake Winnipeg.

(144) *Pyrola uniflora*[2]; in shady woods, with the following of the same genus; abundant on the shores of Lake Winnipeg.

(145) *Pyrola secunda.*

(146) *Pyrola chlorantha.*

(147) *Pyrola rotundifolia.*

(148) *Heuchera* sp. (see Franklin's App.); limestone and granite rocks of Lake Winnipeg; abundant.

(149) *Myrrhis* sp.; perennial; in shady woods; abundant.

(150) *Campanula rotundifolia*, dry grounds and rocks.

(151) *Cardamine* sp.; annual; in moist ground near springs; a low plant.

(151 [*bis*]) *Silene* sp.; annual; viscid; 1 to 2¼ feet high; frequents broken ground; this plant abounds at Fort Vancouver on the Columbia.

(152) *Orchis* sp.; flowers green; in close shady woods; abundant.

(153) *Cypripedium pubescens*; rare.

(154) *Convolvulus* sp.; River Winnipeg and western side of the lake; abundant, Red River.

[1] *Lonicera*, Benth. and Hook. f. Gen. Pl. ii. p. 5.
[2] *Moneses uniflora*, A. Gray, Syn. Fl. N. Am. ii. I. p. 46.

(155) *Ranunculus* sp. ; perennial ; flowers small, yellow ; a very fine species ; margins of pools, springs, and rivulets.

(156) *Potentilla arguta* (? Pursh), fine species ; 2 to 3 feet high ; dry shores of Lake Winnipeg and plains of Red River.

(157) *Apocynum androsaemifolium*, in dry gravelly soils or on rocky places ; plentiful.

(158) *Pentandria, Monogynia,* perennial ; corolla in five segments, rose colour ; leaves opposite, sessile, linear, glaucous ; a low plant, not exceeding 4 to 6 inches high ; on rocky places, Lake Winnipeg.

(159) *Phlox linearis* [*sic*], sparingly on Lake Winnipeg.

(160) *Convallaria bifolia*,[1] in shady woods ; abundant.

(161) *Ranunculus* sp. ; perennial ; small ; on moist grounds ; plentiful.

(162) *Geranium* sp. ; annual ; abundant in broken grounds.

(163) *Potentilla* sp. ; abundant near springs and on the banks of rivers ; this abounds on the Columbia.

(164) *Achillea* sp. ; perennial ; plentiful in all meadows.

(165) *Juncus* sp. ; perennial ; small ; near springs, with the following :

(166–167) *Poa* sp. ; perennial ; a fine strong grass, in fertile soils, 2 feet high.

(168) —— May prove different from the former, in rocky places.

(169) *Gramineae*, perennial ; a low grass ; in elevated situations.

(170) *Gramineae*, perennial ; on moist soils, plentiful on the River Winnipeg.

(171) *Cornus* sp. ; a strong shrub ; abundant, banks of streams and lakes.

(172) *Rubus* sp. ; flowers white ; part of the underwood on the banks of rivers ; in sandy soils.

(173) *Salix* sp. ; female ; a low reclining plant.

(174) *Salix* sp. ; female ; leaves linear, densely tomentose ; a low shrub, 4 to 6 feet high.

(175) *Salix* sp. ; female and male ; a strong species, 10 to 14 feet high ; very plentiful on the shores of Lake Winnipeg.

(176) *Gramineae*, sp. ; on dry soils, Red River.

(177) *Gramineae*, perennial ; a low grass ; on the plains, Red River.

(178) *Asclepias* sp. ; leaves ovate, entire ; flowers faint white ; a small plant, 8 inches to 1 foot high ; in all light soils.

(179) *Solidago* sp. ; small, on all dry elevated places ; common.

(180) *Umbelliferae*, perennial ; flowers yellow ; 1 to 2 feet high ; on the plains.

Tuesday, 17th.—Showery ; thunder and lightning in the evening ; out all day on the plains until dusk. Collected as follows :

(181) *Monarda* sp. ; 18 inches to 2 feet high ; in most dry, light soils ; common ; flowers pink colour.

(182) *Astragalus* sp. ; plant upright, slender ; leaves small, slightly pubescent ; flowers capitate, blue and white ; plentiful on the plains.

(183) *Astragalus* sp. ; creeping and somewhat succulent, smooth ; flowers blue ; a fine species and stronger plant than the former ; grows luxuriantly on moist fertile soils.

---

[1] *Maianthemum bifolium*, Baker, in Journ. Linn. Soc. xiv. p. 563.

(184) *Pentstemon* sp. ; leaves sessile, linear, minutely serrated, smooth ; flowers fine light blue ; 1 to 2 feet high ; in solitary tufts, interspersed over all the dry soils ; this will I think prove one of Nuttall's Missouri plants, perhaps *serrulatus*.[1]

(185) *Helonias* (?) on the plains ; in great abundance.

(186) *Allium* sp. ; flowers rose colour ; on the plains ; very common.

Wednesday, 18*th*.—Having an invitation to dine with the Roman Catholic Bishop at his Mission house, I was prevented from making a long route. In the forenoon collected a few plants :

(187) *Psoralea* sp. ; perennial ; plant hirsute ; flowers blue and white ; this is the Turnip of the Plains of the voyageurs, the roots of which are used by the Indians both boiled and raw ; does not possess any farinaceous substance ascribed to *P. esculenta* by Pursh ; a low, much branching plant ; abundant on the low fertile spots of the plain (endeavour to find seeds of this plant).

(188) *Lobelia inflata* (?) common ; on moist ground.

(189) *Cruciferae*, annual ; a strong-growing plant ; on the banks of rivers ; 2 to 4 feet high.

(190) *Potentilla* sp. ; perennial ; flowers yellow ; a fine plant, 18 inches to 2 feet high ; on dry elevated spots ; rare.

(191) *Galium* sp. ; perennial ; abundant everywhere.

(192) *Polygonum* sp. ; on marshy ground ; a strong plant.

(193) *Spiraea*, a small shrub ; on banks of streams ; common.

(194) *Linum* sp. ; perennial ; flowers blue ; on the plains, rare ; in solitary tufts.

From the Bishop I had much civility and spent the evening in the most agreeable manner. He showed me his garden, farm, church, and Mission establishment, which reflects great credit on its conductor. The aborigines and Bulès have not only a religious education but are taught domestic economy, farming, spinning, and weaving cloth from the wool of the buffalo. This establishment, in common with most others, sustained great injury from the high water during last year, and from the pressing state of the church funds the most rigid economy is required to keep the Mission alive.

Thursday, 19*th*.—Rainy. Shifting papers, &c.

Friday, 20*th*.—Made a short trip to ' The Pines,' a small tract of undulating country a few miles off the banks of the river on the south side. Gathered as follows :

(195) *Gramineae*, annual; in moist fertile soils ; 2 to 3 feet high.

(196) *Gramineae*, perennial ; on the same sort of soils and equally abundant.

(197) *Gramineae*, perennial ; on the meadows, with the following :

(198) *Gramineae*, perennial ;

(199) *Gramineae*, annual ; near the establishments, in all low soils.

(200) *Silene*, annual ; a slender plant, 8 inches to a foot high ; sparingly on the plains, more abundant on hills.

(201) *Apocynum* ; 1 to 2 feet high ; flowers pink colour ; in moist

[1] *Pentstemon diffusus*, A. Gray, Syn. Fl. N. Am. ii. I. p. 271.

low places; this is found over the whole continent, much stronger on the west side.

(202) *Elaeagnus argentea*; abundant on the gravelly or limestone banks of rivers or dry soils on the plains; 2 to 4 feet high; found also on the Saskatchewan River.

(203) *Viburnum* sp.; a strong shrub; on the moist banks of rivers and thin woods; rare.

(204) *Hedysarum*, perennial; 1 to 2 feet high; on low moist plains; seen sparingly in small tufts among the short grass.

(205) *Syngenesia*, perennial; flowers white; a fine low plant; seen on all the dry bare places of the plains.

(206) *Turritis*,[1] annual; abundant on limestone rocks.

(207) *Anemone*, perennial; abundant in the same places with the former one.

(208) *Syngenesia*, perennial; flowers orange; abundant with the former; found on the Columbia.

(209) *Syngenesia*, annual; an inconspicuous weed, found in all inundated grounds; 2 to 3 feet high; in fertile soils.

Saturday, 21*st*.—After the daily shifting and drying papers, made a short turn up the Assiniboine River and added the following to the list:

(210) *Syngenesia*, perennial; flowers yellow; on dry light soils; abundant on limestone soils. Laid in specimens of *Acer* species, the same variety as that on the Saskatchewan at Carlton and Cumberland Houses.

(211) *Quercus* sp.; may prove *Q. alba*; this is one of the few trees that adorn the Red River, which appears to be its most northern range; 40 to 60 feet high, 18 inches to 2 feet diameter; the wood is soft and liable to become shaky.

(212) *Vitis* sp.; this would appear to be the most northern boundary of this genus; all that have come under my observation are enervated stunted plants, and none in fruit, although I am given to believe they bear in favourable seasons.

(213) *Scutellaria*, annual; in moist ground; abundant; *Mentha* sp., in the same situations, has a disagreeable foetid scent.

(214) *Astragalus tenuifolius* (? Nuttall) [*sic*]; flowers fine rose colour; stem reclining and wide-spreading; in large tufts on the light gravelly soils; fragrant; a fine species.

(215) *Asclepias* sp.; perennial; small flowers faint white; in the same places with the preceding; in great profusion.

(216) —— ? annual; an inconspicuous small plant, found abundantly everywhere.

(217) *Convolvulus* sp.; flowers white; leaves somewhat pubescent; a low upright species; in the plains; perhaps *Calystegia*; does not twine.

Sunday, 22*nd*.—At church, morning service. Arranged some plants in the evening.

Monday and Tuesday 23*rd*, 24*th*.—In company with the Rev. Mr. Harper, of the Roman Catholic Mission, who kindly offered to accompany me,

---

[1] *Arabis*, Benth. and Hook. f. Gen. Pl. i. p. 69.

I started for the Whitehorse plain, eighteen miles up the Assiniboine River. The country differs in no wise from the country on the banks of the Red River; put up for the night in the house of Mr. Grant and were very civilly entertained. Returned at sundown on Tuesday, having made some additions:

(218) *Potentilla fruticosa*; frequents elevated dry light soils or limestone rocks.

(219) *Solanum* sp.; annual; rare; flowers yellow; found only one plant near an old camp.

(220) *Psoralea* sp.; leaflets five, silky, particularly underneath; flowers small, blue; this is a much stronger species than the hirsute one; more branching; is never seen in wet soil; abounds on all dry or rocky places; the root is not used, although it is equally large and as well tasted.

(221) *Astragalus* var. *albus*, of 214; found only one plant.

(222) *Hedysarum* sp.; perennial; leaves pinnate; flowers red; a small upright-growing plant; abundant in dry places, with the two following of the same genus:

(223) *Hedysarum*; shrubby; leaves glaucous; flowers faint red;

(224) *Hedysarum*; flowers red; leaves densely tomentose.

(225) *Syngenesia*, perennial; in dry soils; a low reclining plant.

(226) *Oenothera* sp.; suffruticose; flowers yellow; a small plant; in dry light soils and on rocks; abundant.

(227) *Potentilla* sp.; perennial; leaves densely tomentose; peduncle slender; flowers large, yellow; stem reclining; a very beautiful species; abundant on dry clear places.

(228) *Potentilla* sp.; perennial; hirsute; a strong plant; by no means so abundant as the former; 18 inches to 2 feet high.

(229) *Cruciferae*, annual; on the banks of streams; an upright plant 2 to 3 feet high.

(230) *Lysimachia* sp.; perennial; abundant near springs.

(231) *Myosotis* sp.; perennial; abundant in marshy grounds.

(232) *Anemone* sp.; perennial; flowers white; abundant on all fertile grounds.

(233) *Alisma Plantago*, in springs and pools.

(234) *Symphoria*[1] sp.; abundant on all dry places, banks of rivers, &c. *Wednesday, 25th.*—Collected a few specimens around the establishment:

(235) *Triglochin* sp.; in low saline marshes; abundant.

(236) *Aspidium* sp.; in shady places near springs; abundant.

(237) *Oenothera* sp.; annual; flowers yellow; in low fertile soils; plentiful; perhaps *Oenothera fruticosa*.

(238) *Hedysarum* (?) sp.; abundant in all dry light soils; this abounds on the Columbia and its branches.

(239) *Lilium* sp.; flowers fine dark red; 1 foot to 18 inches high; in all fertile soils, abundant; *L. canadense*.

Thursday and Friday, 26th, 27th.

(240) *Cucurbitaceae*, annual; flowers white; leaves partially five-lobed;

[1] *Symphoricarpos*, Benth. and Hook. f. Gen. Pl. ii. p. 4.

banks of streams in thin woody places; abundant; a wide-spreading creeping plant.

(241) *Gramineae*, perennial; in open woods among the underwood; this grass is only seen in tufts, never forming a sward like most others.

(242) *Eupatorium*, perennial; a strong upright plant, frequenting moist meadows on the outskirts of woods and river springs.

(243) *Carex* sp.; a fine strong species; abounds near creeks.

(244) *Gramineae*, perennial; fine delicate slender grass; abounds over all the plains in dry soils; on the Columbia.

(245) *Juncus* sp.; very plentiful in marshy grounds and near springs.

(246) *Hordeum jubatum*; abundant on all the dry plains on the Red and Saskatchewan Rivers.

(247) *Gramineae*, annual; a strong grass, 2 to 3 feet high; abundant everywhere.

(248) *Utricularia* sp.; small slender species, in the low overflowed plains and shallow pools among long grass; this is not so abundant as some other specimens.

(249) *Eriophorum vaginatum* (?), in low moist plains; appears to be rare; seen very sparingly.

(250) *Gramineae*, perennial; in dry light soils; 9 to 18 inches high; abundant.

(251) *Carex* sp.; large root, tuberous, in all moist places.

Saturday, 28th.—Rainy; dried papers and paid a visit to Mr. Logan.

Sunday, 29th.—Very hot and sultry; thunder in the evening. Attended service at the Roman Catholic church and heard a sermon from the Rev. Mr. Harper; music good.

Monday and Tuesday, 30th, 31st.—Much indisposed; violent headache and feverish; had some medicine of Mr. Richard Julian Hamlyn, the Company and Colony surgeon, who has been attentive to me. Unable to go out.

Wednesday, August 1st.—At daylight started on horseback to a small low hill about sixteen or eighteen miles east of the colony, composed of limestone rock with a few low poplars, willow, and birch in the low places. The plains being overflowed four or five miles back from Red River, had to go round by Sturgeon Creek on the Assiniboine River. Returned late at night. Made the following additions:

(252) *Juncus* sp.; dry elevated spots.

(253) *Gramineae*, perennial; 1 to 2½ feet high; common on all dry soils; gives great annoyance to the traveller, the seeds sticking in the trousers and moccasins and accumulating in large masses on the feet.

(254) *Cuscuta americana*; common on all dry soils; prefers syngenesious plants—*Aster, Solidago*—sometimes on others, *Rosa*, &c.

(255) *Umbelliferae*, perennial; in low marshy ground; rare.

(256) *Gramineae*, sp.; perennial; an elegant grass; the strongest of any on the plains; 4 to 6 feet high; in damp low ground; very plentiful.

(257) *Carex* sp.; in low places; abundant.

(258) *Juncus* sp.; this seems to be a variable plant; if I mistake

not, it is seen sometimes not more than 6 or 8 inches high; the present is an instance of the difference; on the plains, abundant.

(259) *Thalictrum*, perennial; flowers white; a tall, strong, much-branched species; plentiful on the banks of rivers and streams and on limestone rocks.

(260) *Gramineae*, perennial; a graceful strong grass; near creeks and springs.

(261) *Polygonum* sp.; in lakes and still streams and pools; strong plant; stem much swelled at the joints; abundant.

(262) *Prunus* sp.; a low shrub; flowers white; fruit bitter and astringent.

(263) *Prunus* sp.; small, 8 inches to a foot high; fruit (not yet ripe) appears to be bitter; the former is seen in abundance on the banks of rivers; the latter only on dry light or gravelly soils and on limestone in the greatest abundance.

(264) *Solidago* sp.; a strong species; outskirts of woods and banks of streams.

(265) *Hopus*[1] sp.; male; banks of the Assiniboine; abundant; I am informed by the Bishop the hops do very well for beer.

(266) *Umbelliferae*, perennial; 4 to 6 feet high; near streams.

(267) *Syngenesia*, perennial; up to 7 feet high; in all fertile soils.

(268) *Pedicularis ;* flowers yellow; on the low meadows; abundant; fine plant.

Thursday to Saturday, 2nd, 3rd, and 4th.—Continued my gleanings daily when the weather permitted:

(269) *Astragalus* sp.; leaves radical; leaflets fifteen to seventeen; stem and calyx woolly, hairs long; leaves silky; flowers fine azure-blue; a foot to 15 inches high; a very fine plant; on the Assiniboine River, in dry places; rare.

(270) *Bartsia* sp.; flowers scarlet; 18 inches to 2½ feet high; in open woody places.

(271) *Monarda* sp., var.; flowers white; found only one plant.

(272) *Pteris* sp.; small; in the dry crevices of limestone rocks; abundant.

(273) *Gramineae*, perennial; on elevated dry places; 6 inches to a foot high; common.

(274) *Gerardia* sp.; annual; on the low damp grounds; abundant.

(275) *Syngenesia*, perennial; flowers yellow; a low plant; in dry open places; common.

(276) *Potentilla* sp.; leaves serrate, upper side hirsute, under pubescent; nerves hirsute; flowers yellow; a fine strong species; on dry elevated grounds; not so abundant as some others of the genus.

(277) *Potentilla* sp.; perennial.

(278) *Potentilla*, perennial; both the latter species are usually seen together in rocky situations on dry soils; abundant; two fine plants.

(279) *Solidago* sp.; small, 4 to 9 inches high; in dry open places; common.

[1] So in MS.—ED.

(280) *Astragalus* sp. ; this appears to differ from (269), the leaves being less hirsute and the flowers fainter blue ; rare.

(281) *Donia* sp. ; annual ; (*glutinosa* [1] ?) ; abundant on the plains.

(282) *Syngenesia*, perennial ; flowers yellow ; abundant on the plains ; 2 to 3 feet high.

(283) *Syngenesia*, perennial ; flowers white ; a slender short plant, frequenting light dry soils ; this may be only a yellow variety of one with white flowers found a few days since.

(284) *Sonchus* sp. ; annual ; plentiful on the plains in all soils.

(285) *Aster* sp. ; perennial ; small ; flowers white ; rare.

(286) *Liatris* sp. ; a plant 1 to 2½ feet high ; abundant on the plains.

(287) *Liatris* sp. ; flowers rose colour ; abundant on dry soils and on limestone rocks.

Sunday, 5th.—Very heavy rain last night, with thunder and lightning ; morning cool and fine. Went to church and heard a sermon from the Rev. David T. Jones ; dined with him and spent the evening at his house.

Monday, 6th.—Collected the following :

(288) *Gramineae*, perennial ; common in all moist grounds near springs.

(289) *Helianthus* sp. ; perennial ; leaves lanceolate, rough ; flowers large, yellow ; a strong plant, 3 to 6 feet high ; abundant in all fertile soils ; this is what is called the 'Indian Potato' by the Canadian voyageurs, a worthless insipid root nothing compared to Jerusalem artichoke.

(290) *Helianthus* sp. ; found in the same place with the preceding number and equally profuse.

(291) *Betula* sp. ; small shrub ; in dry limestone rocks ; abundant ; in fruit.

Tuesday, 7th.—Employed drying and putting in order the collection for my departure. Mr. McKenzie had caused to be put up for me ample stock of provisions and made inquiry about the departure of the boat.

Wednesday, 8th.—Finished packing and putting in order the plants, and made a short turn in the afternoon. Laid in a few plants.

Thursday, 9th.—Severe thunderstorms at daylight, with heavy rain. Dry and windy during the day. Took a walk, but could not venture far from home lest an opportunity should offer of getting away.

Friday, 10th.—Morning fine. Mr. McKenzie having informed me of his intention of sending a boat to Norway House, and lest the others which are to be despatched in a day or two should be delayed in transit by bad weather on the lake too long to meet the ship in Hudson's Bay, I thought it prudent to make my stay no longer. The Rev. Mr. Jones and the Rev. Mr. Cockran called on me and handed me some small packages for Hudson's Bay and for England. Both these gentlemen have shown me much civility. (A few seeds from the Society would be of great benefit to the Missionary establishment and would be thankfully received.) To D. McKenzie (the Governor of the Colony) I am greatly indebted for his polite attentions. After bidding him and the Bishop adieu, I left the establishment in company with Mr. Hamlyn, the surgeon, for Hudson's Bay. In our descent of the river we had to make many short delays, receiving letters, &c., for Hudson's

---

[1] *Grindelia glutinosa*, A. Gray, Syn. Fl. N. Am. i. II. p. 119.

Bay. Had some cheeses presented me, which I could not well refuse. Called on James Bird, Esq., who resides a considerable way down the river, and received a letter from him addressed to Mr. Sabine. I forgot to mention a week ago, being near his house on one of my walks, I called on him and he received me with civility and attention; had tea with him and invited to renew my visit, which I should have done if time permitted. Informed me he had a letter from Mr. Sabine last March, mentioning the probability of my calling on him. Camped a few miles below the rapid. Laid in two species of *Artemisia* and a few other plants.

Saturday, 11*th.*—Thunder and lightning with a few drops of rain at daybreak. At five proceeded down the river with a light air of wind and entered the lake at eight o'clock. Continued our voyage along the south-west side of the lake for fifteen miles, when we came to a small narrow sandy island, where we put ashore to boil the tea-kettle, during which time I picked up two fine *Gramineae*—*Fraxinus* and one species of *Hieracium*, perennial; and *Prunus* sp. (No. 263), found sparingly on gravelly places and limestone rocks on Red River, called by the voyageurs Sand-Cherry—shoots of the present year, upright; of the last prostrate; very luxuriant—the shore is nothing but sand, or more properly very fine lime gravel, not a vestige of earth—exceedingly prolific (more so than any of the Cherry tribe I have seen), fruit not perfectly ripe, large (as large as common Bird Cherry); appears to be astringent, but if only for novelty is worthy of a place in the garden. Continued our route prosperously until three o'clock, when the wind became contrary as we were in a long traverse—that is, crossing a deep bay of the lake. It became suddenly boisterous and much hard labour before we got to shore. The oars were long, and by the heavy swell it was nothing but plunging. Landed on a low thinly wooded island at half-past five; our poor men exhausted, and myself somewhat anxious. From the appearance of this and many others in this lake the water has risen to a considerable degree higher than it must have been at a former period. Trees are buried to the depth of eight to ten feet, and many places are seen with dead poplars standing erect that no doubt were woody islands. Employed in the evening changing papers and drying the same.

Sunday, 12*th.*—Last night the wind increased to a perfect hurricane and the water rose so high as to overflow our camp; so we had to betake ourselves to the boat for the night. Wind more moderate at sunrise. Started at nine o'clock, the sea being nearly calm; kept along the shore. Nothing occurred. Shores of the lake low and not different from yesterday. Camped at seven P.M. Marched twenty-five miles.

Monday, 13*th.*—The wind at two A.M. being favourable, and moonlight, we started under easy sail until daybreak. Morning cloudy and heavy, rain from six to eight; put ashore to replace the things in the boat, and breakfast. The weather being somewhat drier at nine, proceeded and crossed over to the south side of the lake, when the wind veered round to the south-west, which prevented us from going. Put ashore, camped, and remained four hours, when it calmed; proceeded a second time, although the weather was still gloomy. Gained six miles at four, and as the shore to the south-west was steep limestone rocks, affording no harbour, and unfavourable for crossing the lake, we camped. Nothing occurred.

Tuesday, 14th.—Morning dull, cloudy, and drizzly ; rain at eight. Started with a favourable breeze at five o'clock across the lake. Gained the 'Dog's Head' to breakfast at half-past eleven. Delayed forty-six minutes. Continued along the shore and passed 'Rabbit' point at one, and a second at four ; passed it two miles, when the wind shifting to the west we were obliged to run back to a small sandy beach and run the boat on the shore. Ere all the baggage was out, the waves were breaking on the shore with all the violence of a sea hurricane. Shifted plants and dried papers in the evening. Laid in a few plants in fruit— *Pyrola rotundifolia, Cornus canadensis, Artemisia*—finer specimens than before. In the course of the evening the boat had to be hauled up as the surge rose on the shore, all our strength being inadequate to pull it up at once. Blowing with increased violence. Now ten o'clock.

Wednesday, 15th.—Weather same throughout the night ; morning clear and somewhat more moderate at ten, but still unable to proceed. Found and laid in specimens of *Linnaea borealis* (in fruit). This is the first time I have ever seen this plant in this state. Is rare. Mr. Scouler informed me that he found it in 1825, in perfect fruit, in the shady forests of Nootka Sound : here around my tent, in sand that has been thrown on the shore by the lake storms, in partially shady situations. It is not rare in fruit. Employed all the forepart of the day drying papers and shifting plants ; no place that I can walk, being all swamp. More moderate at noon ; started and gained seven miles at three, when the wind sprang up from the same quarter, which obliged us to put to shore on a sandy beach exposed to the weather. Afternoon and evening the same.

Thursday, 16th.—Weather stormy until eleven A.M., when the boat was launched again and pulled off. Calm at two. Called at Banning's River, where we made a stay of a few minutes. Learned from Mr. Spencer that the other boat from Red River had passed the preceding night. On leaving this place at four P.M. a favourable breeze sprang up, and, being anxious to lose no time, did not put ashore to sup, but went on along the shore under easy sail until daylight.

Friday, 17th.—Morning rainy ; took breakfast at six A.M. and continued under a strong breeze till four P.M., the last point of the lake when the wind failed ; pulled over the narrow bay to Norway House, where I found Messrs. John Stuart and Cameron on their way to winter quarters. Both these gentlemen showed me every kindness and informed me that Captain Back had passed two days before for Hudson's Bay.

Saturday, 18th.—Left Norway House at six A.M. in company with Mr. Jos. Bird, with whom I intend to complete the remainder of my journey, as the other boat was to return to Red River. Passed at eight o'clock two canoes in Play Green Lake containing the men belonging to the Land Arctic Expedition on their way to Montreal. Made but little progress, having a strong wind against us. At midday gained the lower establishment on Jack River, where I found my old friend Mr. John McLeod. Learned with regret my Silver-Headed Eagle had died of starvation. I found every other thing safe. The roots, both dry and those hid in the wood, in good condition.

Sunday, 19*th.*—After having everything packed up by ten A.M. embarked in Mr. Bird's boat and descended the river. Camped twenty-five miles below ; country low, swampy ; trees small—*Pinus Banksiana, P. Larix,*[1] *Betula, Populus* ; found nothing different from what I had seen before. In the evening at my usual work changing and drying papers.

Monday, 20*th.*—Started at four A.M. and went along Black Water creek, a narrow natural canal, both sides low swampy ground with low willows. Took breakfast on a small rock on the left-hand side at ten. At midday rainy. Met Mr. Evans, of Red River, from Hudson's Bay. Arrived at the Painted Stone portage at dusk and passed over to White Water creek after dark. Evening fine. Aurora borealis but faintly seen.

Tuesday, 21*st.*—Dull and cloudy. Started a little before day. Passed through a small lake and creek until we arrived at White Fall portage at ten. Took breakfast, and while the men were carrying their property and dragging the boat over, put in order my plants ; picked up one of the *Umbelliferae* [*sic*] (perhaps *Aralia*), *Gentiana, Parnassia caroliniana,* and *Hydrocharis.* Lost overboard a few plants laid in on the upper part of Lake Winnipeg. The White Fall is a small cataract with high rocks on one side adorned by timber of low growth, insignificant, but at the same time worth going a few yards to see. At four left and passed the upper or small Hill Gate, where the boat was lightened previous to running the rapid. Navigation intricate. Country the same as before. Camped on some rocks a mile below. Evening fine with a heavy dew.

Wednesday, 22*nd.*—Started at daylight ; at seven came to Hill Gate, a rocky rapid narrow part of the river where considerable time was lost lowering the boat with the line. Timber gradually becomes smaller as we approach the coast. Continued our route through a chain of small lakes with low marshy and overflowed banks. Shortly after noon entered Oxford Lake, a small narrow but beautiful sheet of water with bold rocky banks and numerous islands. Camped at dusk on one, having had a fine day's journey.

Thursday, 23*rd.*—Thunder and lightning during the night. As usual made an early start and reached Oxford House at ten, where we took breakfast. Received a letter from Mr. Colin Robertson, one of the resident partners of the Company, regarding a few bird-skins he left for me at York. Certainly much obliged, having never seen him nor had any correspondence. Wrote him a note of thanks. Proceeded at eleven and passed some very bad rapids, and launched the boat at Trout Fall portage where the remainder of the day was spent repairing the boat. Laid in a few plants.

Friday, 24*th.*—On leaving Trout Fall we found the boat still made water, but as the wind was favourable for passing through Knee Lake no time was lost. Went prosperously on and breakfasted on Knee near ' Tea Islands,' so named from *Ledum palustre* being abundant on them. The scenery of this, like Oxford Lake, is fine also, with numerous islands.

[1] ? *Larix occidentalis,* Mast. in Journ. R. Hort. Soc. xiv. p. 218.

Entered Lower Jack River at sundown, where we camped ; this is a small narrow stream with low banks.

Saturday 25*th.*—In the course of the morning the boat was considerably injured descending this stream, the water being low. Passed through Swampy lake, a pond a few miles long and then entered ' Hill River.' At noon, while descending a rapid, the boat struck heavily on the rock and shattered seven of the timbers and planking. Just had time to reach a small island when she was filled. My hands tied up—could not get off. Dried papers ; planted in a small box *Erythronium grandiflorum, Lilium pudicum,*[1] and *Claytonia lanceolata,*[2] which I am glad are all fresh. (Why did you not bring *Gaultheria* alive—across the continent—2900 miles ? It could be done.)

Sunday, 26*th.*—Employed all day at the boat until three o'clock, when we set out again, the boat making a great deal of water. Camped a few miles below a low hill on the right.

Monday and Tuesday, 27*th,* 28*th.*—Detained longer than usual, the morning being dull and unfavourable for passing the ' Rock Fall,' which we passed in safety at seven. Entered ' Steel River,' a stream of some magnitude but not so rapid as the last. Breakfasted at its junction with York River. Continued until dusk, when we put to shore, boiled the kettle, and embarked under sail. Aurora borealis beautiful. The idea of finishing my journey, and expectations of hearing from England made the night pass swiftly. At sunrise on Tuesday I had the pleasing scene of beholding York Factory two miles distant, the sun glittering on the roofs of the house (being covered with tin) and in the bay riding at anchor the company's ship from England. The hearty welcome I had to the shores of the Atlantic from Mr. McTavish and all others was to me not a little gratifying. In the most polite manner everything that could add to my comfort was instantly handed ; and I adduce no further proof of this gentleman's goodness than to state that he had, without my knowledge, made for me a new suit of clothing, linen, &c., ready to put on. No letters from England. Regret the death of my Calumet Eagle ; was strangled a few days ago with the cord by which he was tied by the leg : fell over the casing[3] of one of the houses and was found dead in the morning. What can give one more pain ? This animal I carried 2000 miles and now lost him, I might say, at home. Had a note from Governor Simpson. Met Captain Back, Lieutenant Kendall, and Mr. Drummond, who arrived yesterday. It now only remains to state that I have had great assistance, civility, and friendly attentions from the various persons I have formed an acquaintance with during my stay in North America.

[1] *Fritillaria pudica*, Baker, in Journ. Linn. Soc. xiv. p. 267.
[2] *Claytonia caroliniana* var. *sessilifolia*, S. Wats. Bibl. Ind. N. Am. Bot. p. 117.
[3] Word illegible, might possibly be ' eaving.'—ED.

APPENDIX I.

MEMOIR OF DAVID DOUGLAS

[HAVING been requested by the President and Council of the Society to prepare the hitherto unpublished manuscript Journals sent home by Douglas whilst he was travelling in North America on behalf of the Royal Horticultural Society, I feel that it would be a mistake not to attempt to gather together all the facts concerning this eminent Horticulturist which have come to my notice during the researches entailed by the investigation of documents bearing upon his Journal. I make no apology therefore for attempting a very brief Memoir, but only for the inadequacy with which I fear I may have been able to carry it out. I have been unavoidably compelled to present it in the form of four Appendices.—W. W.]

David Douglas was born in Scotland, at Scone near Perth, in the year 1798, and served a seven years' apprenticeship as a gardener in the gardens of the Earl of Mansfield, at Scone. When he was about eighteen years of age he moved to Valleyfield, near Culross, to the garden of Sir Robert Preston, Bart., which was at that time notable for its collection of exotic plants. About two years afterwards he obtained an appointment in the Glasgow Botanic Garden, where his enthusiasm for plants attracted the attention of Dr. William Jackson Hooker, the Professor of Botany at Glasgow University, who made him his companion in his excursions through the Western Highlands and employed him as his assistant in collecting materials for the " Flora Scotica," in the publication of which Dr. Hooker was then engaged. And when Dr. Hooker was consulted by Mr. Joseph Sabine, the secretary of the Royal Horticultural Society, as to a suitable person to be employed by the Society on a Botanical expedition to North America, Dr. Hooker had no hesitation whatever in recommending David Douglas as the most suitable person he knew.

Douglas was accordingly engaged by the Society, and on June 3, 1823, he was despatched to the United States, where he collected a large number of specimens of various Oaks, a detailed descriptive review of which will be found in this volume, pp. 31–49. He also obtained and either sent or brought home a number of other interesting new plants, a collection of fruit trees among them.

Having successfully accomplished this mission he returned to England in the late autumn of 1823, and an opportunity offering next year through the Hudson Bay Company he was again employed by the Society on a similar journey, but spread over a much larger tract of country, extending from California northwards to the Columbia River and as far further north on the western side of the continent as he found he could penetrate. During the interval between his return and starting on his new journey Douglas is related to have worked no less than eighteen hours a day in perfecting himself in various scientific and technical ways in which he felt himself to be at all deficient.

On July 25, 1824, he left England, the narrative of his adventures being recorded in his own handwriting both in a condensed form, pp. 51–76, and also in more or less regular daily entries in his Journal, pp. 77–293.

In the 'Proceedings of the Royal Society' under date April 27, 1837, it is recorded that Mr. Sabine received from Douglas several volumes of lunar, chronometrical, magnetical, meteorological and geographical observations, together with a volume of field sketches. The geographical observations of latitude and longitude refer to two distinct tracts of country : first, the Columbia River, and its tributaries, and the district to the westward of them ; and secondly, California. Douglas very judiciously selected the junctions of rivers, and other well-characterised natural points, as stations for geographical determination. The papers containing the details of his magnetical inquiries comprise records of observation of the dip, and of the intensity, at various stations.

What may have become of these interesting volumes I have not been able to trace, the only books in the Royal Horticultural Society's possession being the Journals reproduced in the present volume. The enthusiasm with which he carried out his instructions and the energy he shewed, coupled with the difficulties he encountered and overcame, are amply apparent in the Journals themselves and do not need to be recapitulated here. It is interesting to notice that he met with Sir John Franklin at Norway House, when he was returning overland from his second Arctic expedition. He sailed from Hudson Bay on September 15, and arrived at Portsmouth on October 11, 1827.

In the seventh volume of the 'Transactions' of the Horticultural Society, December 1830, it is mentioned that Douglas had brought home with him an even far greater number of plants and seeds than he had previously sent home, and that from them 210 distinct species had been raised in the Society's Gardens, 80 of which being then considered to be only " Botanical curiosities " were " abandoned " and 130 species were grown on and distributed to all parts of the world.

The President and Council were so gratified by the result of Douglas's expedition that they persuaded him to return, which he did in October 1829, with the intention of undertaking a wider exploration in the same general districts as before, but the unsettled state of the country and the tribal wars going on amongst the natives made this impossible. He therefore transferred his attention from the Columbia River to California and landed at San Francisco in 1831, whence he journeyed to Monterey, where he was well received by the monks and afforded every facility in their power for exploring the neighbouring country. He remained there during the summer of 1831, intending to return to the Columbia River in the autumn ; but being unable to find any ship or other means of transport he was compelled to spend another season in California making various excursions to the interior, and finally in August 1832 he sailed to the Sandwich Islands, from whence he despatched to this country his Californian collections of seeds and plants and later on returned to the Columbia River. From this collection of seeds more species were raised in the Society's Gardens.

On his way from the Sandwich Islands to the Columbia River he received intelligence that his personal friend Mr. Sabine had resigned the secretaryship of the Society, and through some misconception of the cause Douglas also resigned his appointment of Collector. For more than a year, however, he continued travelling in North Western America.

Of the last eighteen months of his life very little is known. In a letter written from the Columbia River and dated October 24, 1832, he expressed a great desire to become better acquainted with the vegetation of the Sandwich Islands and says " I have made arrangements with Captain Charlton our Consul to aid me should I return there, which I shall earnestly endeavour to do." Accordingly we find him leaving the Columbia River on October 18, 1833, and arriving at Hawaii in the Sandwich Islands on January 2, 1834. On the 7th he started to ascend Mouna Roa, the great volcano of the Sandwich Islands, of which he sent home an account to his brother Mr. John Douglas, which will be found in Appendix II, pp. 298–317. This was the last letter or communication received from David Douglas. All that is known further is contained in Appendix III, pp. 318–323 : a very sad end to the life of one of our greatest and most successful exploring Botanists, to whom the whole world is deeply indebted.

APPENDIX II.

EXTRACT FROM DAVID DOUGLAS'S JOURNAL OF A SUBSEQUENT
EXPEDITION IN 1833–4. WHICH JOURNAL WAS SENT TO HIS
BROTHER MR. JOHN DOUGLAS.[1]

1833

". . . We steered southward for the Sandwich Islands. The island
of Mauai was indistinctly seen, at sun-set, of the 21st of December, 1833,
forty-two miles off; and, on the 22nd, Woahu lay ten miles due West of
us. Having quitted the Harbour of Fair Haven, in Woahu, on Friday,
the 27th, in an American schooner, of sixty tons, she proved too light
for the boisterous winds and heavy seas of these channels, and we were
accordingly obliged to drop anchor in Rahaina Roads, for the purpose
of procuring more ballast. An American Missionary, Mr. Spaulding,
having come on board, I accompanied him on shore, to visit the school,
situated on the hill side, about five hundred feet from the shore, and
returned to the ship at night. On Tuesday, the 31st of December, we
stood in for the island of Hawaii, and saw Mouna Kuāh very clearly, a
few small stripes of snow lying only near its summit, which would seem
to indicate an altitude inferior to that which has been commonly assigned
to this mountain.

1834

" My object being to ascend and explore Mouna Kuāh, as soon as
possible, I started on the 7th January, 1834, and, after passing for rather
more than three miles over plain country, commenced the ascent, which
was however gradual, by entering the wood. Here the scenery was truly
beautiful. Large timber trees were covered with creepers and species
of *Tillandsia*, while the *Tree Ferns* gave a peculiar character to the whole
country. We halted and dined at the Saw Mill, and made some baro-
metrical observations, of which the result is recorded, along with those
that occupied my time daily during the voyage, in my journal. Above
this spot the *Banana* no longer grows, but I observed a species of *Rubus*
among the rocks. We continued our way under such heavy rain, as,
with the already bad state of the path, rendered walking very difficult
and laborious; in the chinks of the lava, the mud was so wet that we
repeatedly sunk in it, above our knees. Encamping at some small huts,
we passed an uncomfortable night, as no dry wood could be obtained

---

[1] See Hooker's Comp. Bot. Mag.

for fuel, and it continued to rain without intermission. The next day we proceeded on our way at eight o'clock, the path becoming worse and worse. The large *Tree Ferns*, and other trees that shadowed it, proved no protection from the incessant rain, and I was drenched to the skin the whole day, besides repeatedly slipping into deep holes, full of soft mud. The number of species of *Filices* is very great, and towards the upper end of the wood, the timber trees, sixty or seventy feet high, and three to ten inches [1] in circumference, are matted with Mosses, which, together with the *Tillandsias* and *Ferns*, betoken an exceedingly humid atmosphere. The wood terminates abruptly; but as the lodge of the cattle-hunter was still about a mile and a-half farther up the clear flank of the mountain, situated on the bank of a craggy lava stream, I delayed ascertaining the exact altitude of the spot where the woody region ends (a point of no small interest to the Botanist), until my return, and sate down to rest myself awhile, in a place where the ground was thickly carpeted with species of *Fragaria*, some of which were in blossom, and a few of them in fruit. Here a Mr. Miles, part owner of the saw-mill that I had passed the day before, came up to me; he was on his way to join his partner, a Mr. Castles, who was engaged in curing the flesh of the wild cattle near the verge of the wood, and his conversation helped to beguile the fatigues of the road, for though the distance I had accomplished this morning was little more than seven miles, still the laborious nature of the path, and the weight of more than 60 lbs. on my back, where I carried my barometer, thermometer, book, and papers, proved so very fatiguing, that I felt myself almost worn out. I reached the lodge at four, wet to the skin, and benumbed with cold, and humble as the shelter was, I hailed it with delight. Here a large fire dried my clothes, and I got something to eat, though, unluckily, my guides all lingered behind, and those who carried my blanket and tea-kettle were the last to make their appearance. These people have no thought or consideration for the morrow; but sit down to their food, smoke and tell stories, and make themselves perfectly happy. The next day my two new acquaintances went out with their guns and shot a young bull, a few rods from the hut, which they kindly gave me for the use of my party. According to report, the grassy flanks of the mountain abound with wild cattle, the offspring of the stock left here by Capt. Vancouver, and which now prove a very great benefit to this island. A slight interval of better weather this afternoon afforded a glimpse of the summit between the clouds, it was covered with snow. At night the sky became quite clear, and the stars, among which I observed Orion, Canis minor, and Canopus, shone with intense brilliancy.

" The next day the atmosphere was perfectly cloudless, and I visited some of the high peaks which were thinly patched with snow. On two of them, which were extinct volcanos, not a blade of grass could be seen, nor any thing save lava, mostly reddish, but in some places of a black colour. Though on the summit of the most elevated peak, the thermo-

---

[1] Probably an error for ' feet.'—ED.

meter under a bright sun stood at 40°, yet when the instrument was
laid at an angle of about fifteen degrees, the quicksilver rose to 63°, and
the blocks of lava felt sensibly warm to the touch. The wind was from
all directions, East and West, for the great altitude and the extensive
mass of heating matter completely destroy the Trade Wind. The last
plant that I saw upon the mountain was a gigantic species of the *Compositae* (*Argyrophyton Douglasii*,[1] Hook. Ic. Plant. t. 75), with a column
of imbricated, sharp-pointed leaves, densely covered with a silky clothing.
I gathered a few seeds of the plants which I met with, among them a
remarkable *Ranunculus*, which grows as high up as there is any soil. One
of my companions killed a young cow just on the edge of the wood, which
he presented me with, for the next day's consumption. Night arrived
only too soon, and we had to walk four miles back to the lodge across
the lava, where we arrived at eight o'clock, hungry, tired, and lame,
but highly gratified with the result of the day's expedition.

" The following morning proved again clear and pleasant, and every
thing being arranged, some of the men were despatched early, but such
are the delays which these people make, that I overtook them all before
eight o'clock. They have no idea of time, but stand still awhile, then
walk a little, stop and eat, smoke and talk, and thus loiter away a whole
day. At noon we came up to the place where we had left the cow, and
having dressed the meat, we took a part and left the rest hanging on
the bushes. We passed to the left of the lowest extinct volcano, and
again encamped on the same peak as the preceding night. It was long
after dark before the men arrived, and as this place afforded no wood,
we had to make a fire of the leaves and dead stems of the species of
Compositae mentioned before, and which, together with a small *Juncus*,
grows higher up the mountain than any other plant. The great difference
produced on vegetation by the agitated and volcanic state of this mountain
is very distinctly marked. Here there is no line between the Phenogamous
and Cryptogamous Plants, but the limits of vegetation itself are defined
with the greatest exactness, and the species do not gradually diminish
in number and stature, as is generally the case on such high elevations.

" The line of what may be called the Woody Country, the upper verge
of which the barometer expresses 21,450 inch.; therm. 46° at 2 P.M.,
is where we immediately enter on a region of broken and uneven ground,
with here and there lumps of lava, rising above the general declivity
to a height of three hundred to four hundred feet, intersected by deep
chasms, which show the course of the lava when in a state of fluidity.
This portion of the mountain is highly picturesque and sublime. Three
kinds of timber, of small growth, are scattered over the low knolls, with
one species of *Rubus* and *Vaccinium*, the genus *Fragaria* and a few
Gramineae, *Filices*, and some alpine species. This region extends to
bar. 20,620 inch.; air 40°, dew-point 30°. There is a third region, which
reaches to the place where we encamped yesterday, and seems to be the
great rise or spring of the lava, the upper part of which, at the foot of
the first extinct peak, is bar. 20,010 inch.; air 39°.

[1] *Argyroxiphium sandwicense*, Hook. Ic. Pl. t. 75.

" At six o'clock the next morning, accompanied by three Islanders and two Americans, I started for the summit of the mountain ; bar. at that hour indicated 20,000 inch., therm. 24°, hygr. 20°, and a keen West wind was blowing off the mountain, which was felt severely by us all, and especially by the natives, whom it was necessary to protect with additional blankets and great coats. We passed over about five miles of gentle ascent, consisting of large blocks of lava, sand, scoriæ, and ashes, of every size, shape, and colour, demonstrating all the gradations of calcination, from the mildest to the most intense. This may be termed the Table Land or Platform, where spring the great vent-holes of the subterranean fire, or numerous volcanos. The general appearance is that of the channel of an immense river, heaped up. In some places the round boulders of lava are so regularly placed, and the sand is so washed in, around them, as to give the appearance of a causeway, while in others, the lava seems to have run like a stream. We commenced the ascent of the Great Peak at nine o'clock, on the N.E. side, over a ridge of tremendously rugged lava, four hundred and seventy feet high, preferring this course to the very steep ascent of the South side, which consists entirely of loose ashes and scoriæ, and we gained the summit soon after ten. Though exhausted with fatigue before leaving the Table Land, and much tried with the increasing cold, yet such was my ardent desire to reach the top, that the last portion of the way seemed the easiest. This is the loftiest of the chimneys ; a lengthened ridge of two hundred and twenty-one yards two feet, running nearly straight N.W. To the North, four feet below the extreme summit of the Peak, the barometer was instantly suspended, the cistern being exactly below, and when the mercury had acquired the temperature of the circumambient air, the following register was entered : at 11 hrs. 20 min. ; bar. 18,362 inch. ; air 33° ; hygr. 0″ 5. At twelve o'clock the horizon displayed some snowy clouds ; until this period, the view was sublime to the greatest degree, but now every appearance of a mountain-storm came on. The whole of the low S.E. point of the island was throughout the day covered like a vast plain of snow, with clouds. The same thermometer, laid on the bare lava, and exposed to the wind at an angle of 27°, expressed at first 37°, and afterwards, at twelve o'clock, 41°, though when held in the hand, exposed to the sun, it did not rise at all. It may well be conjectured that such an immense mass of heating material, combined with the influence of internal fire, and taken in connexion with the insular position of Mouna Kuáh, surrounded with an immense mass of water, will have the effect of raising the snow-line considerably ; except on the northern declivity, or where sheltered by large blocks of lava, there was no snow to be seen ; even on the top of the cairn, where the barometer was fixed, there were only a few handsful. One thing struck me as curious, the apparent non-diminution of sound ; not as respects the rapidity of its transmission, which is, of course, subject to a well-known law. Certain it is, that on mountains of inferior elevation, whose summits are clothed with perpetual snow and ice, we find it needful to roar into one another's ears, and the firing of a gun, at a short distance, does not disturb the timid Antelope on the high snowy peaks of N.W. America. Snow is doubtless a non-conductor of sound, but there may be also something in the mineral substance of Mouna Kuāh which would effect this.

" Until eleven o'clock, the horizon was beautifully defined on the whole N.W. of the island. The great dryness of the air is evident to the senses, without the assistance of the hygrometer. Walking with my trousers rolled up to my knees, and without shoes, I did not know there were holes in my stockings till I was apprised of them by the scorching heat and pain in my feet, which continued throughout the day; the skin also peeled from my face. While on the summit I experienced violent head-ache, and my eyes became blood-shot, accompanied with stiffness in their lids.

" Were the traveller permitted to express the emotions he feels when placed on such an astonishing part of the earth's surface, cold indeed must his heart be to the great operations of Nature, and still colder towards Nature's God, by whose wisdom and power such wonderful scenes were created, if he could behold them without deep humility and reverential awe. Man feels himself as nothing—as if standing on the verge of another world. The death-like stillness of the place, not an animal nor an insect to be seen—far removed from the din and bustle of the world, impresses on his mind with double force the extreme help-lessness of his condition, an object of pity and compassion, utterly un-worthy to stand in the presence of a great and good, and wise and holy God, and to contemplate the diversified works of His hands !

" I made a small collection of geological specimens, to illustrate the nature and quality of the lavas of this mountain, but being only slightly acquainted with this department of Natural History, I could do no more than gather together such materials as seemed likely to be useful to other and more experienced persons. As night was closing and threatening to be very stormy, we hastened towards the camp, descending nearly by the same way as we came, and finding my guide Honori and the other men all in readiness, we all proceeded to the edge of the woody region, and regained the lodge, highly gratified with the result of this very fatiguing day's excursion. Having brought provision from the hill, we fared well.

" January the 13th.—The rain fell fast all night, and continued, accompanied by a dense mist, this morning, only clearing sufficiently to give us a momentary glimpse of the mountain, covered with snow down to the woody region. We also saw Mouna Roa, which was similarly clothed for a great part of its height. Thankful had we cause to be that this heavy rain, wind, and fog did not come on while we were on the summit, as it would have caused us much inconvenience, and perhaps danger.

" The same weather continuing till the 15th, I packed up all the baggage, and prepared to return. It consisted of several bundles of plants, put into paper and large packages tied up in Coa baskets, which are manufactured from a large and beautiful tree, a species of *Acacia*, of which the timber resembles mahogany, though of a lighter colour, and is beautiful, and said to be durable : also some parcels of geological specimens, my instruments, &c.   At seven A.M. I started, having sent the bearers of my luggage before me, but I had hardly entered the wood by the same path as I took on my ascent, when the rain began to fall, which continued the whole day without the least intermission ; but as there was no place suitable for encamping, and the people, as usual, had straggled away from one another, I resolved to proceed.   The path was in a dreadful state, numerous rivulets overflowed it in many places, and, rising above their banks, rushed in foam through the deep glens, the necessity for crossing which impeded my progress in no slight degree. In the low places the water spread into small lakes, and where the road had a considerable declivity, the rushing torrent which flowed down it, gave rather the appearance of a cascade than a path.   The road was so soft that we repeatedly sunk to the knees, and supported ourselves on a lava block, or the roots of the trees.   Still, violent as was the rain and slippery and dangerous the path, I gathered a truly splendid collection of Ferns, of nearly fifty species, with a few other plants, and some seeds, which were tied up in small bundles, to prevent fermentation, and then protected by fresh Coa bark.   Several beautiful species of Mosses and Lichens were also collected ; and spite of all the disadvantages and fatigue that I underwent, still the magnificence of the scenery commanded my frequent attention, and I repeatedly sate down, in the course of the day, under some huge spreading Tree-Fern, which more resembled an individual of the Pine than the Fern tribe, and contemplated with delight the endless variety of form and structure that adorned the objects around me.   On the higher part of the mountain, I gathered a Fern identical with the *Asplenium viride* of my own native country, a circumstance which gave me inexpressible pleasure, and recalled to my mind many of the happiest scenes of my early life.

" In the evening I reached the saw-mill, when the kind welcome of my mountain-friend, Mr. Mills, together with a rousing fire, soon made me forget the rain and fatigues of the day.   Some of the men had arrived before me, others afterwards, and two did not appear till the following day, for having met with some friends, loaded with meat, they preferred a good supper to a dry bed.   My guide, friend, and interpreter, Honori, an intelligent and well-disposed fellow, arrived at seven, in great dismay, having, in the dark, entered the river a short distance above a chain of cataracts, and to avoid these, he had clung to a rock till extricated by the aid of two active young men.   Though he escaped unhurt, he had been exposed to the wet for nearly ten hours.   A night of constant rain succeeded, but I rested well, and after breakfast, having examined all the packages, we quitted the saw-mill for the bay, and arrived there in the afternoon, the arrangement and preservation of my plants affording me occupation for two or three days.   It was no easy matter to dry specimens and papers during such incessantly rainy weather.   I paid the whole of the sixteen men who had accompanied me, not including Honori and the king's man, at the rate of two dollars, some in money and some in goods : the latter consisted of cotton cloth, combs, scissors and thread, &c.; while to those who had acquitted themselves with willingness and activity, I added a small present in addition. Most of them preferred money, especially the lazy fellows. The whole of the number employed in carrying my baggage and provisions, was five men, which left eleven for the conveyance of their own *Tapas* and food. Nor was this unreasonable, for the quantity of *Poe* which a native will consume in a week, nearly equals his own weight! a dreadful drawback on expedition. Still, though the sixteen persons ate two bullocks in a week, besides what they carried, a threatened scarcity of food compelled me to return rather sooner than I should have done, in order that the Calabashes might be replenished. No people in the world can cram themselves to such a degree as the Sandwich Islanders; their food is, however, of a very light kind, and easy of digestion.

" On the 22nd of January, the air being pleasant, and the sun occasionally visible, I had all my packages assorted by nine A.M., and engaged my old guide, Honori, and nine men to accompany me to the volcano and to Mouna Roa. As usual, there was a formidable display of luggage, consisting of *Tapas*, *Calabashes*, *Poe*, *Taro*, &c., while each individual provided himself with the solace of a staff of sugar-cane, which shortens with the distance, for the pedestrian, when tired and thirsty, sits down and bites off an inch or two from the end of his staff. A friend accompanied me as far as his house on the road, where there is a large church, his kind intention being to give me some provision for the excursion, but as he was a stout person, I soon outstripped him. On leaving the bay, we passed through a fertile spot, consisting of Taro patches in ponds, where the ground is purposely overflowed, and afterwards covered with a deep layer of Fern-leaves to keep it damp. Here were fine groves of *Bread-fruit* and ponds of Mullet and Ava-fish; the scenery is beautiful, being studded with dwellings and little plantations of vegetables and of *Morus papyrifera*,[1] of which there are two kinds, one much whiter than the other. The most striking feature in the vegetation consists in the Tree-Fern, some smaller species of the same tribe, and a curious kind of *Compositae*, like an *Eupatorium*. At about four miles and a half from the bay, we entered the wood, through which there is a tolerably cleared path, the muddy spots being rendered passable by the stems or trunks of Tree-Ferns, laid close together crosswise. They seemed to be the same species as I had observed on the ascent to Mouna Kuāh. About an hour's walk brought us through the wood, and we then crossed another open plain of three miles and a half, at the upper end of which, in a most beautiful situation, stands the church, and close to it the chief's house. Some heavy showers had drenched us through; still, as soon as our friend arrived, and the needful arrangements were made, I started and continued the ascent over a very gently rising ground, in a southerly direction, passing through some delightful country, interspersed with low timber. At night we halted at a house, of which the owner was a very civil person, though remarkably talkative. Four old women were inmates of the same dwelling, one of whom, eighty years of age, with hair white

[1] *Broussonetia papyrifera*, Hillebrand, Fl. Hawaiian Isl. p. 407.

as snow, was engaged in feeding two favourite cats with fish. My little terrier disputed the fare with them, to the no small annoyance of their mistress. A well-looking young female amused me with singing, while she was engaged in the process of cooking a dog on heated stones. I also observed a handsome young man, whose very strong stiff black hair was allowed to grow to a great length on the top of his head, while it was cut close over the ears, and falling down on the back of his head and neck, had all the appearance of a Roman helmet.

" January the 23rd.   This morning the old lady was engaged in feeding a dog with fox-like ears, instead of her cats. She compelled the poor animal to swallow *Poe*, by cramming it into his mouth, and what he put out at the sides, she took up and ate herself ; this she did, as she informed me, by way of fattening the dog for food. A little while before day-break my host went to the door of the lodge, and after calling over some extraordinary words which would seem to set orthography at defiance, a loud grunt in response from under the thick shade of some adjoining Tree-Ferns, was followed by the appearance of a fine large black pig, which coming at his master's call, was forthwith caught and killed for the use of myself and my attendants. The meat was cooked on heated stones, and three men were kindly sent to carry it to the volcano, a distance of twenty-three miles, tied up in the large leaves of Banana and Ti-tree. The morning was deliciously cool and clear, with a light breeze. Immediately on passing through a narrow belt of wood, where the timber was large, and its trunks matted with parasitic Ferns, I arrived at a tract of ground, over which there was but a scanty covering of soil above the lava, interspersed with low bushes and Ferns. Here I beheld one of the grandest scenes imaginable ;—Mouna Roa reared his bold front, covered with snow, far above the region of verdure, while Mouna Kuāh was similarly clothed, to the timber region on the South side, while the summit was cleared of the snow that had fallen on the nights of the 12th and two following days. The district of Hido, ' Byron's Bay,' which I had quitted the previous day, presented, from its great moisture, a truly lovely appearance, contrasting in a striking manner with the country where I then stood, and which extended to the sea, whose surface bore evident signs of having been repeatedly ravaged by volcanic fires. In the distance, to the South-West, the dense black cloud which overhangs the great volcano, attests, amid the otherwise unsullied purity of the sky, the mighty operations at present going on in that immense laboratory. The lava, throughout the whole district, appeared to be of every colour and shape, compact, bluish and black, porous or vesicular, heavy and light. In some places it lies in regular lines and masses, resembling narrow horizontal basaltic columns ; in others, in tortuous forms, or gathered into rugged humps of small elevation ; while, scattered over the whole plain, are numerous extinct, abrupt, generally circular craters, varying in height from one hundred to three hundred feet, and with about an equal diameter at their tops. At the distance of five miles from the volcano, the country is more rugged, the fissures in the ground being both larger and more numerous, and the whole tract covered with gravel and lava, &c., ejected at various periods from the crater. The steam that now arose from the cracks bespoke our near approach to the summit, and at two P.M., I arrived at its northern extremity, where finding it nearly level, and observing that water was not far distant, I chose that spot for my encampment. As, however, the people were not likely to arrive before the evening, I took a walk round the West side, now the most active part of the volcano, and sat down there, not correctly speaking, to enjoy, but to gaze with wonder and amazement on this terrific sight, which inspired the beholder with a fearful pleasure.

" From the descriptions of former visitors, I judge that Mouna Roa must now be in a state of comparative tranquillity. A lake of liquid fire, in extent about a thirteenth part of the whole crater, was boiling with furious agitation ; not constantly, however, for at one time it appeared calm and level, the numerous fiery red streaks on its surface alone attesting its state of ebullition, when again, the red hot lava would dart upwards and boil with terrific grandeur, spouting to a height which, from the distance at which I stood, I calculated to be from forty to seventy feet, when it would dash violently against the black ledge, and then subside again for a few moments. Close by the fire was a chimney, above forty feet high, which occasionally discharges its steam, as if all the steam-engines in the world were concentrated in it. This preceded the tranquil state of the lake, which is situated near the South-West, or smaller end of the crater. In the centre of the Great Crater, a second lake of fire, of circular form, but smaller dimensions, was boiling with equal intensity : the noise was dreadful beyond all description. The people having arrived, Honori last, my tent was pitched twenty yards back from the perpendicular wall of the crater ; and as there was an old hut of Ti-leaves on the intermediate bank, only six feet from the extreme verge, my people soon repaired it for their own use. As the sun sunk behind the western flank of Mouna Roa, the splendour of the scene increased ; but when the nearly full moon rose in a cloudless sky, and shed her silvery brightness on the fiery lake, roaring and boiling in fearful majesty, the spectacle became so commanding, that I lost a fine night for making astronomical observations, by gazing on the volcano, the illumination of which was but little diminished by a thick haze that set in at midnight.

" On Friday, January the 24th, the air was delightfully clear, and I was enabled to take the bearings of the volcano and adjoining objects with great exactness. To the North of the crater are numerous cracks and fissures in the ground, varying in size, form, and depth, some long, some straight, round, or twisted, from whence steam constantly issued, which in two of them is rapidly condensed, and collects in small basins or wells, one of which is situated at the immediate edge of the crater, and the other four hundred and eighty yards to the North of it. The latter, fifteen inches deep, and three feet in diameter, about thirteen feet North of a very large fissure, according to my thermometer, compared with that at Greenwich and at the Royal Society, and found without error, maintains a temperature of 65°. The same instrument, suspended freely in the above-mentioned fissure, ten feet from the surface, expressed by repeated trials, 158°; and an equal temperature was maintained when it was nearly level with the surface. When the Islanders visit this mountain, they invariably carry on their cooking operations at this place. Some pork and a fowl that I had brought, together with Taro-roots and Sweet Potatoes were steamed here to a nicety in twenty-seven minutes, having been tied up in leaves of Banana. On the sulphur bank are many fissures, which continually exhale sulphureous vapours, and form beautiful prisms, those deposited in the inside being the most delicate and varied in figure, encrusting the hollows in masses, both large and small, resembling swallows' nests on the wall of a building. When severed from the rock or ground, they emit a crackling noise by the contraction of the parts in the process of cooling. The great thermometer, placed in the holes, showed the temperature to be 195° 5', after repeated trials which all agreed together, the air being then 71°.

" I had furnished shoes for those persons who should descend into the crater with me, but none of them could walk when so equipped, preferring a mat sole, made of tough leaves, and fastened round the heel and between the toes, which seemed indeed to answer the purpose entirely well. Accompanied by three individuals, I proceeded at one P.M. along the North side, and descended the first ledge over such rugged ground as bespoke a long state of repose, the fissures and flanks being clothed with verdure of considerable size : thence we ascended two hundred feet to the level platform that divides the great and small volcanoes. On the left, a perpendicular rock, three hundred feet above the level, shows the extent of the volcano to have been originally much greater than it is at present. The small crater appears to have enjoyed a long period of tranquillity, for down to the very edge of the crust of the lava, particularly on the East side, there are trees of considerable size, on which I counted from sixty to one hundred and twenty-four annual rings or concentric layers. The lava at the bottom flowed from a spot, nearly equidistant from the great and small craters, both uniting into a river, from forty to seventy yards in breadth, and which appears comparatively recent. A little South of this stream, over a dreadfully rugged bank, I descended the first ledge of the crater, and proceeded for three hundred yards over a level space, composed of ashes, scoriæ, and large stones that have been ejected from the mouth of the volcano. The stream formerly described is the only fluid lava here. Hence, to arrive at the black ledge, is another descent of about two hundred and forty feet, more difficult to be passed than any other, and this brings the traveller to the brink of the black ledge, where a scene of all that is terrific to behold presents itself before his eyes. He sees a vast basin, recently in a state of igneous fusion, now, in cooling, broken up, somewhat in the manner of the great American lakes when the ice gives way, in some places level in large sheets, else-where rolled in tremendous masses, and twisted into a thousand different shapes, sometimes even being filamentose, like fine hair, but all displaying the mighty agency still existing in this immense depository of subter-raneous fire. A most uncomfortable feeling is experienced when the traveller becomes aware that the lava is hollow and faithless beneath his tread. Of all sensations in nature, that produced by earthquakes or volcanic agency is the most alarming : the strongest nerves are unstrung and the most courageous mind feels weakened and unhinged, when exposed to either. How insignificant are the operations of man's hands, taken in their vastest extent, when compared with the magnitude of the works of God !

" On the black ledge, the thermometer, held in the hand, five feet from the ground, indicated a temperature of 89°, and when laid on the lava, if in the sun's says, 115° ; and 112° in the shade ; on the brink of the burning lake, at the South end, it rose to 124°. Over some fissures in the lava, where the smoke was of a greyish rather than a blue tinge, the thermometer stood at 94°. I remained for upwards of two hours in the crater, suffering all the time an intense headache, with my pulse strong and irregular,and my tongue parched, together with other symptoms of fever. The intense heat and sulphurous nature of the ground had corroded my shoes so much, that they barely protected my feet from the hot lava. I ascended out of the crater at the South-west, or small and, over two steep banks of scoriæ and two ledges of rock, and returned by the West side to my tent, having thus walked quite round this mighty crater. The evening was foggy ; I took some cooling medicine, and lay down early to rest.

" Saturday, January the 25th.—I slept profoundly till two A.M., when, as not a speck could be seen on the horizon, and the moon was unusually bright, I rose with the intention of making some lunar observations, but though the thermometer stood at 41°, still the keen mountain-breeze affected me so much, of course mainly owing to the fatigue and heat I had suffered the day before, that I was reluctantly obliged to relinquish the attempt, and being unable to settle again to sleep, I replenished my blazing stock of fuel, and sat gazing on the roaring and agitated state of the crater, where three new fires had burst out since ten o'clock the preceding evening. Poor Honori, my guide, who is a martyr to asthma, was so much affected by their exhalations (for they were on the North bank, just below my tent), that he coughed incessantly the whole night, and complained of cold, though he was wrapped in my best blanket, besides his own tapas and some other articles which he had borrowed from my Woahu man. The latter slept with his head toward the fire, coiled up most luxuriously, and neither cold, heat, nor the roaring of the volcano at all disturbed his repose.

" Leaving the charge of my papers and collections under the special care of one individual, and giving plenty of provision for twelve days to the rest, consisting of one quarter of pork, with *poe* and *taro*, I started for Kapupala soon after eight A.M. The path struck off for two miles in a North-West direction, to avoid the rugged lava and ashes on the West flank of Mouna Roa, still it was indescribably difficult in many places, as the lava rose in great masses, some perpendicular, others lying horizontal, in fact with every variation of form and situation. In other parts the walking was pretty good, over grassy undulating plains, clothed with a healthy [1] sward, and studded here and there with *Maurarii* Trees in full blossom, a beautiful tree, much resembling the English Laburnum. As I withdrew from the volcano in order to obtain a good general view of the country lying South and betwixt me and the sea, I ascertained the western ridge or verge of the volcano to be decidedly the most elevated of the table land : and a narrow valley lies to the West of it. A low ridge runs from the mountain, southward, to the sea, terminating at the South end, in a number of craters, of various form and extent. West of this low ridge between the gentle ascent of grassy ground on Mouna Roa there is a space of five to seven miles in breadth to the Grand Discharge from the Great Volcano, where it falls into the ocean at Kapupala. The present aspect of the crater leads me to think that there has been no *overflowing* of the lava for years : the discharge is evidently from the subterranean vaults below. In 1822, the Islanders say there was a great discharge in this direction. Among the grassy undulating ground are numerous caves, some of great magnitude, from forty to sixty-five feet high, and from thirty to forty feet broad, many of them of great length like gigantic arches, and very rugged. These generally run at right angles with the dome of Mouna Roa and the sea. Some of these natural tunnels may be traced for several miles in length, with occasional holes of different sizes in the roofs, screened sometimes with an overgrowth of large Trees and Ferns, which renders walking highly dangerous. At other places, the tops of the vaults have fallen in for the space of one hundred or even three hundred yards, an occurrence which is attributable to the violent earthquakes that sometimes visit this district, and which, as may be readily imagined from the number of these tunnels, is not well supplied with water. The inhabitants convert these caverns to use in various ways ; employing them occasionally as permanent dwellings, but more frequently as cool retreats where they carry on the process of making native cloth from the bark of the *Mulberry Tree*, or where they fabricate and shelter their canoes from the violent rays of the sun. They are also used for goat-folds and pig-styes, and the fallen-in places, where there is a greater depth of decomposed vegetable matter, are frequently planted with *Tobacco, Indian Corn, Melons*, and other choice plants. At a distance of ten miles North of Kapupala, and near the edge of the path, are some fine caverns, above sixty feet deep. The water, dropping from the top of the vault, collected into small pools below, indicated a temperature of 50°, the air of the cave itself 51°, while in the shade on the outside the thermometer stood at 82°. The interiors of the moist caverns are of a most beautiful appearance ; not only from the singularity of their structure, but because they are delightfully fringed with *Ferns, Mosses*, and *Jungermanniae*, thus holding out to the Botanist a most inviting retreat from the overpowering rays of a tropical sun.

" Arrived at Kapupala, at three P.M., I found that the chief or head man had prepared a house for me, a nice and clean dwelling, with abundance of fine mats, &c., but as near it there stood several large canoes filled with water, containing Mulberry Bark in a state of fermentation,

[1] Probably an errror for ' heathy ' or ' heathery.'—ED.

and highly offensive, as also a large pig-fold, surrounded by a lava-wall, and shaded with large bushes of *Ricinus communis*, altogether forming an unsuitable station for making observations, to say nothing of the din and bustle constantly going on when strangers are present, besides the annoyance from fleas, I caused my tent to be pitched one hundred yards behind the house. The chief would have been better pleased if I had occupied his dwelling, but through Honori, I had this matter explained to his satisfaction. He sent me a fowl, cooked on heated stones underground, some baked Taro, and Sweet Potatoes, together with a calabash full of delicious goat's-milk, poured through the husk of a Cocoa-nut in lieu of a sieve.

" As strangers rarely visit this part of the island, a crowd soon assembled for the evening. The vegetation in this district can hardly be compared with that of Hido, nor are the natives so industrious : they have no fish-ponds, and cultivate little else than Taro, which they call *Dry Taro*, no Bananas, and but little Sugar-cane or other vegetables. Flocks of goats brouse over the hills, while fowls, turkeys, and pigs are numerous, and occupy the same dwellings with their owners.

" Honori, my guide, interpreter, purveyor, and, I may say, friend (for in every department of his omnifarious capacity he is a good sort of fellow), preached to-day, Sunday the 26th, in his own language, to an assembly of both sexes, old and young, nearly two hundred in number, both morning and evening. I did not see him, but from my tent-door I could hear him in the School-house, a low small edifice, expounding and exhorting with much warmth. Having made so bold afterwards as to ask him where he took his text, he readily replied, that he ' chose no text, but had taken occasion to say to the people a few good words concerning Paul when at Rome.' He was evidently well pleased himself with his sermon, and seemed to please his audience also. I visited the school in the interval, when Honori had retired to compose his second sermon, and found the assemblage under the direction of the chief, who appears to be a good man, though far from an apt scholar ; they were reading the second chapter of the Epistle to the Galatians, and proceeded to the third, reading verse and verse, all round. The females were by far the most attentive, and proved themselves the readiest learners. It is most gratifying to see, far beyond the pale of what is called civilization, this proper sanctification of the Lord's Day, not only consisting in a cessation from the ordinary duties, but in reading and reflecting upon the purifying and consolatory doctrines of Christianity. The women were all neatly dressed in the native fashion, except the chief's wife, and some few others who wore very clean garments of calico. The hair was either arranged in curls or braided on the temples, and adorned with tortoise-shell combs of their own making, and chaplets of balsamic flowers, the pea-flowering racemes of the Maurarii-Tree, and feathers, &c. The men were all in the national attire, and looked tolerably well dressed, except a few of the old gentlemen.

" The schoolmaster, a little hump-backed man, about thirty years old, little more than three feet high, with disproportionately long legs, and having a most peculiar cast in his right eye, failed not to prompt and reprove his scholars when necessary, in a remarkably powerful tone of voice, which when he read, produced a trumpet-like sound, resembling the voice of a person bawling into a cask.

"Honori ' had the people called together ' by the sound of a conch-shell, blown by a little imp of a lad, perched on a block of lava, in front of the school-house, when as in the morning, he ' lectured ' on the third chapter of St. John. The congregation was thinner than in the morning, many who lived at a distance having retired to their homes.

"I spent the Monday (January the 27th) in making observations and arranging matters for returning to Mouna Roa : my men cooked a stock of Taro, and I purchased a fine large goat for their use.

"Tuesday, January the 28th.—I hired two guides, the elder of whom a short stout man, was particularly recommended to me by the chief, for his knowledge of the mountain. By profession he is a bird-catcher, going in quest of that particular kind of bird which furnishes the feathers of which the ancient cloaks, used by the natives of these islands, are made. The other guide was a young man. Three volunteers offered to accompany me ; one a very stout, fat dame, apparently about thirty, another not much more than half that age, a really well-looking girl, tall and athletic : but to the first, the bird-catcher gave such an awful account of the perils to be undergone, that both the females finally declined the attempt, and only the third person, a young man, went with me. My original party of ten, besides Honori and the two guides, set out at eight, with, as usual, a terrible array. of Taro, calabashes full of Poe, Sweet Potatoes, dry Poe tied up in Ti-leaves, and goat's flesh, each bearing a pole on his shoulder with a bundle at either end. Of their vegetable food, a Sandwich Islander cannot carry more than a week's consumption, besides what he may pick up on the way. One, whose office it was to convey five quires of paper for me, was so strangely attired, in a double-milled grey great coat, with a spencer of still thicker materials above it, that he lamented to his companions that his load was too great, and begged their help to lift it on his back. I had to show the fellow, who was blind of one eye, the unreasonableness of his grumbling by hanging the parcel, by the cord, on my little finger. He said, ' Ah ! the stranger is strong,' and walked off. Among my attendants was one singular-looking personage, a stripling, who carried a small packet of instruments, and trotted away, arrayed in ' a Cutty-Sark,' of most ' scanty longitude,' the upper portion of which had been once of white, and the lower of red flannel. Honori brought up the rear, with a small telescope slung over his shoulder, and an umbrella, which, owing perhaps to his asthmatic complaint, he never fails to carry with him, both in fair and foul weather. We returned for about a mile and a half along the road that led to the Great Volcano, and then struck off to the left in a small path that wound in a northerly direction up the green grassy flank of Mouna Roa. I soon found that Honori's cough would not allow him to keep up with the rest of the party, so leaving one guide with him, and making the bird-catcher take the lead, I proceeded at a quicker rate. This part of the island is very beautiful; the ground, though hilly, is covered with a tolerably thick coating of soil, which supports a fine sward of Grass, Ferns, climbing plants, and in some places, timber of considerable size, Coa, Tutui, and Mamme trees. Though fallen trees and brushwood occasionally intercepted the path, still it was by no means so difficult as that by which I had ascended Mouna Kuāh. To avoid a woody point of steep ascent, we turned a little eastward, after having travelled about five miles and a half, and passed several deserted dwellings, apparently only intended as the temporary abodes of bird-catchers and sandal-wood-cutters. Calabashes and Pumpkins, with Tobacco, were the only plants that I observed growing near them. At eleven A.M. we came to a small pool of fresh water, collected in the lava, the temperature of which was 55°, here my people halted for a few minutes to smoke. The barometer stood at 26 inch., the air 62°, and the dew point at 58°. The wind was from the South, with a gentle fanning breeze and a clear sky. Hence the path turns North-West, for a mile and half, becoming a little steeper, till it leads to a beautiful circular well, three feet deep, flowing in the lava, its banks fringed with Strawberry Vines, and shaded by an Acacia Tree grove. Here we again rested for half an hour. We might be said here to have ascended above the woody country; the ground became more steep and broken, with a thinner soil and trees of humbler growth, leading towards the South-East ridge of Mouna Roa, which, judging from a distance, appeared the part to which there is the easiest access. I would recommend to any Naturalists who may in future visit this mountain, to have their canteens filled at the well just mentioned, for my guide, trusting to one which existed in a cave further up, and which he was unable to find, declined to provide himself with this indispensable article at the lower well, and we were consequently put to the greatest inconvenience. Among the brushwood was a strong kind of Raspberry-bush, destitute of leaves; the fruit, I am told, is white. At four P.M. we arrived at a place where the lava suddenly became very rugged, and the brushwood low, where we rested and chewed sugar-cane, of which we carried a large supply, and where the guides were anxious to remain all night. As this was not very desirable, since we had no water, I proceeded for an hour longer, to what might be called the Line of Shrubs, and at two miles and a half further on, encamped for the night. We collected some small stems of a heath-like plant, which, with the dried stalks of the same species of *Compositae* which I observed on Mouna Kuāh, afforded a tolerably good fire. The man who carried the provisions did not make his appearance—indeed it is very difficult, except by literally driving them before you, to make the natives keep up with an active traveller. Thus I had to sup upon Taro-roots. Honori, as I expected, did not come up. I had no view of the surrounding country, for the region below, especially over the land, was covered with a thick layer of fleecy mist, and the cloud which always hovers above the great volcano, overhung the horizon and rose into the air, like a great tower. Sunset gave a totally different aspect to the whole, the fleecy clouds changed their hue to a vapoury tint, and the volume of mist above the volcano, which is silvery bright during the prevalence of sunshine, assumed a fiery aspect, and illumined the sky for many miles around. A strong North-West mountain-breeze sprung up, and the stars, especially Canopus and Sirius, shone with unusual brilliancy. Never, even under a tropical sky, did I behold so many stars. Sheltered by a little brush-wood, I lay down on the lava beside the fire, and enjoyed a good night's rest, while my attendants swarmed together in a small cave, which they literally converted into an oven by the immense fire they kindled in it.

" Wednesday, January the 29th.—The morning rose bright and clear, but cold, from the influence of a keen mountain-breeze. As the man who carried the provisions was still missing, the preparation of breakfast occupied but little time, so that, accompanied by the bird-catcher and Cutty-Sark, I started at half-past six for the summit of the mountain, leaving the others to collect fuel and to look for water. Shortly before daybreak the sky was exceedingly clear and beautiful, especially that part of the horizon where the sun rose, and above which the upper limb of his disc was visible like a thread of gold, soon to be quenched in a thick haze, which was extended over the horizon. It were difficult, nay, almost impossible, to describe the beauty of the sky and the glorious scenes of this day. The lava is terrible beyond description, and our track lay over ledges of the roughest kind, in some places glassy and smooth like slag from the furnace, compact and heavy like basalt ; in others, tumbled into enormous mounds, or sunk in deep valleys, or rent into fissures, ridges, and clefts. This was at the verge of the snow—not twenty yards of the whole space could be called level or even. In every direction vast holes or mouths are seen, varying in size, form, and colour, from ten to seventy feet high. The lava that has been vomited forth from these openings presents a truly novel spectacle. From some, and occasionally indeed from the same mouth, the streams may be seen, pressed forward transversely, or in curved segments, while other channels present a floating appearance ; occasionally the circular tortuous masses resemble gigantic cables, or are drawn into cords, or even capillary threads, finer than any silken thread, and carried to a great distance by the wind. The activity of these funnels may be inferred from the quantity of slag lying round them, its size, and the distance to which it has been thrown. Walking was rendered dangerous by the multitude of fissures, many of which are but slightly covered with a thin crust, and everywhere our progress was exceedingly laborious and fatiguing. As we continued to ascend, the cold and fatigue disheartened the Islanders, who required all the encouragement I could give to induce them to proceed. As I took the lead, it was needful for me to look behind me continually, for when once out of sight, they would pop themselves down, and neither rise nor answer to my call. After resting for a few moments at the last station, I proceeded about seven miles further, over a similar kind of formation, till I came to a sort of low ridge, the top of which I gained soon after eleven P.M.,[1] the thermometer indicating 37°, and the sky very clear. This part was of gradual ascent, and its summit might be considered the

---

[1] Should evidently be A.M. not P.M.—ED.

southern part of the dome. The snow became very deep, and the influence of the sun melting its crust, which concealed the sharp points of the lava, was very unfavourable to my progress. From this place to the North towards the centre of the dome, the hill is more flattened. Rested a short time, and a few moments before noon, halted near the highest black shaggy chimney to observe the sun's passage. In recording the following observations, I particularly note the places, in order that future visitors may be able to verify them. To the S.W of this chimney, at the distance of one hundred and seventy yards, stands a knoll of lava, about seventy feet above the gradual rise of the place. The altitude was 104° 52' 45". This observation was made under highly favourable circumstances, on a horizon of mercury, without a roof, it being protected from the wind by a small oil-cloth :—bar. 18° 953' ; therm. 41° ; in the sun's rays 43° 5' ; and when buried in the snow, 31° ; the dew-point at 7° ! ! wind S.W. The summit of this extraordinary mountain is so flat, that from this point no part of the island can be seen, not even the high peaks of Mouna Kuāh, nor the distant horizon of the sea, though the sky was remarkably clear. It is a horizon of itself, and about seven miles in diameter. I ought, ere now, to have said that the bird-catcher's knowledge of the volcano did not rise above the woody region, and now he and my two other followers were unable to proceed further. Leaving these three behind, and accompanied by only Calipio, I went on about two miles and a half, when the Great Terminal Volcano or Cone of Mouna Roa burst on my view : all my attempts to scale the black ledge here were ineffectual, as the fissures in the lava were so much concealed, though not protected by the snow, that the undertaking was accompanied with great danger. Most reluctantly was I obliged to return, without being able to measure accurately its extraordinary depth. From this point I walked along upon the brink of the high ledge, along the East side, to the hump, so to speak, of the mountain, the point which, as seen from Mouna Kuāh, appears the highest. As I stood on the brink of the ledge, the wind whirled up from the cavity with such furious violence that I could hardly keep my footing within twenty paces of it. The circumference of the black ledge of the nearly circular crater, described as nearly as my cir- cumstances would allow me to ascertain, is six miles and a quarter. The ancient crater has an extent of about twenty-four miles. The depth of the ledge, from the highest part (perpendicular station on the East side) by an accurate measurement with a line and plummet, is twelve hundred and seventy feet. It appears to have filled up considerably all round ; that part to the North of the circle seeming to have, at no very remote period, undergone the most violent activity, not by boiling and over- flowing, nor by discharging underground, but by throwing out stones of immense size to the distance of miles around its opening, together with ashes and sand. Terrible chasms exist at the bottom, appearing, in some places, as if the mountain had been rent to its very roots : no termi- nation can be seen to their depth, even when the eye is aided with a good glass, and the sky is clear of smoke, and the sun shining brightly. Fearful indeed must the spectacle have been, when this volcano was in a state of activity. The part to the South of the circle, where the outlet of the lava has evidently been, must have enjoyed a long period of repose. Were it not for the dykes on the West end, which show the extent of the ancient cauldron, and the direction of the lava, together with its proximity to the existing volcano, there is little to arrest the eye of the Naturalist over the greater portion of this huge dome, which is a gigantic mass of slag, scoriæ, and ashes. The barometer remained stationary during the whole period spent on the summit, nor was there any change in the temperature nor in the dew-point to-day. While passing, from eight to nine o'clock, over the ledges of lava of a more compact texture, with small but numerous vesicles, the temperature of the air being 36°, 37°, and the sun shining powerfully, a sweet musical sound was heard, proceeding from the cracks and small fissures, like the faint sound of musical glasses, but having at the same time, a kind of hissing sound, like a swarm of bees. This may perhaps be owing to some great internal fire escaping—or is it rather attributable to the heated air on the surface of the rocks, rarefied by the sun's rays ? In a lower region, this sound might be overlooked, and considered to proceed, by possibility, from the sweet harmony of insects, but in this high altitude it is too powerful and remarkable not to attract attention. Though this day was more tranquil than the 12th, when I ascended Mouna Kuāh, I could perceive a great difference in sound ; I could not now hear half so far as I did on that day when the wind was blowing strong. This might be owing to this mountain being covered with snow, whereas, on the 12th, Mouna Kuāh was clear of it. Near the top I saw one small bird, about the size of a common sparrow, of a light mixed grey colour, with a faintly yellow beak—no other living creature met my view above the woody region. This little creature, which was perched on a block of lava, was so tame as to permit me to catch it with my hand, when I instantly restored it its liberty. I also saw a dead hawk in one of the caves. On the East side of the black ledge of the Great Terminal Crater, is a small conical funnel of scoriæ, the only vent-hole of that substance that I observed in the crater. This mountain appears to be differently formed from Mouna Kuāh ; it seems to be an endless number of layers of lava, from different overflowings of the great crater. In the deep caves at Kapupala, two thousand feet above the level of the sea, the several strata are well defined, and may be accurately traced, varying in thickness with the intensity of the action, and of the discharge that has taken place. Between many of these strata are layers of earth, containing vegetable substances, some from two feet to two feet seven inches in thickness, which bespeak a long state of repose between the periods of activity in the volcano. It is worthy of notice that the thickest strata are generally the lowest, and they become thinner towards the surface. In some places I counted twenty-seven of these layers, horizontal, and preserving the declination of the mountain. In the caves which I explored near my camp, which are from forty to seventy feet deep, thin strata of earth intervene between the successive beds of lava, but none is found nearer the surface than thirteen layers. No trace of animal, shell or fish, could I detect in any of the craters or caves , either in this mountain or Mouna Kuāh. At four P.M. I returned to the centre of the dome, where I found the three men whom I had left all huddling together to keep themselves warm. After collecting a few specimens of lava, no time was to be lost in quitting this dreary and terrific scene. The descent was even more fatiguing, dangerous, and distressing than the ascent had proved, and required great caution in us to escape unhurt ; for the natives, benumbed with cold, could not walk fast. Darkness came on all too quickly, and though the twilight is of considerable duration, I was obliged to halt, as I feared, for the night, in a small cave. Here, though sheltered from the N.W. breeze, which set in more and more strongly as the sun sunk below the horizon, the thermometer fell to 19°, and as I was yet far above the line of vegetation, unable to obtain any materials for a fire, and destitute of clothing except the thin garments soaked in perspiration in which I had travelled all day, and which rendered the cold most intense to my feelings, I ventured, between ten and eleven P.M. to make an effort to proceed to the camp. Never shall I forget the joy I felt when the welcome moon, for whose appearance I had long been watching, first showed herself above the volcano. The singular form which this luminary presented, was most striking. The darkened limb was uppermost, and as I was sitting in darkness, eagerly looking for her appearance on the horizon, I descried a narrow silvery belt, 4° to 5° high, emerging from the lurid fiery cloud of the volcano. This I conceived to be a portion of the light from the fire, but a few moments showed me the lovely moon shining in splendour in a cloudless sky, and casting a guiding beam over my rugged path. Her pale face actually threw a glow of warmth into my whole frame, and I joyfully and thankfully rose to scramble over the rough way, in the solitude of the night, rather than await the approach of day in this comfortless place. Not so thought my followers. The bird-catcher and his two companions would not stir ; so with my trusty man Calipio, who follows me like a shadow, I proceeded in the descent. Of necessity we walked slowly, stepping cautiously from ledge to ledge, but still having exercise enough to excite a genial heat. The splendid constellation of Orion, which had so often attracted my admiring gaze in my own native land, and which had shortly passed the meridian, was my guide. I continued in a South-East direction till two o'clock, when all at once I came to a low place, full of stunted shrubs, of more robust habit, however, than those at the camp. I instantly struck a light, and found by the examination of my barometer that I was nearly five hundred feet below the camp. No response was given to our repeated calls—it was evident that no human being was near, so by the help of the moon's light, we shortly collected plenty of fuel, and kindled a fine fire. No sooner did its warmth and light begin to diffuse themselves over my frame, than I found myself instantly seized with violent pain and inflammation in my eyes, which had been rather painful on the mountain, from the effect of the sun's rays shining on the snow ; a slight discharge of blood from both eyes followed, which gave me some relief, and which proved that the attack was as much attributable to violent fatigue as any other cause

Having tasted neither food nor water since an early hour in the morning, I suffered severely with thirst; still I slept for a few hours, dreaming the while of gurgling cascades, overhung with sparkling rainbows, of which the dewy spray moistened my whole body, while my lips were all the time glued together with thirst, and my parched tongue almost rattled in my mouth. My poor man, Calipio, was also attacked with inflammation in his eyes, and gladly did we hail the approach of day. The sun rose brightly on the morning of Thursday, January 30th, and gilding the snow over which we had passed, showed our way to have been infinitely more rugged and precarious than it had appeared by moon-light. I discovered that by keeping about a mile and a half too much to the East, we had left the camp nearly five hundred feet above our present situation; and returning thither over the rocks, we found Honori engaged in preparing breakfast. He had himself reached the camp about noon on the second day. He gave me a Calabash full of water, with a large piece of ice in it, which refreshed me greatly. A few drops of opium in the eyes afforded instant relief both to Calipio and myself. The man with the provision was here also, so we shortly made a comfortable meal, and immediately after, leaving one man behind with some food for the bird-catcher and his two companions, we prepared to descend, and started at nine A.M. to retrace the path by which we had come. Gratified though one may be at witnessing the wonderful works of God in such a place as the summit of this mountain presents, still it is with thankfulness that we again approach a climate more congenial to our natures, and welcome the habitations of our fellow-men, where we are refreshed with the scent of vegetation, and soothed by the melody of birds. When about three miles below the camp, my three companions of yesterday appeared like mawkins, on the craggy lava, just at the very spot where I had come down. A signal was made them to proceed to the camp, which was seen and obeyed, and we proceeded onwards, collecting a good many plants by the way. Arriving at Strawberry Well, we made a short halt to dine, and ascertained the barometer to be 25° 750'; air 57°, and the well 51°; dew 56°. There were vapoury light clouds in the sky, and a S.W. wind. We arrived at Kapupala at four P.M. The three other men came up at seven, much fatigued, like myself. Bar. at Kapupala at eight P.M. 27° 936'; air 57°; and the sky clear."

APPENDIX III

LETTER FROM THE MISSIONARIES OF HAWAII TO RICHARD CHARLTON, ESQ., HIS BRITANNIC MAJESTY'S CONSUL AT THE SANDWICH ISLANDS [1]

Hido, Hawaii, July 15th, 1834.

" DEAR SIR,—Our hearts almost fail us when we undertake to perform the melancholy duty which devolves upon us, to communicate the painful intelligence of the death of our friend Mr. Douglas, and such particulars as we have been able to gather respecting this distressing providence. The tidings reached us when we were every moment awaiting his arrival, and expecting to greet him with a cordial welcome : but alas ! He whose thoughts and ways are not as our's, saw fit to order it otherwise ; and instead of being permitted to hail the *living* friend, our hearts have been made to bleed while performing the offices of humanity to his mangled corpse. Truly we must say, that the ' ways of the Lord are mysterious, and His judgments past finding out ! ' but it is our unspeakable consolation to know, that those ways are directed by infinite wisdom and mercy, and that though ' clouds and darkness are round about Him, yet righteousness and judgment are the habitation of His throne ! ' But we proceed to lay before you as full information as it is in our power to do at the present time, concerning this distressing event. As Mr. Diell was standing in the door of Mr. Goodrich's house yesterday morning, about eight o'clock, a native came up, and with an expression of countenance which indicated but too faithfully that he was the bearer of sad tidings, inquired for Mr. Goodrich. On seeing him, he communicated the dreadful intelligence, that the body of Mr. Douglas had been found on the mountains, in a pit excavated for the purpose of taking wild cattle, and that he was supposed to have been killed by the bullock which was in the pit, when the animal fell in.[2] Never were our feelings so shocked, nor could we credit the report, till it was painfully confirmed as we proceeded to the beach, whither his body had been conveyed in a canoe, by the native who informed us of his death. As we walked down with the native, and made further inquiries of him, he gave for substance the following relation :—that on the evening of the 13th instant, the natives who brought the body down from the mountain came to his house at Laupashoohoi, about twenty-five or thirty miles distant from Hido, and employed him to bring it to this place in his canoe—the particulars which he learned

---

[1] See Hooker's Comp. Bot. Mag.

[2] It is not quite clear whether the bullock is thought to have fallen in upon Douglas or to have been already in the pit when Douglas fell in, as suggested in the explanation of the plan on page 322.—ED.

from them were as follows :—that Mr. D. left Rohala Point last week, in company with a foreigner (an Englishman), as a guide, and proceeded to cross Mouna Roa on the North side—that on the 12th he dismissed his guide, who cautioned him, on parting, to be very careful lest he should fall into the pits excavated for the purpose mentioned above ; describing them as near the place where the cattle resorted to drink—that soon after Mr. D. had dismissed his guide, he went back a short distance to get some bundle which he had forgotten, and that as he was retracing his steps, at some fatal moment he tumbled into one of the pits in which a bullock had previously fallen—that he there was found dead by these same natives, who, ignorant of the time of his passing, were in pursuit of cattle, and observed a small hole in one end of the covering of the pit. At first they conjectured that a calf had fallen in ; but on further examination they discerned traces of a man's footsteps, and then saw his feet, the rest of the body being covered with dust and rubbish. They went in pursuit of the guide, who returned, shot the beast in the hole, took out the corpse, and hired the natives at the price of four bullocks, which he killed immediately, to convey the body to the sea-shore. He himself accompanied them, and procured the native who related the affair to bring the corpse to this place, promising to come himself immediately, and that he would bring the compass, watch (which was somewhat broken, but still going), some money found in Mr. D.'s pocket, and the little dog, that faithful companion of our departed friend.—Thus far the report of the native, who brought the corpse in his canoe, and who professes to relate the facts to us, as he learned them from the natives who came down from the mountain. We do not stop, at present, to examine how far it is consistent or inconsistent with itself, as we have not the means of making full investigation into the matter. On reaching the canoe, our first care was to have the remains conveyed to some suitable place, where we could take proper care of them, and Mr. Dibble's family being absent, it was determined to carry the body to his house. But what an affecting spectacle was presented, as we removed the bullock's hide in which he had been conveyed !—we will not attempt to describe the agony of feeling which we experienced at that moment : can it be he ? can it be he ? we each exclaimed—can it be the man with whom we parted but a few days before, and who then was borne up with so high spirits and expectations, and whom, but an hour previously, we were fondly anticipating to welcome to our little circle. The answer was but too faithfully contained in the familiar articles of dress—in the features, and in the noble person before us. They were those of our friend. The body, clothes, &c. appeared to be in the same state they were in when taken from the pit : the face was covered with dirt, the hair filled with blood and dust ; the coat, pantaloons, and shirt, considerably torn. The hat was missing. On washing the corpse, we found it in a shocking state : there were ten to twelve gashes on the head—a long one over the left eye, another, rather deep, just above the left temple, and a deep one behind the right ear ; the left cheek-bone appeared to be broken, and also the ribs on the left side. The abdomen was also much bruised, and also the lower parts of the legs. After laying him out, our first thought was to bury him within Mr. Goodrich's premises ; but after we had selected a spot, and commenced clearing away the ground, doubts were suggested by a foreigner who was assisting us, and who has for some time been engaged in taking wild cattle, whether the wounds on the head could have been inflicted by a bullock. Mr. G. said that doubts had similarly arisen in his mind while examining the body. The matter did not seem clear—many parts of the story were left in obscurity. How had Mr. Douglas been left alone—without any guide, foreign or native ?—Where was John, Mr. Diell's coloured man, who left Honolulu with Mr. Diell, and who, on missing a passage with him from Lahaina, embarked with Douglas, as we are informed by the captain of the vessel in which Mr. D. sailed from Lahaina to Rohala Point, and then left the vessel with Mr. D. on the morning of the 9th instant, in order to accompany him across the mountain to Hido ? How was it that Mr. D. should fall into a pit when retracing his steps, after having once passed it in safety ? And if a bullock had already tumbled in, how was it that he did not see the hole necessarily made in its covering ?—These difficulties occurred to our minds, and we deemed it due to the friends of Mr. D., and to the public, whom he had so zealously and so usefully served, that an examination should be made of the body by medical men. The only way by which this could be effected, was by preserving his body, and either sending it to Oahu or keeping it till it could be examined. The former method seemed most advisable ; accordingly we had the contents of the abdomen removed, the cavity filled with salt, and placed in a coffin, which was then filled with salt, and the whole enclosed in a box of brine. Some fears are entertained whether the captain of the native vessel will convey the body : this can be determined in the morning. After the corpse was laid in the coffin, the members of the Mission family and several foreigners assembled at the house of Mr. Dibble, to pay their tribute of respect to the mortal remains of the deceased, and to improve this affecting providence to their own good. Prayers were offered, and a brief address made ; and we trust that the occasion may prove a lasting blessing to all who were present. After the services were concluded, the body was removed to a cool native house, where it was enclosed in the box.

" 16th. As neither the guide nor any natives have arrived, we have employed two foreigners to proceed to the place where the body was received on the sea-shore, with directions to find the persons who discovered it, and go with them to the pit, and after making as full inquiries as possible, to report to us immediately. So far as we can ascertain, the guide is an Englishman, a convict from Botany Bay, who left a vessel at these islands some years ago. He has a wife and one child with him, and to this circumstance in part may be attributed his delay. There are two native vessels in port, besides the one about to sail to-day. By these vessels we shall apprise you of all the information we can obtain, and yet hope that the darkness which involves the subject may be removed. Mr. G. has just returned from the vessel about to sail to-day. The application to convey the remains of Mr. D. to Honolulu will, we fear, prove unsuccessful, as the cargo is already taken in, consisting of wood, canoes, food, &c. It is barely possible that a consent may yet be obtained ; but if not, you must be so kind as to dictate what course is to be pursued. Should you deem it advisable to come up in person, we think that the body will be in such a state of preservation, as will admit of its being examined upon your arrival. Meanwhile, we shall take all possible pains to procure information. The principal part of Mr. D.'s baggage, his trunks, instruments, &c. are in Mr. Goodrich's possession, who will take care of them, subject to your order.—*Three o'clock p.m.* Edward Gurney, the Englishman spoken of before, has arrived, and our minds are greatly relieved, as to the probable way in which the fatal event was brought about. He states that on the 12th instant, about ten minutes before six in the morning, Mr. D. arrived at his house on the mountain, and wished him to point out the road, and go a short distance with him. Mr. D. was then alone, but said that his man had gone out the day before (this man was probably John, Mr. Diell's coloured man). After taking breakfast, Ned accompanied Mr. D. about three quarters of a mile, and after directing him in the path, and warning him of the traps, went on about half a mile further with him. Mr. D. then dismissed him, after expressing an anxious wish to reach Hido by evening, thinking he could find out the way himself. Just before Ned left him, he warned him particularly of three bullock-traps, about two miles and a half a-head, two of them lying directly in the road, the other on one side, as exhibited in the following rude sketch [p. 322].

" Ned then parted with Mr. D. and went back to skin some bullocks which he had previously killed. About eleven o'clock, two natives came in pursuit of him, and said that the European was dead ; that they had found him in the pit where a bullock was. They mentioned, that as they were approaching this pit, one of them, observing some of the clothing on the side, exclaimed *Lole*, but in a moment afterwards discovered Mr. D. in the cave, trampled under the beast's feet. They immediately hastened back for Ned, who, leaving his work, ran into the house for a musket, ball, and hide ; and on arriving at the pit, found the bullock standing upon poor Douglas' body, which was lying on the right side. He shot the animal, and after drawing it to one side of the pit, succeeded in extricating the corpse. Douglas' cane was there, but not his dog and bundle : Ned knowing that he had the latter with him, asked for it. After a few moments' search, the dog was heard to bark, at a little distance a-head on the road to Hido. On coming up to the spot, indicated by No. 4, the dog and bundle were found. On further scrutiny, it appeared that Mr. D. had stopped for a moment and looked at the empty pit, No. 1,—and also at that where the cow was ; and that after proceeding about fifteen fathoms up the hill, he had laid down his bundle and returned to the side of the pit where the bullock was entrapped, No. 3, and which was situated on the side of the pond opposite to that along which the road runs ; and that whilst looking in, by making a false step, or some other fatal accident, he fell into the power of the infuriated animal, which speedily executed the work of death. The body was covered in part with stones, which probably prevented its being entirely crushed. After removing the corpse, Ned took charge of the dog and bundle, also of his watch and chronometer (which is injured in some way), his pocket compass, keys, and money, and after hiring the natives to convey the body to the shore, a distance of about twenty-seven miles, came directly to this place. This narrative clears up many of the difficulties which rested upon the whole affair, and perhaps affords a satisfactory account of the manner in which Mr. D. met with his awful death. We presume that it would be agreeable to you that the body should be sent down, and as the vessel is still delayed by a calm, we hope to receive a favourable

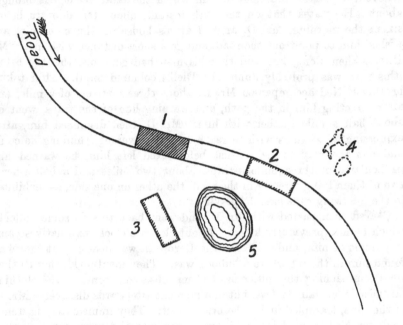

1. Trap empty, covered.
2. Ditto, cow in, open.
3. Ditto, bullock in, open.
4. The place where Mr. Douglas' dog and bundle were found.
5. Water.

answer from the captain. If we should not, it may perhaps be well to inter the body, which can easily be disinterred for exami tion, if desirable.

   " We have thus, dear Sir, endeavoured to furnish you with all the particulars we have been able to gather concerning this distressing event. It is no common death which has thus called forth our tears and sympathies : it presents a most affecting comment on the truth, that ' in the midst of life we are in death ! ' How forcible then is the admonition to all of us, whose privilege it was to be acquainted with him who is thus snatched from us, to ' prepare to meet our God,' ' for the Son of Man cometh at an hour that we know not of.' You will be pleased, dear Sir, to accept for yourself and family, the expression of our kindest sympathies under this afflicting dispensation, and allow us to subscribe ourselves, with sincere regard, your friends and obedient servants,

JOSEPH GOODRICH,
JOHN DIELL.

"P.S. The bearer, Mr. Martin, will take charge of the little dog. There are several matters of expenses, incurred for conveying the body to this place, paying the natives, &c., which Mr. Goodrich will meet, so far as can be done, with the clothes, &c.—of these and of Mr. D.'s other things, he will present a full statement."

COPY OF A LETTER FROM MR. CHARLTON TO JAMES BANDINEL, ESQ.

Woahoo, August 6th, 1834.

"My dear Sir,—It has devolved on me to inform you of the melancholy death of our friend, poor Douglas. On his arrival at this island from the Columbia River, he took the first opportunity of visiting Hawaii, where he remained for some time, with great satisfaction to himself, and usefulness to the public. After his return to this island, he suffered much from rheumatism; but on the 3rd ultimo,[1] finding himself quite recovered, he re-embarked for Hawaii. On the 19th ult. I received the accompanying letter from Messrs. Diell and Goodrich, two gentlemen belonging to the Mission: from it you will learn the particulars relative to his melancholy fate. On the 3rd instant, the body was brought here in an American vessel. I immediately had it examined by the medical gentlemen, who gave it as their opinion that the several wounds were inflicted by the bullock. I assure you that I scarcely ever received such a shock in my life. On opening the coffin, the features of our poor friend were easily traced, but mangled in a shocking manner, and in a most offensive state. The next day, I had his remains deposited in their last resting-place; the funeral was attended by Captain Seymour and several of the officers of His Majesty's Ship Challenger, and the whole of the foreign Residents. I have caused his grave to be built over with brick, and perhaps his friends may send a stone to be placed (with an inscription) upon it. As I am about to embark in the Challenger to-morrow for Otaheite, I have left all his effects in the hands of my friend, Mr. Rooke, with a request to sell his clothing, and forward his collections, books, papers, and instruments, to the Secretary of the Horticultural Society. One of his chronometers, reflecting circle, and dipping needle, are on board the Challenger, in charge of Capt. Seymour. As I do not know the address of the friends or relations of Mr. Douglas, I shall feel very much obliged to you to forward the copy of Messrs. Goodrich's and Diell's letter to them.

I remain, my dear Sir,
Yours, &c.
RICHARD CHARLTON."

[1] July 3rd is evidently intended.—ED.

APPENDIX IV

A CALIFORNIAN newspaper of 1856, under the heading of 'The Sandwich Islands,' says:

"On a reçu de San Francisco un monument en marbre blanc, érigé par M. Julius L. Brenchley à la mémoire d'un illustre voyageur, l'infortuné David Douglas, qui mourut en 1834 au pied du Maunakea, dans l'île de Havaii, assassiné, suivant les uns, par un convict échappé de Botany Bay ; massacré, suivant les autres, par un bœuf sauvage et furieux. Ce monument, qui fait honneur au patriotisme et à la générosité dont M. Brenchley a laissé tant de traces dans nos îles, porte l'inscription suivante :—

Hîc jacet
D. DAVID DOUGLAS,
Scotiâ, anno 1799, natus ;
Qui,
Indefessus viator,
A Londinensi Regiâ Societate Horticulturali missus,
In Havaii saltibus
Die 12ᵃ Julii, A.D. 1834,
Victima scientiae
Interiit.
'Sunt lacrymae rerum et mentem mortalia tangunt.'—Virg.

"Douglas est enterré dans le Cimetière de la grande église à Honolulu. Sa tombe, qui était confondue avec celles de kanaks obscurs, va recevoir enfin une distinction méritée et trop longtemps attendue. "

APPENDIX V

LIST OF PAPERS WRITTEN BY DAVID DOUGLAS.

1. An Account of a new Species of *Pinus*, native of California.—*Trans. Linn. Soc.* xv. 1827, pp. 497–500.

2. Observations on the *Vultur californianus* of Shaw.—*Zool. Journ.* iv. 1829, pp. 328–330.

3. Observations on two undescribed Species of North American Mammalia (*Cervus leucurus* et *Ovis californianus*).—*Zool. Journ.* iv. 1829, pp. 330–332.

4. An Account of the Species of *Calochortus*; a Genus of American Plants. [1828].—*Trans. Hort. Soc.* vii. (1830) pp. 275–280.

5. An Account of some new, and little known Species of the Genus *Ribes*. [1829].—*Trans. Hort. Soc.* vii. (1830) pp. 508–518.

6. Observations on some Species of the Genera *Tetrao* and *Ortyx*, natives of North America, with descriptions of four new species of the former, and two of the latter genus [1828].—*Trans. Linn. Soc.* xvi. 1833, pp. 133–150.

7. Description of a new Species of the Genus *Pinus* (*P. Sabiniana*) [1832].—*Trans. Linn. Soc.* xvi. 1833, pp. 747–750.

8. Volcanoes in the Sandwich Islands.—*Geogr. Soc. Journ.* iv. 1834, pp. 333–343.

APPENDIX VI

PLANTS INTRODUCED BY DAVID DOUGLAS DURING THE YEARS 1826-34.

Names in black type are those now used by the best authorities. Names in italic type have been superseded by those which follow them; the authority for the change being appended to the black-letter names; as *Amelanchier florida*, Lindl. is referred by S. Watson to **Amelanchier alnifolia**; and so in all similar cases throughout this volume.

Abronia mellifera, Dougl. in *Hook. Bot. Mag.* t. 2879—Northern California.

Acer circinnatum, Pursh, *Fl. Am. Sept.* i. p. 267—North-West America.

A. macrophyllum, Pursh, *l.c.*; *Hook. Fl. Bor. Am.* i. p. 112, t. 38— North-West America.

Amelanchier florida, Lindl. *Bot. Reg.* t. 1589—North-West America.

= **A. alnifolia,** S. Wats. *Bibl. Ind. N. Am. Bot.* p. 272.

Anemone Hudsoniana, Richards. in *Frankl. 1st Journ.* ed. II. App. p. 22 —North America.

= **A. multifida,** S. Wats. *Bibl. Ind. N. Am. Bot.* p. 4.

Antirrhinum glandulosum, Lindl. *Bot. Reg.* t. 1893—California.

Arbutus procera, Dougl. ex Lindl. *Bot. Reg.* t. 1753—North-West America.

= **A. Menziesii,** A. Gray, *Syn. Fl. N. Am.* ii. i. p. 27.

A. tomentosa, Pursh, *Fl. Am. Sept.* i. p. 282—North-West America.

= **Arctostaphylos tomentosa,** A. Gray, *Syn. Fl. N. Am.* ii. i. p. 28.

Astragalus succulentus, Richards. in *Frankl. 1st Journ.* App. p. 746; Lindl. *Bot. Reg.* t. 1324—North America.

= **A. caryocarpus,** S. Wats. *Bibl. Ind. N. Am. Bot.* p. 192.

Audibertia incana, Benth. in *Bot. Reg.* t. 1469—North-West America.

Bartonia aurea, Lindl. in *Bot. Reg.* t. 1831—California.

= **Mentzelia Lindleyi,** Torr & Gray, *Fl. N. Am.* i. p. 533.

B. conferta, See Hook. *Comp. Bot. Mag.* ii. (1836) p. 153. *quid?*

Benthamia lycopsoides, Lindl. ex A. DC. *Prod.* x. p. 118—California.

= **Amsinckia lycopsoides,** A. Gray, *Syn. Fl. N. Am.* ii. i. p. 198.

Berberis Aquifolium, Pursh, *Fl. Am. Sept.* i. p. 219, t. 4; Lindl. *Bot. Reg.* t. 1425—North-West America.

B. glumacea, Spreng. *Syst.* ii. p. 120; Lindl. *Bot. Reg.* t. 1426—River Columbia.

= **B. nervosa,** S. Wats. *Bibl. Ind. N. Am. Bot.* p. 34.

Brassavola nodosa, Lindl. *Gen. et Sp. Orch.* p. 114; *Bot. Reg.* t. 1465— Rio.

Brodiaea congesta, Sm. in *Trans. Linn. Soc.* x. (1811) p. 3, t. 1—North-West America.

B. grandiflora, Sm. *l.c.* p. 2; Lindl. *Bot. Reg.* t. 1183 — North-West America.

Calandrinia speciosa, Lindl. *Bot. Reg.* t. 1598—Northern California.
  = **C. Menziesii,** S. Wats. *Bibl. Ind. N. Am. Bot.* p. 116.
Calliprora lutea, Lindl. *Bot. Reg.* t. 1590—North-West America.
  = **Brodiaea ixioides,** S. Wats. in *Proc. Am. Acad.* xiv. p. 238.
Calochortus luteus, Dougl.; *Bot. Reg.* t. 1567—California.
C. macrocarpus, Dougl. in *Trans. Hort. Soc.* vii. (1830) p. 276, t. 8; Lindl. *Bot. Reg.* t. 1152—North-West America.
C. splendens, Dougl. ex Benth. in *Trans. Hort. Soc.* Ser. II. i. p. 411, t. 15, f. 1; *Bot. Reg.* t. 1676—California.
C. venustus, Dougl. ex Benth. *l.c.* p. 412, t. 15, f. 3; *Bot. Reg.* t. 1669—California.
Camassia esculenta, Lindl. *Bot. Reg.* t. 1486—North-West America.
Caprifolium ciliosum, Pursh, *Fl. Am. Sept.* i. p. 160—North-West America.
  = **Lonicera ciliosa,** A. Gray, *Syn. Fl. N. Am.* i. ii. p. 16.
C. Douglasii, Lindl. in *Trans. Hort. Soc.* vii. (1830) p. 244—North America.
  = **Lonicera hirsuta,** A. Gray, *l.c.* p. 17.
C. hispidulum, Dougl. ex Lindl. *Bot. Reg.* t. 1761—North-West America.
  = **Lonicera hispidula,** A. Gray, *l.c.* p. 18.
C. occidentale, Lindl. *Bot. Reg.* t. 1457—North-West America.
  = **Lonicera ciliosa,** A. Gray, *Syn. Fl. N. Am.* i. ii. p. 16.
Castilleja coccinea, Lindl. *Bot. Reg.* t. 1136—North-West America.
  = **C. parviflora,** A. Gray, *l.c.* ii. i. p. 296.
Chelone centranthifolia, Benth.; *Bot. Reg.* t. 1737—California and Western Arizona.
  = **Pentstemon centranthifolius,** A. Gray, *Syn. Fl. N. Am.* ii. i. p. 264.
C. nemorosa, Dougl. ex Lindl. *Bot. Reg.* t. 1211—North-West America.
Chryseis compacta, Lindl. *Bot. Reg.* t. 1948—California.
  = **Eschscholzia californica,** S. Wats. *Bibl. Ind. N. Am. Bot.* p. 41.
C. crocea, Lindl. *l.c.*—California.
  = **Eschscholzia californica,** S. Wats. *l.c.*
Clarkia elegans, Dougl. in Lindl. *Bot. Reg.* t. 1575—North-West America.
C. gauroides, Dougl. ex Sweet, *Brit. Flow. Gard.* Ser. II. t. 379—California.
  = **C. rhomboidea,** S. Wats. *Bibl. Ind. N. Am. Bot.* p. 364.
C. pulchella, Pursh, *Fl. Am. Sept.* i. p. 260, t. 11; Lindl. *Bot. Reg.* t. 1100 —North-West America.
Clematis virginiana, Linn. *Cent.* i. p. 15; *Amoen. Acad.* iv. p. 275—North America.
Clintonia elegans, Dougl. ex Lindl. *Bot. Reg.* t. 1241—North-West America.
  = **Downingia elegans,** A. Gray, *Syn. Fl. N. Am.* ii. i. p. 8.
C. pulchella, Lindl., *l.c.* t. 1909—California.
  = **Downingia pulchella,** A. Gray, *l.c.* p. 9.

Collinsia bicolor, Benth. in *Trans. Hort. Soc.* N.S. i. p. 480 ; *Bot. Reg.* t. 1734—California.

C. grandiflora, Dougl. ex Lindl. *Bot. Reg.* sub t. 1166 et t. 1107—North-West America.

C. parviflora, Lindl. *Bot. Reg.* t. 1082—North-West America.

Collomia bellidifolia, Dougl. ex Hook. *Fl. Bor. Am.* ii. p. 76—North-West America.

= **C. gracilis,** Hook, *l.c.*

C. gracilis, Dougl. ex Benth. in *Bot. Reg.* sub t. 1622—North-West America.

C. grandiflora, Dougl. ex Lindl. *Bot. Reg.* t. 1174 — North-West America.

C. linearis, Nutt. *Gen. Am.* i. p. 126 ; Lindl. *Bot. Reg.* t. 1166—North-West America.

C. *pinnatifida,* See Hook. *Comp. Bot. Mag.* ii. (1836) p. 140. *quid ?*

Coreopsis Atkinsoniana, Dougl. ex Lindl. *Bot. Reg.* t. 1376—North-West America.

Cornus alba, quid ? Lam. *Encyc.* ii. p. 115 *partim* (and American authors) —North America.

? = **C. stolonifera,** Torr. & Gray, *Fl. N. Am.* i. p. 650.

Crataegus Douglasii, Lindl. *Bot. Reg.* t. 1810—North-West America.

Cyclobothra alba, Benth. in *Trans. Hort. Soc.* Ser. II. i. p. 413, t. 14, f. 3 —California.

= **Calochortus albus,** Baker, in *Journ. Linn. Soc.* xiv. p. 304.

C. pulchella, Benth., *l.c.* p. 412, t. 14, f. 1—California.

= **Calochortus pulchellus,** Baker, *l.c.* p. 303.

C. pusilla, See *Trans. Hort. Soc.* Ser. II. ii. p. 376. *quid ?*

Delphinium Menziesii, DC. *Syst.* i. p. 355 ; Lindl. *Bot. Reg.* t. 1192— North-West America.

Dendromecon rigidum, Benth. *Bot. Mag.* t. 5134—California.

Diplopappus incanus, Lindl. *Bot. Reg.* t. 1693—California.

= **Aster canescens** var. **viscosus,** A. Gray, *Syn. Fl. N. Am.* i. II. p. 206.

Donia villosa, See Hook, *Comp. Bot. Mag.* ii. (1836) p. 141. *quid ?*

Douglasia nivalis, Lindl. ; *Bot. Reg.* t. 1886—Rocky Mountains.

Epilobium minimum, See Hook. *Comp. Bot. Mag.* ii. (1836) p. 141. Is this a slip of the pen for **E. minutum,** Hook. *Fl. Bor. Am.* i. p. 207 ?—North-West America

Eriogonum compositum, Dougl. ex Benth. in *Bot. Reg.* t. 1774—North-West America.

E. nudum, Dougl. ex Benth. in *Trans. Linn. Soc.* xvii. (1837) p. 143— North-West America.

Eriophyllum caespitosum, Dougl. ex Lindl. *Bot. Reg.* t. 1167—North-West America.

Erythronium grandiflorum, Pursh, *Fl. Am. Sept.* i. p. 231 ; Lindl. *Bot. Reg.* t. 1786—North-West America.

Eschscholzia caespitosa, Benth. in *Trans. Hort. Soc.* Ser. II. i. (1835)
p. 408—California.
= **E. californica** var. **caespitosa,** S. Wats. *Bibl. Ind. N. Am. Bot.*
p. 42.
E. californica, Lindl. *Bot. Reg.* t. 1168—California.
This is **E. californica** var. **Douglasii** according to S. Wats. *Bibl. Ind.
N. Am. Bot.* p. 42.
E. crocea, Benth.; *Bot. Reg.* t. 1677—California.
= **E. californica,** S. Wats. *Bibl. Ind. N. Am. Bot.* p. 41.
E. hypecoides, Benth. in *Trans. Hort. Soc.* Ser. II. i. (1835) p. 408—
California.
= **E. californica** var. **hypecoides,** S. Wats. *l.c.* 42.
E. tenuifolia, Benth.; *Bot. Mag.* t. 4812—California.
= **E. californica** var. **caespitosa,** S. Wats. *l.c.* 42.
Eutoca divaricata, Benth. in *Trans. Linn. Soc.* xvii. (1837) p. 278 ; Lindl.
Bot. Reg. t. 1784—California.
= **Phacelia divaricata,** A. Gray, *Syn. Fl. N. Am.* ii. I. p. 168.
E. multiflora, Dougl. ex Lehm. *Pugill.* ii. p. 19 ; Lindl. *Bot. Reg.* t. 1180
—North-West America.
= **Phacelia Menziesii,** A. Gray, *l.c.* p. 166.
E. viscida, Benth. in. *Bot. Reg.* t. 1808—California.
= **Phacelia viscida,** A. Gray, *l.c.* p. 163.

Gaillardia aristata, Pursh, *Fl. Am. Sept.* ii. p. 573 ; Lindl. *Bot. Reg.*
t. 1186—North America.
Garrya elliptica, Dougl. ex Lindl. *Bot. Reg.* t. 1686—California.
Gaultheria Shallon, Pursh, *Fl. Am. Sept.* i. p. 283, t. 12 ; Hook. *Bot.
Mag.* t. 2843—North-West America.
Geranium carolinianum, Linn. *Sp. Pl.* p. 682 ; *Cav. Diss.* p. 206, t. 84,
f. 1, t. 124, f. 2—North America.
Gesneria Douglasii, Lindl. in *Trans. Hort. Soc.* vii. (1826), p. 62 ; *Bot.
Reg.* t. 1110—Brazil.
Gilia achilleaefolia, Benth. in *Bot. Reg.* sub t. 1622 ; Hook. *Bot. Mag.*
t. 5939—California.
G. achilleaefolia, Lindl. *Bot. Reg.* t. 1682—California.
= **G. multicaulis,** A. Gray, *Syn. Fl. N. Am.* ii. I. p. 147.
G. capitata, Sims, *Bot. Mag.* t. 2698—North-West America.
G. coronopifolia, Pers. ; *Bot. Reg.* t. 1691—North-West America.
G. pharnaceoides, Benth. in *Bot. Reg.* sub t. 1622 ; Hook. *Fl. Bor. Am.*
ii. p. 74, t. 161—California.
= **G. liniflora,** var. **pharnaceoides,** A. Gray, *Syn. Fl. N. Am.* ii. I. p. 137.
G. pungens, Benth. in DC. *Prod.* ix. p. 316—North-West America.
G. splendens, See Hook. *Comp. Bot. Mag.* ii. (1836) p. 141. *quid ?*
G. tenuiflora, Benth.; *Bot. Reg.* t. 1888—California.
G. tenuifolia, err. typ. See Hook. *Comp. Bot. Mag.* ii. (1836) p. 153.
= **G. tenuiflora.**
G. tricolor, Benth. : in *Bot. Reg.* sub t. 1622 ; Lindl. *Bot. Reg.* t. 1704—
California.

Godetia densiflora, See *Trans. Hort. Soc.* Ser. II. ii. p. 377—California
? = **Oenothera densiflora.**

G. lepida, Lindl. *Bot. Reg.* t. 1849—California.
= **Oenothera lepida,** Torr. & Gray, *Fl. N. Am.* i. p. 504.

G. rubicunda, Lindl. *Bot. Reg.* t. 1856—California.
= **Oenothera amoena,** S. Wats. *Bibl. Ind. N. Am. Bot.* p. 371.

G. vinosa, Lindl. *Bot. Reg.* t. 1880—California.
= **Oenothera amoena,** S. Wats. *Bibl. Ind. N. Am. Bot.* p. 371.

Helianthus lenticularis, Dougl. in *Bot. Reg.* t. 1265—North-West America.
= **H. annuus,** A. Gray, *Syn. Fl. N. Am.* i. ii. p. 272.

Helonias tenax, Pursh, *Fl. Am. Sept.* i. p. 243—California.
= **Xerophyllum asphodeloides** var. Baker, in *Journ. Linn. Soc.* xvii.
p. 467.

Hesperoscordum lacteum, Lindl. *Bot. Reg.* t. 1639—California.
= **Brodiaea lactea,** S. Wats. in *Proc. Am. Acad.* xiv. p. 238.

Heuchera cylindracea, Lindl. *Bot. Reg.* t. 1924 = *seq.*

H. cylindrica, Dougl. in *Hook. Fl. Bor. Am.* i. p. 236—North-West
America.

H. micrantha, Dougl. in *Lindl. Bot. Reg.* t. 1302—North-West America.

Horkelia congesta, Dougl. in *Hook. Bot. Mag.* t. 2880—California.
= **Potentilla congesta,** *Ind. Kew.* fasc. ii. p. 1174.

H. fusca, Lindl. *Bot. Reg.* t. 1997—California.

Hosackia bicolor, Dougl. ex Benth. in *Lindl. Bot. Reg.* t. 1257—North-
West America.

H. stolonifera, Lindl. *Bot. Reg.* t. 1977—California.
= **H. crassifolia,** S. Wats. *Bibl. Ind. N. Am. Bot.* p. 225.

Hyssopus urticifolius, Dougl. ex Benth. *l.c.* sub t. 1282—North-West
America.
= **Lophanthus urticifolius,** Benth. *l.c.* sub t. 1282.

Ipomopsis elegans, Lindl. *Bot. Reg.* t. 1281—North-West America.
= **Gilia aggregata,** A. Gray, *Syn. Fl. N. Am.* ii. i. p. 145.

Iris tenax, Dougl. ex Lindl. *Bot. Reg.* t. 1218—North-West America.

Lasthenia californica, DC.; *Bot. Reg.* t. 1823—California.
= **L. glabrata,** A. Gray, *Syn. Fl. N. Am.* i. ii. p. 324.

L. glabrata, Lindl. *Bot. Reg.* t. 1780—California.

Lathyrus californicus, Dougl. ex Lindl. *l.c.* t. 1144—North-West America.
= **L. maritimus,** S. Wats. *Bibl. Ind. N. Am. Bot.* p. 229.

Leptosiphon androsaceus, Benth.; *Bot. Mag.* t. 3491—California.
= **Gilia androsacea,** A. Gray, *Syn. Fl. N. Am.* ii. i. p. 139.

L. densiflorus, Benth.; *Bot. Reg.* t. 1725—California.
= **Gilia densiflora,** A. Gray, *l.c.* p. 139.

Lespedeza capitata, Michx. *Fl. Bor. Am.* ii. p. 71—North America.

Lilium pudicum, Pursh, *Fl. Am. Sept.* i. p. 228, t. 8—North-West
America.
= **Fritillaria pudica,** Baker in *Journ. Linn. Soc.* xiv. p. 267.

Limnanthes Douglasii, R. Br.; *Bot. Reg.* t. 1673—California.

Linum Lewisii, Pursh, *Fl. Am. Sept.* i. p. 210—North America.

= **L. perenne**, S. Wats. *Bibl. Ind. N. Am. Bot.* p. 146.

L. sibiricum, see Hook. *Comp. Bot. Mag.* ii. (1836) p. 141. This undoubtedly refers to *L. sibiricum* var. *Lewisii*, Lindl. *Bot. Reg.* t. 1163. —North-West America.

= **L. perenne**, S. Wats. *l.c.* p. 147.

Lupinus albifrons, Benth. in *Trans. Hort. Soc.* Ser. II. i. (1835) p. 410 ; Lindl. *Bot. Reg.* t. 1642—California.

= **L. Chamissonis**, S. Wats. *l.c.* p. 236.

L. arbustus, Dougl. ex Lindl. *Bot. Reg.* t. 1230—North California and Fort Vancouver.

= **L. laxiflorus**, S. Wats. *l.c.* p. 237.

L. aridus, Dougl. in Lindl. *Bot. Reg.* t. 1242—North-West America.

L. bicolor, Lindl. *Bot. Reg.* t. 1109—River Columbia.

= **L. micranthus** var. **bicolor**, S. Wats. *l.c.* p. 238.

L. densiflorus, Benth. in *Trans. Hort. Soc.* Ser. II. i. (1835) p. 410 ; Lindl. *Bot. Reg.* t. 1689—California.

L. flexuosus, Lindl. ex Agardh, *Syn. Gen. Lupin.* p. 34—North-West America.

L. grandiflorus, see Hook. *Comp. Bot. Mag.* ii. (1836) p. 141. Evidently a slip for *L. grandifolius*, Lindl. ex Agardh, *Syn. Gen. Lupin.* p. 18—California.

= **L. polyphyllus** var. **grandifolius**, Torr. & Gray, *Fl. N. Am.* i. p. 375.

L. hirsutissimus, Benth. in *Trans. Hort. Soc.* Ser. II. i. (1835) p. 411 —California.

L. latifolius, Agardh ; *Bot. Reg.* t. 1891—California.

= **L. rivularis** var. **latifolius**, S. Wats. *Bibl. Ind. N. Am. Bot.* p. 240.

L. laxiflorus, Dougl. ex Lindl. *Bot. Reg.* t. 1140—River Columbia.

L. lepidus, Dougl. ex Lindl. *Bot. Reg.* t. 1149—North-West America.

L. leptophyllus, Benth. in *Trans. Hort. Soc.* Ser. II. i. (1835) p. 409 ; Lindl. *Bot. Reg.* t. 1670—California.

L. leucophyllus, Dougl. *l.c.* t. 1124—North-West America.

L. littoralis, Dougl. *l.c.* t. 1198—North-West America.

L. lucidus, Hook. *Comp. Bot. Mag.* ii. (1836) p. 141. *quid ?*

L. micranthus, Dougl. ex Lindl. *Bot. Reg.* t. 1251—North-West America.

L. nanus, Dougl. ex Benth. in *Trans. Hort. Soc.* Ser. II. i. (1835) p. 409 ; Lindl. *Bot. Reg.* t. 1705—California.

L. ornatus, Dougl. ex Lindl. *Bot. Reg.* t. 1216—North-West America.

L. plumosus, Dougl. ex Lindl. *Bot. Reg.* t. 1217—North-West America.

= **L. leucophyllus**, S. Wats. *Bibl. Ind. N. Am. Bot.* p. 238.

L. polyphyllus, Lindl. *Bot. Reg.* t. 1096—California.

L. polyphyllus var. **albiflorus**, Lindl. *Bot. Reg.* t. 1377—North-West America.

L. rivularis, Dougl. ex Lindl. *Bot. Reg.* t. 1595—California.

L. Sabinianus, Dougl. ex Lindl. *Bot. Reg.* t. 1435—North-West America.

= **L. Sabinii**, Dougl. ex Hook. *Fl. Bor. Am.* i. p. 166.

Lupinus succulentus. The first publication of this name is in *Hook. Comp.*
Bot. Mag. ii. (1836) p. 141, but it is thus referred to by S. Watson.
L. succulentus, Dougl. ex C. Koch, *Wochenschr.* iv. (1861) p. 277.

= **L. densiflorus,** S. Wats. *Bibl. Ind. N. Am. Bot.* p. 236.

L. sulphureus, Dougl. in Hook. *Fl. Bor. Am.* i. p. 166—North-West
America.

L. tristis, Hook. *Comp. Bot. Mag.* ii. (1836) p. 141. *quid?*

L. versicolor, Lindl. *Bot. Reg.* t. 1979—California.

= **L. littoralis,** S. Wats. *Bibl. Ind. N. Am. Bot.* p. 238.

Madia elegans, D. Don, in *Lindl. Bot. Reg.* t. 1458—North-West America.

M. floribunda, *Trans. Hort. Soc.* Ser. II. ii. p. 377. *quid?*

Malva Munroana, Dougl. ex Lindl. *Bot. Reg.* t. 1306—North-West America.

= **Malvastrum Munroanum,** S. Wats. *Bibl. Ind. N. Am. Bot.* p. 138.

Meconopsis crassifolia, Benth. in *Trans. Hort. Soc.* Ser. II. i. (1835)
p. 408—California.

= **M. heterophylla,** S. Wats. *l.c.* p. 42.

M. heterophylla, Benth.; Hook. *Ic. Pl.* t. 732—California.

Mikania scandens, Willd., *Sp. Pl.* iii. p. 1743—North America to South
Brazil.

Figured in Desc. *Fl. Pittoresque Antilles,* vii. t. 484, under the name
of *Eupatorium scandens.*

Mimulus cardinalis, Dougl. ex Benth. *Scroph. Ind.* p. 28; Hook. *Bot.*
Mag. t. 3560—North-West America.

M. floribundus, Dougl. in *Lindl. Bot. Reg.* t. 1125—North-West America.

M. guttatus, [Fisch.] *Hort. Gorenk.* ed. II. (1812) p. 25; DC. *Cat. Hort.*
Monsp. p. 127—North-West America.

= **M. luteus,** A. Gray, *Syn. Fl. N. Am.* ii. i. p. 277.

M. moschatus, Dougl. ex Lindl. *Bot. Reg.* t. 1118—North-West America.

M. roseus, Dougl. *l.c.* t. 1591—California.

= **M. Lewisii,** A. Gray, *Syn. Fl. N. Am.* ii. i. p. 276.

Nemophila aurita, Lindl. *Bot. Reg.* t. 1601—California.

N. insignis, Dougl. ex Benth. in *Trans. Linn. Soc.* xvii. (1835) p. 275;
Lindl. *Bot. Reg.* t. 1713—California.

Nicotiana multivalvis, Lindl. *Bot. Reg.* t. 1057—North-West America.

= **N. quadrivalvis** var. **multivalvis,** A. Gray, *Syn. Fl. N. Am.* ii. i. p. 243.

Oenothera albicaulis, Fras. *Cat.* (1913) ex Nutt. *Gen. Am.* i. p. 245—
North America.

Syn. *O. pallida,* Lindl. *Bot. Reg.* t. 1142.

O. decumbens, Dougl. ex Hook. *Bot. Mag.* t. 2889—California.

O. densiflora, Lindl. *Bot. Reg.* t. 1593—California.

O. dentata, Cav. *Ic.* iv. p. 67, t. 398—North-West America.

O. lepida, D. Dietr. *Syn. Pl.* ii. p. 1287; Hook. et Arn. *Bot. Beech.*
Voy. p. 342—California.

= **O. decumbens,** *Ind. Kew.* fasc. iii. p. 334.

O. Lindleyi, Dougl. in Hook. *Bot. Mag.* t. 2832—North-West America.

= **O. amoena,** *Ind. Kew.* fasc. iii. p. 334.

Oenothera muricata, *Linn. Syst.* ed. XII. p. 263—North-West America.
= **O. biennis** var. **muricata**, S. Wats. *Bibl. Ind. N. Am. Bot.* p. 378.
O. quadrivulnera, Dougl. ex Lindl. *Bot. Reg.* t. 1119—North-West America.
O. rubicunda, Torr. & Gray, *Fl. N. Am.* i. p. 502—California.
= **O. amoena,** Bailey, *Cyc. Am. Hort.* iii. p. 1121.
O. speciosa, Nutt. in *Journ. Acad. Philad.* ii. (1821) p. 119 ; Hook. *Exotic Flora*, ii. t. 80—N. America.
O. tenella var. *albiflora*, Hook. *Comp. Bot. Mag.* ii. (1836) p. 153. *quid ?*
O. triloba, Nutt. ; Sims, *Bot. Mag.* t. 2566—North-West America.
O. viminea, Dougl. ex Hook. *Bot. Mag.* t. 2873—California.
O. vinosa, Torr. & Gray, *Fl. N. Am.* i. p. 503—California.
= **O. amoena,** Bailey, *Cyc. Am. Hort.* iii. p. 1121.
Oncidium Pubes, Lindl. *Bot. Reg.* t. 1007—Brazil.
Oxyura chrysanthemoides, Lindl. *Bot. Reg.* t. 1850—California.
= **Layia calliglossa,** A. Gray, *Syn. Fl. N. Am.* i. ii. p. 316.

Paeonia Brownii, Dougl. in Hook. *Fl. Bor. Am.* i. p. 27 ; Lindl. *Bot. Reg.* xxv. t. 30—North-West America.
Pedicularis canadensis, Linn. *Mant.* i. p. 86 ; *Bot. Mag.* t. 2506—North America.
Pentstemon acuminatus, Dougl. ex Lindl. *Bot. Reg.* t. 1285—North-West America.
P. attenuatus, Dougl. *l.c.* t. 1295—North-West America.
P. breviflorus, Lindl. *Bot. Reg.* t. 1946—California.
P. confertus, Dougl. ex Lindl. *l.c.* t. 1260—North-West America.
P. crassifolius, Lindl. *Bot. Reg.* (1838) t. 16—North-West America.
= **P. Menziesii,** var. **Douglasii,** A. Gray, *Syn. Fl. N. Am.* ii. i. p. 260.
P. deustus, Dougl. ex Lindl. *l.c.* 1318—North-West America.
P. diffusus, Dougl. *l.c.* t. 1132—North-West America.
P. digitalifolius, Hook. *Comp. Bot. Mag.* ii. (1836) p. 153. *quid ?*
P. glandulosus, Dougl. ex Lindl. *Bot. Reg.* t. 1262—North-West America.
P. gracilis, Nutt, *Gen. Am.* ii. p. 52 ; Graham in *Bot. Mag.* t. 2945— North-West America.
P. heterophyllus, Lindl. *Bot. Reg.* t. 1899—California.
P. ovatus, Dougl. in *Bot. Mag.* t. 2903—North-West America.
P. pruinosus, Dougl. ex Lindl. *Bot. Reg.* t. 1280—North-West America.
P. Richardsonii, Dougl. *l.c.* t. 1121—North-West America.
P. Scouleri, Lindl. *l.c.* t. 1277—North-West America.
= **P. Menziesii** var. **Scouleri,** A. Gray, *Syn. Fl. N. Am.* ii. i. p. 260.
P. speciosus, Dougl. ex Lindl. *l.c.* t. 1270—North-West America.
= **P. glaber,** A. Gray, *l.c.* p. 262.
P. staticifolius, Lindl. *Bot. Reg.* t. 1770—California.
= **P. glandulosus,** A. Gray, *Syn. Fl. N. Am.* ii. i. p. 271.
P. triphyllus, Dougl. ex Lindl. *Bot. Reg.* t. 1245—North-West America.
P. venustus, Dougl. ex Lindl. *Bot. Reg.* t. 1309—North-West America.
Phacelia tanacetifolia, Benth. ; *Bot. Reg.* t. 1696—California.
Phlox speciosa, Lindl. *Bot. Reg.* t. 1351—River Columbia.
= **P. linearifolia,** A. Gray, *Syn. Fl. N. Am.* ii. i. p. 133.

Pinus amabilis, Dougl. in Hook. *Comp. Bot. Mag.* ii. (1836) p. 93—North-West America.

= **Abies amabilis,** Mast. in *Journ. R. Hort. Soc.* xiv. p. 189.

P. Douglasii, Lamb. *Gen. Pin.* ed. II. iii. t. 90—North-West America.

= **Pseudotsuga Douglasii,** Mast. in *Journ. R. Hort. Soc.* xiv. p. 245.

P. grandis, Dougl. in Hook. *Comp. Bot. Mag.* ii. (1836) p. 147—North-West America.

= **Abies grandis,** Mast. in *Journ. R. Hort. Soc.* xiv. p. 192.

P. insignis, Dougl. ex Loud. *Arboret.* iv. p. 2243, f. 2132–37—California.

P. Lambertiana, Dougl. in *Trans. Linn. Soc.* xv. (1827) p. 500—California.

P. macrocarpa, Lindl. *Bot. Reg.* (1840), Misc. p. 62—California.

= **P. Coulteri,** Mast. in *Journ. R. Hort. Soc.* xiv. p. 227.

P. Menziesii, Dougl. ex Lamb. *Gen. Pin.* ed. II. iii. p. 161, t. 71—California.

= **Picea sitchensis,** Mast. *l.c.* p. 224.

P. monticola, Dougl. ex Lamb. *Gen. Pin.* ed. I. iii. p. 149, t. 87—California.

P. nobilis, Dougl. ex Loud. *Encyc. Pl. Suppl.* i. p. 1276—California.

= **Abies nobilis,** Mast. in *Journ. R. Hort. Soc.* xiv. p. 193.

P. ponderosa, Dougl. in Loud. *Arboret.* iv. p. 2243, f. 2132–2137—North-West America.

P. Sabiniana, Dougl. in *Trans. Linn. Soc.* xvi. (1833) p. 749—California.

Platystemon californicus, Benth.; *Bot. Reg.* t. 1679—California.

Platystigma lineare, Benth.; *Bot. Mag.* t. 3575—California.

Pogonia pendula, Lindl. in *Bot. Reg.* t. 908—North America.

Potentilla arachnoidea, appears to have been first published in Hook. *Comp. Bot. Mag.* ii. (1836) p. 141.

= **P. pensylvanica,** *Ind. Kew.* fasc. iii. p. 611.

P. arguta, Pursh, *Fl. Am. Sept.* ii. p. 636 ; Lindl. *Bot. Reg.* t. 1379—North America.

P. effusa, Dougl. ex Lehm. *Pugill.* ii. p. 8—North-West America.

P. glandulosa, Lindl. *Bot. Reg.* t. 1583—North-West America.

P. obscura, Hook. *Comp. Bot. Mag.* ii. (1836) p. 141. *quid ?*

P. pectinata, Fisch. ex DC. *Prod.* ii. p. 581—North America.

= **P. pensylvanica** var. **strigosa,** Hook. *Fl. Bor. Am.* i. p. 188.

Prunus depressa, Pursh, *Fl. Am. Sept.* i. p. 332—North America.

= **P. pumila,** S. Wats. *Bibl. Ind. N. Am. Bot.* p. 306.

Psoralea macrostachya, DC.; *Bot. Reg.* t. 1769—California.

P. orbicularis, Lindl. *Bot. Reg.* t. 1971—California.

Purshia tridentata, DC. in *Trans. Linn. Soc.* xii. (1817) p. 158 ; Lindl. *Bot. Reg.* t. 1446—North-West America.

Pyrus rivularis, Dougl. ex Hook. *Fl. Bor. Am.* i. p. 203, t. 68—North-West America.

Ribes cereum, Dougl. in *Trans. Hort. Soc.* vii. (1830) p. 512 ; Lindl. *Bot. Reg.* t. 1263—North-West America.

R. divaricatum, Dougl. *l.c.* p. 515 ; Lindl. *Bot. Reg.* t. 1359—North-West America.

R. echinatum, Dougl. ex Lindl. *Bot. Reg.* sub t. 1349—California.

= **R. lacustre,** S. Wats. *Bibl. Ind. N. Am. Bot.* p. 334.

Ribes glutinosum, Benth. in *Trans. Hort. Soc.* Ser. II. i. (1835) p. 476
    —California.
   = **R. sanguineum** var. **glutinosum,** S. Wats. *Bibl. Ind. N. Am. Bot.*
       p. 337.
R. irriguum, Dougl. in *Trans. Hort. Soc.* vii. (1830) p. 516—North-West
    America.
   = **R. divaricatum** var. **irriguum,** S. Wats. *l.c.* p. 333.
R. lacustre, Poir. *Encyc. Suppl.* ii. p. 856 ; Lodd. *Bot. Cab.* t. 884—
    North America.
R. malvaceum, Sm. in Rees, *Cycl.* xxx. n. 13 ; Sweet, *Brit. Fl. Gard.*
    Ser. II. t. 340—California.
   = **R. sanguineum** var. **malvaceum,** S. Wats. *l.c.* p. 337.
R. Menziesii, Pursh, *Fl. Am. Sept.* ii. p. 732 ; Lindl. *Bot. Reg.* xxxiii. 56
    —North-West America.
R. niveum, Lindl. *Bot. Reg.* t. 1692—North-West America.
   = **R. gracile,** S. Wats. *l.c.* p. 333.
R. petiolare, Dougl. in *Trans. Hort. Soc.* vii. (1830) p. 514—North-West
    America.
   = **R. Hudsonianum** var. *β*, S. Wats. *l.c.* p. 334.
R. sanguineum, Pursh, *Fl. Am. Sept.* i. p. 164 ; Hook. *Bot. Mag.* t. 3335
    —North-West America.
R. setosum, Lindl. *Bot. Reg.* t. 1237—North America.
R. speciosum, Pursh, *Fl. Am. Sept.* ii. p. 731 ; Lindl. *Bot. Reg.* t. 1557—
    North-West America.
    (" First, however, introduced by Mr. Collie.")
R. tenuiflorum, Lindl. in *Trans. Hort. Soc.* vii. (1830) p. 242 ; *Bot. Reg.*
    t. 1274—North-West America.
   = **R. aureum** var. **tenuiflorum,** S. Wats. *Bibl. Ind. N. Am. Bot.*
       p. 332.
R. viscosissimum, Pursh, *Fl. Am. Sept.* i. p. 163 ; Hook. *Fl. Bor. Am.* i.
    p. 234, t. 76—North-West America.
Rodriguezia planifolia, Lindl. ; *Bot. Mag.* t. 3504—Brazil.
   = **Gomesa planifolia,** Cogn. in *Mart. Fl. Bras.* iii. vi. p. 244.
Rubus leucodermis, Dougl. ex Hook. *Fl. Bor. Am.* i. p. 178 et ex Torr. &
    Gray, *Fl. N. Am.* i. p. 454—North-West America.
R. leucostachys, Hook. *Comp. Bot. Mag.* ii. (1836) p. 142.
   = **R. leucodermis,** *Ind. Kew.* fasc. iv. p. 752.
R. longipetalus, Hook. *Comp. Bot. Mag.* ii. (1836) p. 142. *quid ?*
R. nutkanus, Moç. ex Scr. in DC. *Prod.* ii. p. 566 ; Lindl. *Bot. Reg.*
    t. 1368—North-West America.
R. spectabilis, Pursh, *Fl. Am. Sept.* i. p. 348, t. 16 ; Lindl. *Bot. Reg.* t. 1424
    —North America.

Scilla esculenta, *Bot. Mag.* t. 2774 = **Camassia esculenta.**
Sida malvaeflora, [Moç. and Sesse], ex DC. *Prod.* i. p. 474 ; Lindl. *Bot.*
    Reg. t. 1036—North-West America.
   = **Sidalcea malvaeflora,** S. Wats. *Bibl. Ind. N. Am. Bot.* p. 143.
Silene inamoena, Hook. *Comp. Bot. Mag.* ii. (1836) p. 142. *quid ?*
Sinningia Helleri, Nees ; Lindl. *Bot. Reg.* t. 997—Brazil.

Spergula ramosissima, Dougl. ex Hook. *Fl. Bor. Am.* i. p. 93—North-
West America.
= **S. arvensis** var. Hook. *l.c.*

Spiraea americana, Steud. *Nom.* ed. I. p. 805—North America.
= **S. Aruncus,** *Ind. Kew.* fasc. iv. p. 964.

S. ariaefolia, Sm. in Rees, *Cycl.* xxxiii. n. 16 ; Lindl. *Bot. Reg.* t. 1365
—North-West America.
= **S. discolor** var. **ariaefolia,** S. Wats. *Bibl. Ind. N. Am. Bot.* p. 321.

S. Aruncus, Linn. *Sp. Pl.* p. 490—North Temperate Region.

Stenactis speciosa, Lindl. *Bot. Reg.* t. 1577—California.
= **Erigeron speciosus,** A. Gray, *Syn. Fl. N. Am.* i. ii. p. 209.

Symphoria racemosa, Pursh, *Fl. Am. Sept.* i. p. 162—North America.
= **Symphoricarpos racemosus,** A. Gray, *Syn. Fl. N. Am.* i. ii. p. 13.

Tanacetum boreale, Nutt. in *Trans. Am. Phil. Soc.* N.S. vii. (1841)
p. 401—North America.
= **T. huronense,** A. Gray, *l.c.* p. 366.

Trifolium fucatum, Lindl. *Bot. Reg.* t. 1883—California.

Triteleia laxa, Benth. in *Trans. Hort. Soc.* Ser. II. i. (1835) p. 413,
t. 15, f. 2 ; *Bot. Reg.* t. 1685—California.
= **Brodiaea laxa,** S. Wats. in *Proc. Am. Acad.* xiv. p. 237.

Vaccinium ovatum, Pursh, *Fl. Am. Sept.* i. p. 290 ; Lindl. *Bot. Reg.*
t. 1354—North-West America.

APPENDIX VII

ICE LETTUCE

THIS lettuce is described in the 'Transactions of the Horticultural Society,' vol. vi. (1826), p. 575, as follows :—

"Seeds of this lettuce were brought from the United States, under the above name, by Mr. David Douglas, in 1823, and it was raised the following year. It belongs to the division of Silesian or Batavian Lettuces, and must not be confounded with the Ice Lettuce of Scotland, which is our White Cos Lettuce. The leaves are of a light shining green, blistered on the surface, very undulated, and slightly jagged round the edges ; they grow nearly erect, being eight inches long, and five or six broad. The outer spread a little at the top, but grow very close at the heart. It blanches without tying up, and becomes very white, crisp, and tender. It comes into use with the White Silesian, from which it differs, as it also does from any other of its class, in being much more curled, having a lucid sparkling surface, whence probably its name, and not turning in so much at the heart. It lasts as long in crop as the White Silesian."

APPENDIX VIII

SOME AMERICAN PINES

[At the time when this volume had already been pronounced ready for the press, one of the Society's officers, engaged in turning out a very old packing-case, which had apparently not been opened since the Society moved from South Kensington in 1887, came upon two manuscripts, which, on careful examination, proved to be in David Douglas' handwriting, and, as far as can be judged, are of about the same date as the other manuscripts from which the rest of this volume has been printed. It was at once decided to delay the issue of the volume a few weeks in order that these newly discovered manuscripts might be included.

As is the case with much of the previous part of this Volume there are two distinct manuscripts covering precisely the same ground, but in which one will from time to time contain a few words or a sentence which does not occur in the other. These Manuscripts have been most carefully compared and collated with the following result. As with the preceding parts of the Volume the original expressions have been scrupulously retained even at the cost, in places, of grammatical accuracy and clearness of meaning.—W. W., ED.]

1. *Pinus Douglasii.*[1] *Foliis solitariis planis subdistichis, strobilis ovatis pendulis, bracteolis exsertis, 3-cuspidatis.* Sabine in Trans. Hort. Soc. Vol.[2]

Flowers in April and May, fruit ripe in September.

Leaves solitary, flat, entire, imperfectly two-ranked, blunt at the apex, dark shining green above, glaucous underneath, about an inch long. Common filament erect, shorter than the bractea. Another reniform, inflated, destitute of a crest, having instead a blunt, short entire point. Bractea nearly round, concave, densely ciliated or fringed. Female catkin erect, sessile, oblong or elliptic, one inch long, of a bright pink colour. Bractea linear-oblong, ciliate, tricuspidate, persistent, very long. Cone sessile, ovate, pointed, pendulous in clusters at the extremities of the twigs, two to two and a half inches long, one and a half of an inch in diameter. Scales orbicular, ciliate, slightly notched near the base, entire at the apex, soft and velvety to the touch, fuscous, the bractea of a glossy reddish tint and exserted beyond the scale five-eighths of an inch.

Seeds small, pointed at the base, widening upwards, brown, wing pointed, broad and large in proportion to the seed.

Tree remarkably tall, unusually straight, having the pyramidal form peculiar to the *Abies* tribe of Pines. The trees which are interspersed in groups or standing solitary in dry upland, thin, gravelly soils or on rocky situations, are thickly clad to the very ground with widespreading pendent branches, and from the gigantic size which they attain in such places and

---

[1] *Pseudotsuga Douglasii*, Mast. in Journ. R. Hort. Soc. xiv. p. 245.
[2] Reference is several times made in these MSS. to 'Sabine in Trans. Hort. Soc. Vol.,' but in none of the volumes of the Society's Transactions does any such entry of Mr. Sabine's occur, the inference being that it was intended to publish it, but that the intention was never carried out.—ED.

from the compact habit uniformly preserved they form one of the most striking and truly graceful objects in Nature. Those on the other hand which are in the dense gloomy forests, two-thirds of which are composed of this species, are more than usually straight, the trunks being destitute of branches to the height of 100 to 140 feet, being in many places so close together that they naturally prune themselves, and in the almost impenetrable parts where they stand at an average distance of five square feet, they frequently attain a greater height and do not exceed even 18 inches in diameter close to the ground. In such places some arrive at a magnitude exceeded by few if any trees in the world generally 20 or 30 feet apart. The actual measurement of the largest was of the following dimensions : entire length 227 feet, 48 feet in circumference 3 feet above the ground, $7\frac{1}{2}$ feet in circumference 159 feet from the ground.

Some few even exceed that girth, but such trees do not carry their proportionate thickness to such a vast height as that above mentioned. Behind Fort George, near the confluence of the Columbia River, the old establishment of the Honourable the Hudson's Bay Company, there stands a *stump* of this species which measures in circumference 48 feet, 3 feet above the ground, without its bark. The tree was burned down to give place to a more useful vegetable, namely potatos.

On a low estimation the average size may be given at 6 feet diameter, and 160 high. The young trees have a thin, smooth, pale whitish-green bark covered with a profusion of small blisters like *P. balsamea*[1] or Balm of Gilead Fir, which, when broken, yield a limpid oily fluid possessing a fragrant and very peculiar odour, and which, after a few days' exposure to the action of the atmosphere, acquires a hard brittle consistence like other rosins, assuming a pale amber colour. The bark of the aged trees is rough, rotten, and corky, the pores smaller and containing less rosin, and in the most aged, 4 to 12 inches thick, greatly divided by deep fissures.

Often in the space or vacuity between the bark and the timber of standing dead trees [2] is found and may be flayed off in large pieces of several square yards and from its texture and colour might without examination be taken for sheepskin. There is no doubt but this curious species of Fungus hastens the decay of the timber like dry rot in Oak, though perhaps not in the same degree, and, as I have observed it only on erect trees or those dead on the stump, I infer it will not exist in the seasoned wood, consequently cannot detract any merit from it.

Wood straight and regular in the grain, fine, heavy, and easily split ; the layers or rings of a darker tint, closely resembling the timber of the well-known Larch. Whether it will prove durable or not remains yet to be known.

If we judge from the quantity of charcoal produced it will not prove so durable as the Larch ; the coal is moderately hard, bulky, brown, which might be expected from the great quantity of gaseous matter it contains. What might be the exact age of one of the largest dimensions could not be ascertained, not having the means of preparing a transverse section

[1] *Abies balsamea*, Mast. in Journ. R. Hort. Soc. xiv. p. 189.
[2] Blank space left in MS.

sufficiently well polished to be able to determine with accuracy the number of annual layers. One tree 14 feet in diameter, counting from the centre, gave 167 rings or layers to within four and three-fourths of an inch of the outside, where they became so thin that they could no longer be exactly ascertained ; although with sufficient accuracy upon which to ground a moderate calculation that fifty years added only nine and a quarter inches to the diameter of the trunk.

This is a common tree from Cape Blanco, situated in 43°, to the Straits of Juan de Fuca in 49° North Latitude, abounding in all the mountainous parts of the coast, preferring light, dry, thin, or gravelly soils, on a substratum of sand and clay or on rocky places. A few straggling trees are seen at Cape Mendocino in 40° which may be regarded as its most southern range, and likely it will extend much farther north than the habitat above given.

The mountains of the Grand Rapids of the Columbia, situated in Lat. 46°, are clothed to the top, some peaks of which exceed 5000 feet above the level of the sea. On the Blue Mountains of the interior it is also found, and clothes also the subalpine range or base of Mount Hood, Mount St. Helens, Mount Baker, and Mount Vancouver as well as the western base of the Rocky Mountains in 52° 7′ 9″ N. Lat., 115° N. Long., where it maintains a place, and arrives at a considerable size at an altitude of 9000 feet above the level of the sea, 1000 from the verge of perpetual snow.

It is not a little surprising the vast change of climate and of soil experienced between the western and eastern base of the last-mentioned ridge in the same parallel of latitude, and is beautifully exemplified by the growth of the present tree as well as others of the same tribe. On the west side it is enormously large, on the east a low scrubby tree, and without the recesses of the mountains on the same side it ceases to exist.

Being an inhabitant of a country nearly in the same parallel of Latitude with Great Britain, where the winter (even judging from the immense covering afforded it by Nature in its bark) is more severe, gives every reason to hope that it is in every respect well calculated to endure our climate and that it will prove a beautiful acquisition to English Sylva if not an important addition to the number of useful timbers.

The wood may be found very useful for a variety of domestic purposes: the young slender ones exceedingly well adapted for making ladders and scaffold poles, not being liable to cast ; the larger timber for more important purposes ; while at the same time the rosin may be found deserving attention.

In the memorable journey of Lewis and Clarke (p. 455), in their interesting account of the timber of that country, we find that they " measured some 42 feet in circumference, at a point beyond the reach of an ordinary man. This trunk for the distance of two hundred feet was destitute of limbs ; the tree was perfectly sound, and, at a moderate calculation, its size may be estimated at three hundred feet." I am most willing to bear testimony to the correctness of their statements as respects the girth of the timber, but after a two years' residence, during which time

I measured any tree that appeared from its magnitude as interesting, I was unable to find any from actual measurement exceeding the height I have mentioned.

I am unable to gather from their description whether the largest mentioned by them can be the same as the *present*. It would appear not, for in the fifth species mentioned we find that they mention (p. 457): "A thin leaf is inserted in the pith of the cone which overlays the centre of, and extends half an inch beyond, the point of each scale"—an important character in the species now observed, but from what is stated of it generally it belongs to neither.

2. *Pinus Menziesii*.[1] *Foliis solitariis subtriangularis incurvis, strobilis subcylindricis, bracteolis exsertis integris undulatis.* Sabine in Trans. Hort. Soc. Vol.

Flowers in April, May. Fruit ripe in October.

Leaves solitary, nearly triangular, flat on the upper side, angular or keeled below, rigid, incurved, acute, very glaucous, nearly an inch long. Common filament erect, shorter than the bractea. Crest of the anther orbicular, slightly notched at the apex. Bractea obovate-spathulate, concave, entire. Female catkin unknown. Cone pendulous, nearly cylindrical, blunt at the point, two and a half to three inches long with entire obovate scales rounded at the apex with a pale or dun colour. Bractea persistent linear-oblong, entire, waved, exserted one eighth of an inch beyond the scale. Seeds small, rusty-brown colour. Wings hatchet-shaped, acute at the point. Bark thin, smooth, light grey or brown, peeling off in thin flakes.

The appearance of this species closely resembles *P. Douglasii*[2]; although neither so large nor so plentiful as that species, it may nevertheless become of equal if not of greater importance, as it possesses one great advantage over that one by growing to a very large size on the Northern declivities of the mountain in apparently poor, thin, damp soils; and even in rocky places, where there is scarcely a sufficiency of earth to cover the horizontal wide-spreading roots, their growth is so far from being retarded that they exceed one hundred feet high and eight feet in circumference. This unquestionably has great claims on our consideration as it would thrive in such places in Britain where even *P. sylvestris* finds no shelter. It would become a useful and large tree.

The wood is remarkably fine, white, smooth, and regularly grained, lighter than the wood of *P. Douglasii*,[2] in texture and quality not far removed from the wood of *P. Abies*.[3]

So far as my observation went, this species is only found on the bleak cold mountainous parts of the Coast from Arguilar River in 43° to 49° N. Lat., and was not seen to the east of 121° W. Long. Is not, as far as I know, in the Rocky Mountains.

It is to be regretted that all the seeds of this truly magnificent tree were lost and could not then be replaced.

[1] *Picea sitchensis*, Mast. in Journ. R. Hort. Soc. xiv. p. 224.
[2] *Pseudotsuga Douglasii*, Mast. loc. cit., p. 245.
[3] ? *Picea excelsa*, Mast. loc. cit., p. 221.

3. *P. alba*,[1] Ait. Hort. Kew. ed. I. 3, p. 371 ; Lamb. Gen. Pin. ed. I. i. p. 39, t. 26. Pursh, Fl. Am. Sept. 2, p. 641 ; Richards. in Frankl. Journ. App. p. 752.

Common on the North-West coast and in the Rocky Mountains, also in Hudson's Bay ; grows even to a large tree in the 54° at an elevation of 7800 feet above the level of the sea.

4. *Pinus rubra*,[2] Lamb. Gen. Pin. ed. II. ii. p. 43, t. 28 ; Pursh, Fl. Am. Sept. 2, p. 640.

Near the Boat Encampment of the Columbia River in 52° 17′ 19″ ; more plentiful on the aqueous flats of Cedar River in 48° 23′ 00″. Sparingly seen and of diminutive growth in the 46° on the North-West coast.

5. *P. nigra*,[2] Ait. Hort. Kew. ed.I. 3, p. 370 ; Lamb. Gen. Pin. ed. I. t. 27 ; Pursh, Fl. Am. Sept. 2, p. 640 ; Richards. in Frankl. Journ. App. p. 752.

Plentiful on low swampy ground about Hudson's Bay, Lake Winnipeg, and Lesser Slave Lake, and on low aqueous flats of the Rocky Mountains, on the low marshy islands of the Columbia River above Tongue Point, and not uncommon in similar situations along the coast to Nootka Sound.

6. *P. canadensis*,[3] Linn. Sp. Pl. ed. II. p. 1421 ; Lamb. Gen. Pin. ed. I. t. 32 ; Pursh, Fl. Am. Sept. 2, p. 640 ; *Abies canadensis*, Michx. f. Hist. Arb. Am. p. 137, t. 13.

Abundant on the woody parts of the North-West coast, fully larger than any found on the Atlantic side of the Continent from the 43° to the 49° ; not in the interior nor in the valleys of the Rocky Mountains.

7. *P. nobilis*.[4] *Foliis solitariis planis secundis imbricatis glaucis, strobilis cylindricis erectis, squamis rotundatis.* Sabine in Trans. Hort. Soc. Vol.

Flowers in May ?

Leaves solitary, flat, linear, entire, stiff, blunt, with a minute concavity at the apex, ascending, somewhat two-ranked, so densely imbricated that the lower side of the branch is completely hid, very glaucous, seven-eighths of an inch long. Flower unknown. Cone solitary, cylindrical, erect, six to eight inches long, one and a quarter in diameter, dark brown. Scales round. Bractea scarcely exceeding the scale, oblong, obtuse.

The present tree, among the many highly interesting species by which it is surrounded in its native woods, in point of elegance justly claims the pre-eminence. The trees are straight, one hundred and seventy feet high, two to six feet in diameter with a white smooth polished bark. An open-growing tree, sparingly clad with wide-spreading horizontal branches placed in regular whorls round the tree, the distance between the whorls diminishing towards the top. The cones are always on the highest branches, near the top like *P. Picea*[5] and *P. balsamea*.[6]

The wood is soft, white and very light, containing but little rosin. An inhabitant only of the mountains, seldom if ever seen to arrive at any considerable size lower down on the hills than 5000 feet above the level of the sea in the 46° and 48° N. Lat.

Common on a chain of mountains that run nearly parallel with the

---

[1] *Picea alba*, Mast. in Journ. R. Hort. Soc. xiv. p. 220.

[2] *Picea nigra*, Mast. *loc. cit.*, p. 222.   [3] *Tsuga canadensis*, Mast. *loc. cit.*, p. 255.

[4] *Abies nobilis*, Mast. *loc. cit.*, p. 193.   [5] ? *Abies pectinata*, Mast. *loc. cit.*, p. 194.

[6] *Abies balsamea*, Mast. *loc. cit.*, p. 189.

coast, from Arguilar River on the South to the base of Mount St. Helens on the North and in the intermediate distance intersecting Mount Hood and Mount Vancouver. Not on the Rocky Mountains ? This grows more luxuriantly and more abundantly in thin poor dry soil on a rocky bottom; near springs or rivulets it entirely disappears.

This if introduced would profitably clothe the bleak barren hilly parts of Scotland, Ireland, and Cumberland, besides increasing the beauty of the country.

8. *P. amabilis.*[1] *Foliis solitariis planis pectinato-distichis emarginatis subtus glaucis, strobilis cylindricis erectis, squamis rotundatis.* Sabine in Trans. Hort. Soc. Vol.

Flowers in May ?

Leaves solitary, flat, pectinate, two-ranked, emarginate, glaucous underneath, shining green above, one and an eighth of an inch long, the upper rank only three-eighths of an inch, bent forward, lying close to the wood, less glaucous than the others. Flower unknown. Cones cylindrical, erect, mostly solitary, four to five inches long, one and a half in diameter, brown with rounded scales. Seeds large, dark brown. Wing short. Bractea pointed, somewhat longer than the scales.

This is another tree of singular beauty, not quite so tall but exceeding in girth the preceding one. This does not much exceed one hundred and thirty feet, but is sometimes seen eight in diameter, a very graceful and more compact tree than the former. The branches are very long, drooping, and flat, the leaves maintaining the same character. Is well contrasted with the distinct and varied foliage of those that surround it. The bark of the full-grown timber is partly green with large whitish blotches, smooth, producing little or no rosin. That of the young trees is smooth and polished green, with minute round or oval scattered blisters, yielding a limpid fluid, which possesses a less pungent taste and has a less aromatic odour than Balm of Gilead Fir *P. balsamea,*[2] to which in some instances it is related. The wood is soft, white, very light, but heavier than the last-mentioned, clean-grained, and takes a good polish, and under the microscope shows a multitude of minute angular reservoirs filled with the like fluid contained in the blisters but somewhat more viscid and less fragrant.

This inhabits the same range, and is equally as plentiful as the last-mentioned species, and, though not quite so fine, justly merits our further consideration.

9. *P balsamea,*[2] [Linn. Sp. Pl. p. 1002]; Willd. Sp. Pl. 4, p. 504 ; Lamb. Gen. Pin. ed. I. t. 31 ; Pursh, Fl. Am. Sept. 2, p. 639. *Abies balsamifera,* Michx. f. Hist. Arb. Am. 1, p. 145, t. 14.

Common on the mountainous parts of the North-West Coast from the 43° to the 50° N. Lat., and abundant on the western side of the Rocky Mountains.

10. *P. Lambertiana. Foliis quinis rigidis scabriusculis, vaginis brevissimis, strobilis crassis longissimis cylindricis; squamis laxis rotundatis.* Dougl. in Trans. Linn. Soc. Vol. 15 (1827), p. 500.

---

[1] *Abies amabilis,* Mast. in Journ. R. Hort. Soc. xiv. p. 189.
[2] *Abies balsamea,* Mast. *loc. cit.,* p. 189.

Common in Northern California, between the parallels of 40° and 43°, in dry, barren sandy soils. Trunks 150 to 200 feet high, varying from 20 to nearly 60 feet in circumference. Bark smooth, light brown colour bleached like on the north side. Cones 12 to 17 inches long, 9 to 11 round, erect the first year, pendulous the second. Seeds ripe in September, eaten by the native tribes on river.

11. *P. monticola. Foliis quinis rigidis, vaginis brevissimis, strobilis sessilibus crassis cylindricis, squamis rotundatis.*

Flowers in April and May ? Fruit ripe in September.

Leaves five together, rigid, finely serrated, somewhat glaucous, two and a half inches long with a very short sheath. Flowers male and female unknown. Cone, sessile, cylindrical, nearly straight, obtuse at the apex, five to six inches long, one and a half to two inches in diameter, light brown when ripe. Scales rounded, flat, entire. Seeds four lines long and two broad, oval. Wing falcate, two and a half times the length of the seed, fuliginous, with minute dark veins; cotyledons 11.

This species is intermediate between *P. Lambertiana* and *P. Strobus.* Like the former the leaves are more rigid, and less glaucous than *P. Strobus,* and are somewhat shorter than either species. The cones are considerably shorter, nearly double the thickness, straight, obtuse pointed, with larger and flatter scales, more rounded at the apex, lying more compactly over each other and yielding a less copious rosin, which, instead of being whitish with a blue tint and clammy during the intense heat of summer, the present yields a hard, candied, bright amber-coloured rosin; the seeds never, as it were, glued to the inside of the scales, which is constantly the case in *P. Strobus.*

The bark of the young trees is smooth and nearly olive-green; that of the full-grown trees, smooth, light grey, peeling off in flakes. The wood is very light, white, soft, even-grained, smooth, easily wrought, containing fewer and more minute resinous reservoirs than *P. Strobus,* and on this account may be less valuable. A handsome tree of large dimensions, particularly in aqueous deposits consisting of decayed vegetable substances in the mountain valleys which are washed by the torrents from higher altitudes, also in rocky, bare, thin soils it particularly abounds. They grow immensely large, frequently 5 feet in diameter, and a hundred and sixty feet high, and, as is the case with many others, they naturally prune themselves, leaving a clean straight trunk, without a single branch, exceeding one hundred feet.

A common tree in the mountainous districts of the Columbia, from its confluence with the sea in 46° 19' N. Lat. to its source in 52° 30'; also on the banks of Flathead River, and the western base of the Rocky Mountains.

A truly distinct and beautiful species intermediate between the present and *Pinus Lambertiana,* exists with cones as long as, if not longer than it, of an elegant slender taper form about 2 inches in diameter, it was only seen on one of my journeys to the mountains near the base of Mount St. Helens, I never had it in my power to procure perfect specimens of this desirable tree.

12. *P. ponderosa.* *Foliis ternis elongatis, strobilis subcylindraceis deflexis ; squamarum spinis reflexis vagina foliorum abbreviata.* Sabine, in Trans. Hort. Soc. Vol.

Flowers in April and May; fruit perfect in September of the second year.

Leaves three in a sheath, slender, acute, rounded on the upper surface, furrowed below, finely serrated, six to nine inches long. Vagina or sheath half an inch long, entire at the end. Flower unknown. Cone sessile, partly cylindrical, tapering to the point, slightly curved and bent downwards, six to eight inches long, dark brown. Scale dilated in the middle. Seeds about five lines long and three broad, oval, black, encircled by the base of the wing, the inner side of the seed being nearly covered. Wing elongated, blunt at the apex, fuliginous.

Trees tall, straight, seldom divided by large branches, very elegant, ninety to one hundred and thirty feet high ; sometimes exceeding four feet in diameter, three feet above the ground, carrying their thickness to a very great height, frequently measuring eighteen inches in diameter at seventy feet, and not uncommon to see this without so much as a branch of any description whatever.

The wood is remarkably clean-grained, though somewhat coarse in texture, smooth, heavy, reddish, works fine, and is impregnated with a copious rosin.

The bark is very smooth, tawny-red. This appears to be rapid-growing, is highly ornamental, and may, though not so valuable as some, be of importance. Like all the species of this genus which have plural leaves, that inhabit the western parts of the continent of America, it never grows in nor composes thick forests like the *Abies* section, but is found on declivities of low hills and undulating grounds in unproductive sandy soils in clumps, belts, or forming open woods, and in low, fertile, moist soils totally disappears. This may have greater claims on our attention than merely its beauty, for, in addition to its timber, a great portion of turpentine could be extracted.

The young trees are liable to injury by   [1] which in many places appears in such abundance on some trees as completely to destroy them ere they reach 20 feet high. It is not confined to the cleft or between the limbs and branches like Viscum on its various supports, but spreads spontaneously over the young shoots, springing from the base of the sheath, which with the bud scales and rosin sufficiently protect the seeds until the insertion of the roots in the bark takes place. Before the first season goes past the branch affected becomes distorted, swollen to six or eight times the natural size, afterwards rough and rotten, the rosin copiously flowing from the part affected until it exhausts its supporter. Then, being deprived of its natural aliment, it does not long outlive the tree.

[1] A blank is left here in the principal MS. and the other runs thus :—" A singular species of *Viscum* or mistletoe is found on this tree which grows so rapidly and in such abundance that it in time completely destroys it. A second species of the same genus is found on *Pinus Banksiana* west of the Rocky Mountains and more sparingly in the valleys of that ridge."—ED.

This species like *P. Lambertiana* only grows in the poorest and apparently barren, dry, sandy or gravelly soils, which it almost exclusively occupies. In low fertile moist soils it totally disappears. Very abundant on the lower parts of the Blue Mountains in 46° 02′ 37″ N. Lat., 118° 25′ 07″ W Long., frequenting rocky places or such soils as noticed above. Equally if not more abundant throughout the chain of Spokane River to its junction with the Columbia in 48° 12′ 43″ North Lat., also on the banks of Flathead River, near the Kettle Falls 48° 37′ 40″, decorating the country on the North towards Fraser River 49° 50′ 00″ and on the south to Salmon River in 47°.

To *P. Taeda*, Linn. Sp. Pl. 1000, Willd. Sp. Pl. 4, p. 498 ; Lamb. Gen. Pin. ed. I. p. 23 tt. 16, 17 ; Ait. Hort. Kew. ed. I. 3, p. 368 ; Michx. f. Hist. Arb. Am. 1, p. 97, t. 9 ; Pursh, Fl. Am. Sept. 2, p. 644, this in general habit agrees. The principal differences are in the form and length of the cone, the reflected spine, and in the shortness of the sheaths which are neither torn nor dilated at the apex, giving us essential characters to form a species.

To *P. rigida*, Mill. Gard. Dict. ed. 8, n. 10 ; Lamb. Gen. Pin. ed. I. tt. 18, 19, it in some respects approaches in habit but differs materially in station ; this species occupies large tracts of country on the western parts of America, being confined to the coast in the northern boundaries of California between the parallels of 40° and 45° N. Lat., while *P. ponderosa* is an inhabitant only of the interior of the country.

The seeds of this as well as those of *P Lambertiana* are eaten by several of the native tribes raw but more generally dried or roasted in the embers. Vegetates readily and has 9–11 cotyledons. This species vegetated quickly in the Society's garden and I hear also that most of the seeds distributed amongst the Fellows grew.

13. *Pinus contorta. Foliis geminis rigidis, strobilis. sessilibus ovatis recurvis confertis ; aculeis squamarum reflexis.* Sabine in Trans. Hort. Soc. Vol.

Flowers in April and May ; fruit ripe in September and October.

Leaves in pairs, rounded on the back, concave on the inner side, rigid, acute, 2 to 2½ inches long, having a very short ragged or ciliated sheath. Flower, male terminal cylindrical, numerous, on a short peduncle. Crest of the anther reniform, minutely toothed ; female round or nearly globular, with a profusion of small scaly pointed bractea, erect. Cone, sessile, rounded, bent downwards, shorter than the leaves, in clusters round the shoots, remaining on the twigs for several years. Scales oblong, slightly dilated in the middle, of a hard woody tough texture with a short sharp slender reflected spine, brown. Seed small, sharp at the base, wing short and acute.

Trunk 20 to 30 feet high, one foot to eighteen inches in diameter, with a rough, white bark, seldom divided by its branches. Wood soft, spongy, coarse-grained, brown-coloured, with abundance of rosin. Branches drooping, greatly twisted in every direction, remarkably tough, the younger ones covered by acuminate chaffy brown scales. As far as my opportunities of observing this species went, it is exclusively the inhabitant of the sea coast from the 44° to the 49° of North Latitude, at all times confined to damp boggy soils where *Vaccinium uliginosum, Oxycoccus macrocarpus* delight to grow and where there is uniformly a dense carpet of *Sphagnum obtusifolium* and tufts of *Bartramia.* Not uncommon from "Cape Look Out" to the confluence of the Columbia ; more abundant towards Puget Sound, and very likely will be found to increase in number to the 60°.

Little can be said in favour of this tree either for ornament or as a useful wood.

14. *P Banksiana,* Lamb. Gen. Pin. ed. I. t. 3 ; Pursh, Fl. Am. Sept. 2, p. 642 ; Richards. in Frankl. Journ. App. p. 752.

P. rupestris, Michx. f. Hist. Arb. Am. 1, p. 49. t. 2.

Flowers in May.

A common tree in the mountainous districts of the Columbia River in 48° Lat., 118° West Long., and in the valleys of the Rocky Mountains and on the banks of streams that flow into Hudson's Bay.

This species occupies higher altitudes than any other natives of the continent of America. In the 53° North Latitude it is not uncommon, though of diminutive growth, at the height of 11,000 feet above the level of the sea, and exists to within 700 feet of the confines of perpetual snow.

As a proof of their hardiness, such trees as spring in sheltered coves or recesses that are screened from the blast, grow vigorously till such time as their tops outgrow the height of the rocks which protect them, when, by the severity of the climate and the keenness of the atmosphere, the tops become horizontal as if cut with hedge shears.

In the valleys where the wind is at all seasons of the year by projecting places of the rocks directed to the same point, the trees have their branches literally all blown to one side, the one side being perfectly smooth as if effected by art.

In the valleys of the central ridge of the continent and west of that ridge it is found and exclusively confined to this species.

15. *P. rigida,* [Mill. Gard. Dict. ed. 8, n. 10] ; Marshall. Arb. Am. p. 101 ; Lamb. Gen. Pin. ed. I. tt. 18, 19 ; Michx. f. Arb. Hist. Am. 1, p. 89, t. 8 ; Pursh, Fl. Am. Sept. 2, p. 643.

P. Taeda rigida β, Ait. Hort. Kew. ed. I. v. 3, p. 368.

Common in Northern California in barren soils extending as far north as the 45° about a hundred miles from the sea, forming small clumps or thinly scattered over the ground with *P. Lambertiana.*

16. *P microcarpa,*[1] Lamb. Gen. Pin. ed. II. 2, p. 56, t. 37 ; *P laricina,* Du Roi, Obs. Bot. p. 49 ; Harbk. Baumz., v. 2, p. 83, t. 3, f. 5–7.

Near the source of River Athabasca in the eastern valleys of the Rocky Mountains in the 53° North Lat. the present is of very humble growth, never exceeding two or three feet high and six to ten inches in diameter. Abundant on the shores of Lesser Slave Lake, Lake Winnipeg, and the streams that flow into Hudson's Bay. Has not yet been seen on the North-West coast.

[1] *Larix americana,* Sargent, Silva N. Am. xii. p. 7.

17. *P. pendula*?[1] [Soland in] Ait. Hort. Kew. ed. I. 3, p. 369 ; **Lamb.** Gen. Pin. ed. I. i. p. 56, t. 36 ; Pursh, Fl. Am. Sept. 2, p. 645.

Flowers in April and May.

For want of perfect cones I am unable to decide whether this tree found west of the Rocky Mountains be truly distinct from *P. pendula*.

The trees are one to two hundred feet high, five to twenty-five feet in circumference. Beautifully straight and remarkably singular, being spirally twisted, which is advantageously seen in dead trees that have lost the bark. The bark is smooth and reddish. The branches are horizontal, short and twiggy, springing from the trunk in regular whorls and nearly at equal distances from each other, which gives it an aspect very different from any of the species of this Section. From recollection I state that the cones will be found larger. The wood is very durable, judging from the drift-wood usually found on the rocks of the river. I have seen some which had lain thirty years and still comparatively sound. The native tribes do not use it, its hard tough texture being too great for the slender means they have of working it. Common on the west side of the Rocky Mountains in Lat. 52° 07′ 09″ and at the Kettle Falls of the Columbia in 48° 37′ 40″ in aqueous deposits composed of decayed vegetables ; also in rocky places where the soil is thin, composed of decayed granite and earth, it arrives at the greatest size.

[1] *Larix americana*, Sargent, Silva N. Am. xii. p. 7.

INDEX

All scientific names are printed in italics. The pages on which the species of *Pinus* and *Quercus* are described are printed in thick type.

2 A

Printed in the United States
By Bookmasters